ADVANCED STRUCTURAL DAMAGE DETECTION

ADVANCED STRUCTURAL DAMAGE DETECTION: FROM THEORY TO ENGINEERING APPLICATIONS

Tadeusz Stepinski
AGH University of Science and Technology, Poland
and Uppsala University, Sweden

Tadeusz Uhl and Wieslaw Staszewski
AGH University of Science and Technology, Poland

Registered Office
John Wiley & Sons, Ltd, The Atrium, Southern Gate, Chichester, West Sussex, PO19 8SQ,
United Kingdom

For details of our global editorial offices, for customer services and for information about how to apply for permission to reuse the copyright material in this book please see our website at www.wiley.com.

Library of Congress Cataloging-in-Publication Data applied for

A catalogue record for this book is available from the British Library.

Print ISBN: 9781118422984

Set in 10/12pt Times by SPi Publisher Services, Pondicherry, India
Printed and bound in Malaysia by Vivar Printing Sdn Bhd

Contents

List of Contributors

Łukasz Ambroziński
Alberto Gallina
Krzysztof Holak
Andrzej Klepka
Piotr Kohut
Michał Mańka
Adam Martowicz
Krzysztof Mendrok
Paweł Paćko
Łukasz Pieczonka
Mateusz Rosiek
Mariusz Szwedo

The above authors are affiliated with the Department of Mechatronics and Robotics, Faculty of Mechanical Engineering and Robotics, AGH University of Science and Technology, Poland.

Preface

Structural Health Monitoring (SHM) is an interdisciplinary engineering field that deals with innovative methods of monitoring structural safety, integrity and performance without affecting the structure itself or particularly impairing its operation. SHM utilizes several types of sensors – embedded in or attached to – a structure to detect the presence, location, severity and consequence of damage.

SHM technology integrates nondestructive evaluation (NDE) techniques using remote sensing and smart materials to create smart self-monitoring structures characterized by an increased reliability and long life. SHM primarily applies to systems with critical demands concerning performance where classical onsite assessment is related to high costs, is difficult or even impossible.

Written by academic experts in the field, this book will provide students, engineers and other interested technical specialists with a comprehensive review of recent developments in various monitoring techniques and their applications to SHM. By providing a comprehensive review of the main SHM techniques it contributes to an area which is the subject of intensive research and development. This book offers both theoretical principles and feasibility studies for a number of SHM techniques. It also presents a number of novel data processing algorithms and demonstrates real operating prototypes.

This book reports results of the research project MONIT conducted at the AGH University of Science and Technology (AGH-UST) in Kraków and supported by The European Regional Development Fund in the framework of the programme Innovative Economy. The book includes 10 chapters written by researchers active in the research team at the Department of Robotics and Mechatronics at the AGH-UST where the Editors are faculty members.

Chapter 1 contains an introduction that briefly presents SHM philosophy and explains its relationship to the traditional NDE techniques. The interdisciplinary character of SHM is outlined in the context of NDE and condition monitoring (CM). Structural damage and structural damage detection aspects are discussed and the main levels of SHM procedures are outlined.

The structure of the SHM systems using global and local SHM methods is presented. Aspects related to the design process of SHM systems are discussed in the final part of the chapter.

Chapter 2 is devoted to the numerical simulation of elastic wave propagation in planar structures. A brief overview of the available numerical methods for elastic wave propagation is presented, which include: finite element methods (both implicit and explicit formulations), spectral element method, finite volume method and finite difference methods. Both theoretical background and practical aspects are considered in the overview. The main focus in this chapter is on the new implementations of LISA (Local Interaction Simulation Approach) which belongs to the finite difference methods. Recently, LISA has been rediscovered and found attractive on account of rapidly evolving techniques based on graphical processing units (GPUs). The GPU implementation of LISA developed at AGH-UST is presented and its performance is illustrated using benchmark modelling cases.

Chapter 3 presents an application of the model assisted probability of detection (MAPOD) for SHM systems under uncertain crack configuration and system variability. The phenomena related to a SHM system are studied using numerical simulations of planar structures monitored by transducer arrays. The monitoring reliability is evaluated using a three-dimensional model including both Lamb wave propagation in a plate-like structure and scattering from cracks with different configurations. Computer models are implemented using parallel processing technology, presented in Chapter 2, which significantly speeds up the simulation time. The configurations considered in the chapter account for the variations of the relative crack position, orientation and size.

Chapter 4 provides the state of the art in the area of applications of nonlinear acoustics to SHM. An extensive literature review dealing with physical mechanisms related to nonlinearities encountered in elastic materials in the presence of damage is provided as an introduction. Different stress–strain characteristics are considered and two particular effects, contact acoustics nonlinearity and nonlinear resonance, are presented in some detail. The principles of frequency mixing that occurs if two waves with different amplitude and frequency are introduced into a damaged structure are explained. The main part of the chapter contains a review of damage detection methods and their applications to metallic and composite structures as well as glass, plexiglas, concrete and rocks.

Chapter 5 discusses piezoelectric transducers used for generation and sensing surface and Lamb waves. After a short overview of the transducer designs used for that purpose, the focus is on the flexible transducers made of piezocomposite materials. A novel type of piezoelectric transducer, based on a macro-fibre composite (MFC), is presented. Contrary to the commonly used conventional MFC transducers with dense electrodes, the transducers presented here are provided with sparse interdigital electrodes matched to a certain wavelength. Two different designs of interdigital transducers (IDTs) made of MFC substrate are presented and compared with the classical MFC transducers with dense electrodes. The comparison includes the results of numerical simulations and experimental tests performed using a scanning laser vibrometer for the conventional MFC and the proposed IDT attached to an aluminium plate.

The electromechanical impedance (EMI) method is presented in Chapter 6. After providing a theoretical background, the measurement setups and signal processing algorithms used for damage detection are introduced. The EMI is illustrated with the results of numerical finite element simulations performed for simple mechanical structures, such as beams and plates. Experimental results obtained in the laboratory conditions for simple structures (e.g., aluminium plate and pipeline section) are also presented. Finally, results of the EMI measurements performed on two aircraft structures (bolted joint in the main undercarriage bay and riveted fuselage panel) are presented and discussed in terms of EMI feasibility for SHM.

Chapter 7 reviews methods used for selective focusing of Lamb waves using transducer arrays. After a brief introduction to the field of phased arrays, conventional beamforming techniques based on delay and sum (DAS) operation are discussed. First, a few example two-dimensional (2D) array topologies are compared in terms of spatial resolution based on the simulation results performed using a simplified model with point-like transducers and dispersion-free medium. In the second part, a new beamforming technique is presented, which is an extension of the DORT method (a French acronym for the decomposition of time reversal operator) where the continuous wavelet transform (CWT) is used for the time–frequency representation (TFR) of nonstationary snapshots. An application of the proposed technique to self focusing of Lamb waves in an aluminium plate is demonstrated both for a linear- and a 2D star-shaped array. It is shown that the decomposition of the time reversal operator obtained with the proposed method enables separation of point-like scatterers in the aluminium plate and allows the focusing of Lamb waves on a number of individual damages present in the plate.

Chapter 8 presents the theory and applications of modal filters with focus on their potential in SHM systems. Modal filters can be successfully used for global monitoring of large engineering structures. Structural modification of a mechanical system (e.g. drop in stiffness due to a crack) results in the appearance of peaks at the output of the modal filter. These peaks result from the imperfect modal filtration due to the system's local structural changes. An SHM system based on modal filters is theoretically insensitive to environmental changes, such as temperature or humidity variation (global structural changes do not cause a drop in modal filtering accuracy). A number of practical implementations of the presented technique are provided including a description of the developed SHM system as well as the results of its extensive simulation, laboratory and operational testing. It is demonstrated that modal filters can create a valuable damage detection indicator characterized by low computational effort (due to the data reduction ability); moreover, an SHM system utilizing this concept is easy to automate.

Different ways of using thermographic measurements in NDE and SHM are discussed in Chapter 9. First, classification of the thermographic techniques focused on active thermography with internal excitation, including vibrothermography (VT), is presented. An overview of the measurement equipment including thermographic cameras and excitation sources is also provided. Numerical simulations of coupled thermo mechanical phenomena using explicit finite elements are presented. Hardware and software components of the VT measurement system developed at AGH-UST are provided and illustrated by the description of laboratory measurements performed using this system on composite and metallic samples. A parametric study of the influence of measurement parameters on the resultant thermal response is also provided. Finally, measurements performed on a military aircraft fuselage and wing panels are reported and the practical issues related to field measurements are discussed.

Chapter 10 is devoted to vision based monitoring systems. It starts from an overview of the background and related work in the field of vision based measurement methods; the systems available in the market are presented and their advantages and shortcomings are discussed. The following section focuses on the main steps of the developed method, such as camera calibration and scale coefficient calculation algorithms as well as image rectification using homographic mapping. Simulation tests of the developed method carried out using the developed programming tool as well as using the virtual model of the construction are reported. Results of the numerical investigations of the uncertainty propagation in the proposed algorithms are presented and discussed; the probability of damage detection is considered as well. Laboratory tests, carried out using a setup consisting of a steel frame loaded by a point force and two high resolution digital single-lens reflex (SLR) cameras, are reported. Results of the vision based method are compared with those obtained using other contact and non contact measurement techniques. Finally, an evaluation of the method on civil engineering construction is presented.

Acknowledgments

The research presented in this book has been supported by funding from the research project MONIT (No. POIG.01.01.02-00-013/08-00) co-financed by the European Regional Development Fund under the operational programme Innovative Economy and research project N N501158640 sponsored by the Polish National Science Center.

1

Introduction

Tadeusz Uhl[1], Tadeusz Stepinski[1,2] and Wieslaw Staszewski[1]

[1]*Department of Mechatronics and Robotics, Faculty of Mechanical Engineering and Robotics, AGH University of Science and Technology, Poland*
[2]*Signals and Systems, Department of Engineering Sciences, Uppsala University, Sweden*

1.1 Introduction

It is widely accepted that maintenance of engineering structures is important to ensure structural integrity and safety. This is particularly relevant to civil engineering and transportation. Aerospace structures for example are inspected regularly. Airframes are monitored for possible fatigue cracks.

A variety of different Nondestructive Testing and Evaluation (NDT/E) methods have been developed for damage detection. Ultrasonic inspection and eddy current technique are good examples of mature, well-established technologies that are widely used for crack detection. NDT/E techniques are often limited to single-point measurements and require scanning when large areas need to be monitored.

In recent years there have been a range of new damage detection techniques and sensing technologies. These methods allow for global, online monitoring of large structures and fall into the area of Structural Health Monitoring (SHM). They are capable of achieving continuous monitoring for damage involving the application of new sensors. Damage monitoring systems, which often use advanced sensor technologies, are concerned with a new design philosophy. Actuators, sensors, and signal processing are integrated to offer progress in this area.

SHM involves integrating sensors and actuators, possibly smart materials, data transmission and computational power within a structure in order to detect, localize, assess and predict damage which can be a cause of structure malfunction now or in the future (Adams 2007; Balageas *et al.* 2006). A typical SHM system is associated with an online global damage

Advanced Structural Damage Detection: From Theory to Engineering Applications, First Edition.
Edited by Tadeusz Stepinski, Tadeusz Uhl and Wieslaw Staszewski.
© 2013 John Wiley & Sons, Ltd. Published 2013 by John Wiley & Sons, Ltd.

identification in structures; such systems are most often applied in aerospace (Staszewski *et al.* 2004) and civil engineering (Wenzel 2005).

Although SHM systems utilize NDT/E methods as tools there are many differences in SHM and NDT/E operation principles. NDT/E techniques are commonly applied offline and locally in regions of expected damage while SHM methods should provide real time monitoring of a whole structure during its operation. SHM is a successive step in the evolution of structure diagnostics, which historically evolved from the damage detection concept implemented in the form of condition monitoring (CM) systems. CM systems should be capable of detecting damage based on a global assessment of technical structures during their operation. SHM is expected to go further by the ability of detecting damages in early stages of their development or, in an ideal case, of predicting their occurrence before they really take place (Inman *et al.* 2005).

1.2 Structural Damage and Structural Damage Detection

There are many different connotations of the term *damage* in mechanical structures which in the area of SHM damage can be understood intuitively as an imperfection, defect or failing which impairs functional and working conditions of engineering structures. A more precise definition of damage can be offered when system analysis is used. Structures can be modelled as systems with input excitations and output measurable signals. In this context damage can be considered as an additional excitation that results in energy flow and transformation, leading to modifications of output signals. Therefore damage detection is an inverse problem; measurable outputs are used to detect damage. Damage can also be regarded as a modification to material properties and/or structural physical parameters. These properties and parameters can be modified due to sedimentation and plasticity of material or fatigue and corrosion. In this context damage detection is an identification problem. Material properties and physical parameters need to be extracted to assess damage.

Many different damage detection methods have been developed in the last few decades. Altogether these methods can be classified into model based and signal based approaches. Vibration based methods often utilize physical and/or modal parameters, obtained from physical models, for damage detection. Models are also essential when loads are monitored to obtain information about structural usage. Signal based methods rely on various types of direct measurements such as noise, vibration, ultrasound or temperature.

Both, i.e. model and signal based, approaches require signal processing techniques; the former to develop appropriate models and to analyse changes in these models that are relevant to damage; and the latter to extract features and establish a relationship between these features and possible damage.

The majority of signal based methods rely on a relationship between structural condition and signal features or symptoms. This condition – symptom relationship is often not easy to analyse due to the complexity of engineering structures, sophistication of design and use of advanced materials. Different signal processing methods need to be used for the analysis. This includes signature and advanced signature analysis. The former is based on simple features such as statistical spectral moments or physical/modal parameters. The

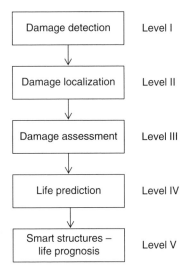

Figure 1.1 Main levels of SHM procedure

latter uses multidimensional features (e.g. vectors, matrices, images) such as spectra, signal instantaneous characteristics or time – frequency distributions.

In this context, damage detection can be regarded as a problem of pattern recognition. Pattern recognition requires feature selection procedures for training and is usually based on statistical, syntactic or neural approaches. Many recent studies in this area are based on new developments related to signal processing (Staszewski and Worden 2009) and machine learning (Worden *et al.* 2011). It is clear that these developments are essential for implementation of any SHM system.

Damage detection forms the primary objective of the overall problem of damage identification. SHM systems' tasks can be classified as a process consisting of five activities that form five important elements or levels (Balageas *et al.* 2006; Cempel 1991; Rytter 1993), as shown in Figure 1.1.

These are: I, damage detection; II, damage localization; III, assessment of damage size; IV, remaining life prediction; and V, smart structures with self-evaluating, self-healing or control capabilities. In this context detection gives a qualitative indication that damage might be present, localization gives information about the probable position of damage, assessment estimates its severity by providing information about damage type and size and finally, prognosis estimates the residual structural life and predicts possible breakdown or failure. The first three levels (i.e. detection, localization and assessment) are mostly related to system identification, modelling and signal processing aspects. The level of prognosis falls into the field of fatigue analysis, fracture mechanics, design assessment, reliability and statistical analysis. This level is very intensively investigated in many laboratories but there are currently no commercially available solutions. All these levels require various elements of data, signal and/or information processing.

1.3   SHM as an Evolutionary Step of NDT

Damage detection/monitoring, NDT/E and SHM are often misunderstood as synonyms and may have the same meaning in many engineering areas. Damage, health and monitoring of structures can be described using various definitions. In general, health is the ability to function/perform and maintain structural integrity throughout the entire lifetime of the structure; monitoring is the process of diagnosis and prognosis and damage is a material, structural or functional failure. Also, in this context, structural integrity is the boundary condition between safety and failure of engineering components and structures. In aircraft maintenance, damage detection and direct monitoring of damage accumulation offers an alternative approach to loads monitoring.

Recent developments in SHM are related either to modifications of well-established techniques, new equipment and sensor technologies or new monitoring principles. This can be illustrated using three examples. First, Acoustic Emission (AE) is a well-established NDT technique used for damage detection for many years. However, when optical fibre sensors – that can be integrated with monitoring structures – are used, AE (passive NDT approach) can be combined with Lamb wave based damage detection (active SHM approach). Secondly, the first NDT/E application of Lamb waves goes back to the 1950s although significant progress was achieved when low profile, smart transducers (e.g. piezo-ceramic, polymer, discs, paints, fibres) were introduced in the early 1990s allowing a real SHM approach. Thirdly, new damage detection methods based on a nonclassical approach to nonlinear acoustics have been proposed recently offering good damage detection sensitivity.

NDT techniques are often limited to single point measurements and require scanning when large areas need to be monitored. There have been a range of new damage detection techniques and sensing technologies in recent years. SHM methods allow for global, online monitoring of large structures and also offer damage localization. These methods are capable of achieving continuous monitoring for damage with the application of new sensors. Damage monitoring systems that use smart sensor technologies are concerned with a design philosophy directed to the integration of actuators, sensors, and signal processing. There has been an enormous research effort in this area in the last 20–30 years.

SHM, damage detection/monitoring and NDT are often used replaceably to describe the process of nondestructively evaluating structural condition. However, only SHM defines the entire process of implementing a strategy that includes five important identification elements (or levels), as discussed above. What distinguishes SHM from NDT is the global and online implementation of various damage detection technologies which require periodically spaced measurements (or observations), as accurately pointed out in Adams (2007). This process of online implementation needs more advanced signal processing for reliable damage detection than classical NDT techniques.

NDT involves comparing the known input of a measured signal with a known model – it does not require sacrificing the physical system, as disassembly or failure testing would. NDT is usually carried out offline in a local manner, after the damage has been located, or periodically, to improve performance of a structure. NDT techniques are mainly used to characterize damages and assess their severity, if their location is known.

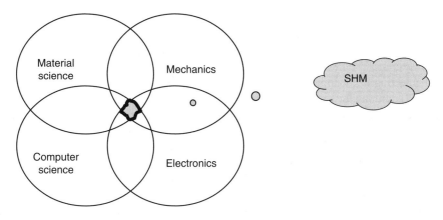

Figure 1.2 Interdisciplinary nature of SHM

1.4 Interdisciplinary Nature of SHM

SHM is an interdisciplinary area of research that integrates such basic sciences as materials science, mechanics, electronics and computer science, and which is strongly related to structures and their life cycle. This is illustrated schematically in Figure 1.2.

SHM approaches require three important elements for damage detection implementation. These are: (1) knowledge of monitored structures and possible damage scenarios; (2) sensors and instrumentation used to obtain signals/data that can be used for damage detection; and (3) relevant analysis which can extract information about possible damage and overall structural integrity. Modelling, numerical simulations and signal processing are important activities of all these three elements.

The interdisciplinary nature of SHM requires a dedicated approach during design, manufacturing and operation, therefore SHM systems are mostly installed on new structures and rarely on old ones that have been operating for a long time. Installation of SHM systems on structures with unknown history of operation is very difficult and the probability of correct damage assessment is much lower than for the new structures.

The design of an SHM system depends on the type of damages which can occur, type of materials applied for the design and physical phenomena employed for damage detection. The complexity of an SHM system design depends on the local nature of material damages that are most likely to occur in the assessed structure and that may not significantly influence the structure's response measured normally during its operation, e.g. its low frequency vibration spectrum.

Another factor that makes SHM data from a damaged structure difficult to acquire is limited accessibility of its particular components during operation. This often requires an in-depth study of local structure behaviour with the application of analytical and simulation tools that are widely used for understanding damaged structure behaviour and characteristics of the related signals. Due to the relatively high cost of an SHM system implementation of a very careful design and optimization process is recommended, taking into account minimization of the influence of the system on the structure's performance, minimization of the hardware costs (e.g. number of sensors and actuators), and maximization of the correctness of damage detection and assessment.

Nowadays, multi-physics and multiscale simulation are extremely valuable tools in designing SHM systems. The design process consists of several steps, the most challenging of which are: (1) selecting a phenomenon which is sensitive enough to the damages that are to be detected; (2) defining the required sensing system with self-validation capability; (3) selecting data acquisition and processing architecture; (4) defining feature extraction and information reduction procedures; (5) formulating and implementing procedure of damage detection; and last but not least, (6) damage localization and its size assessment.

There are no general rules on how to solve all these design and implementation problems for any structure. The design methods are dedicated to a given structure, used materials and chosen physical phenomena employed for health monitoring. SHM technology helps to achieve better operational safety and has an economic impact on decreasing maintenance and operating costs because it allows the prediction of possible damage a long time before its appearance and, consequently, gives operators enough time to plan a proactive service and maintenance action.

There are several disciplines that are very closely related to SHM:

- CM – condition monitoring (Inman *et al.* 2005);
- NDT/E – Nondestructive Testing/Evaluation (Staszewski *et al.* 2004);
- SPC – statistical process control (Inman *et al.* 2005);
- DP – damage prognosis (Inman *et al.* 2005);
- MP – Maintenance Planning (Pietrzyk and Uhl 2005), for instance, RCM (Reliability Centred Maintenance).

CM is in many aspects very similar to SHM but in practical use it is dedicated to rotating and reciprocating machinery. The main features of the CM approach are: damage localization is approximately known as well as the type of damage – the number of possible damages is limited, databases with damage symptoms are available, the influence of environmental conditions on the measurement results is very slight, and the economic benefits from employment of CM procedures are well defined. An essential advantage of CM over SHM methods is the well defined economic benefits from the use of CM procedures as well as the fact that many standards used worldwide require the monitoring of rotating machinery.

On the other hand, SHM also has disadvantages: localization of damage is not known, there are difficulties in measurements, due to the limited admittance to the monitored structural components, the type of damage is often difficult to identify, the influence of environmental conditions on measurement results is significant, and the cost of SHM systems is relatively high, which is the reason for their application only on critical structures.

CM systems rely on measurements of structural responses during operation, but they do not use dedicated actuators to excite or trigger effects which can help to detect damage. The differences between CM, NDT and SHM systems in terms of integration of hardware with a structure are illustrated in Figure 1.3.

The main difference between NDT and SHM systems can be noticed in the hardware architecture. In the case of a SHM system, sensors and actuators are built into (or integrated with) the structure, while NDT is an external system with an independent (not integrated with the structure) set of sensors and actuators [cf. Figure 1.3(c)]. Integration of SHM and NDT systems with the other tools is shown schematically in Figure 1.4.

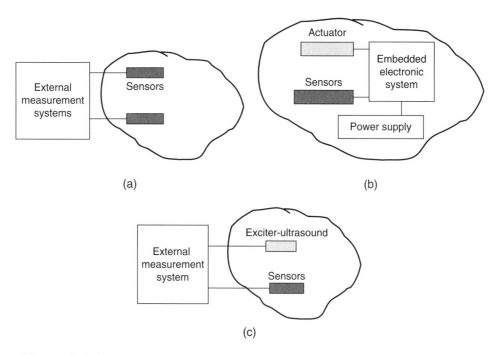

Figure 1.3 Schematic diagram of a typical (a) CM system, (b) SHM system and (c) NDT system

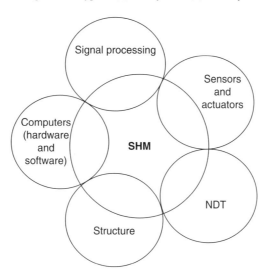

Figure 1.4 Main components of an SHM system

The main difference between NDT/E and SHM is their implementation – NDT/E techniques are implemented offline while SHM ones are implemented online, which makes SHM tasks much more complex than the autonomous NDT/E applications.

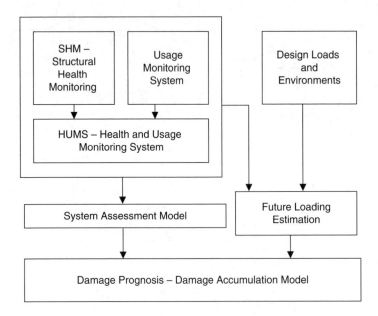

Figure 1.5   Schematic diagram of a typical DP procedure. Adapted from Inman D, Farrar C, Lopes V and Steffen J. *Damage Prognosis for Aerospace, Civil and Mechanical Systems*, © 2005 John Wiley and Sons Ltd.

SPC has similar aims to SHM but the final aim of SPC systems is not only detecting structural damages but process diagnostics – they use a variety of sensors to monitor changes in the process parameters. The process parameters can change due to structural failure and in this respect SHM and SPC are comparable.

DP is used to predict the remaining lifetime of operating structures during which their performance will remain above a given threshold. DP systems use the knowledge about damage size and location as well as expected operational loads. The remaining life prediction is based on a predictive model that acquires information from a usage monitoring system (the system that monitors loading cycle during the structure's operation), the SHM system as well as the past, current and future environmental conditions and expected load levels.

Today's DP systems give only a very rough estimation of remaining life prognosis, owing to the very complex physics of structure destruction if material level is to be considered. Multi-scale simulation including molecular dynamics (Packo and Uhl 2011) methods can be helpful to solve this problem in the future. Chapter 2 contains recent results concerning modern numerical simulation methods.

The interaction between different types of monitoring systems in DP is shown diagrammatically in Figure 1.5 (Farrar *et al.* 2005). The most general is the MP system which defines requirements and tasks that are to be accomplished for achieving, restoring and maintaining the operational capability for the whole life of a structure.

MP systems use data from the installed SHM system but also help to analyse historical data in order to detect events that could have been the reason for performance loss. This approach enables preventive service action before damage occurs. Several approaches can be distinguished within this discipline, one of the most useful for mechanical structures is RCM that helps minimize maintenance costs and minimize the risk of structural failure (Pietrzyk and Uhl 2005).

1.5 Structure of SHM Systems

A typical SHM system includes two main parts – a hardware section and an algorithmic section with software. The hardware part includes sensors and optional actuators and the units performing signal conditioning and acquisition, communication, and power supply. These components work autonomously and very often are embedded into the structure. Communication and power supply problems can be often encountered in this type of architecture, which calls for miniaturization of the applied hardware components. The problem with power supply can be solved by energy harvesting units that are currently the subject of many research projects.

Examples of already available technologies are shown in (Priya and Inman 2010). A feasible solution to the miniaturization of SHM hardware is designing and manufacturing MEMS chips dedicated to the SHM purpose, which include sensors, actuators, communication units and processors integrated on board (Staszewski *et al.* 2004). Wireless communication is one of the most commonly used data transfer type in SHM systems, there are dedicated solutions but some of the commercially available, such as zigbee or radio-frequency identification (RFID), can also be applied (Dargie and Poellabauer 2010; Lynch and Loh 2006; Uhl *et al.* 2007).

The software part contains basic procedures for signal processing, signal fusion, hardware control, structure health detection and remaining life prognosis. More advanced systems contain also some procedures related to structural health management.

1.5.1 Local SHM Methods

Modern SHM approaches can be classified into two main groups, the global methods (Adams and Farrar 2002; Doebling *et al.* 1996; Uhl 2004) and local methods (Grimberg *et al.* 2001; Maldague 2007; Raghavan and Cesnik 2007). Local methods monitor a small area of structure surrounding the sensor (sensors) using measurements of a structural response to certain applied excitation. Ultrasonic waves (Raghavan and Cesnik 2007), eddy currents (Grimberg *et al.* 2001), thermal field (Maldague 2007; Uhl *et al.* 2008) and acoustic emission (Pao 1978) are examples of phenomena that are most commonly employed for local SHM. The methods that are most often used for the design of SHM systems are guided waves (GWs; Raghavan and Cesnik 2007), those based on FBG sensors (strain, temperature measurements and ultrasound sensing) (Betz 2003), vibrothermography (Uhl *et al.* 2008) and electromechanical impedance (Bhalla and Soh 2004; Park *et al.* 2003). The state-of-art of the vibrothermography and electromechanical impedance methods can be found, respectively, in Chapters 9 and 6.

There are many other methods that can also be classified as local SHM (Chung 2001) but those methods are mostly used for more specific applications. Classical NDT methods that rely on the characteristics of ultrasound waves propagating in solid bodies can be used in combination with different signal processing and damage imaging techniques.

In the context of SHM, however, the waves are generated by permanently installed actuators that are integrated within a structure. The response is measured by a built-in set of piezoelectric sensors. There are many different techniques employed for excitation and sensing GWs. Recent results in the area of piezoelectric transducers made of macro-fibre composite (MFC) are presented in Chapter 5.

In thin plate-like structures elastic waves can propagate in the form of Lamb waves and the methods making use of these type of GWs are one of the most often proposed local methods in SHM. GW based techniques can be applied for both metallic and composite structures. Generally, local GW based methods require dense sensor networks (to provide a large number of measurement points distributed in space) which generate a large amount of data that have to be processed in order to detect, localize and assess structural damage. The cost of installation of such sensor networks is usually much higher than that for the global methods. Using two-dimensional (2D) phased arrays, capable of electronic beamforming, instead for the dense sensor networks has been proposed as a feasible solution to this problem; details are given in Chapter 7.

Local GW based methods have a number of advantages, the most important is their ability to monitor structural parts without the need for disassembly. Also, since the wavelength of GWs is in the same range as damage dimension they are sensitive to small damages. However, they also have an essential disadvantage – they require dense sensor networks or sophisticated phased arrays that need to be located in proximity to the potential damage, which means that knowledge of the critical damage location (hot spot) is of primary importance. Therefore, local methods are applied when critical structures are to be tested and early phase of damage has to be detected, and the high cost of the SHM system is acceptable.

1.5.2 Global SHM Methods

The global methods are performed if global motion of the structure is induced during its operation (Adams 2007; Balageas *et al.* 2006). Vibration based methods belong to this class. The global methods make use of the fact that local damage, for instance, local stiffness reduction, has an influence on the global structure's behaviour in terms of time and space.

In comparison with local methods, global methods have a number of essential advantages: they can monitor the whole structure using a rough sensor network, sensors do not necessarily have to be located close to damage, and only a limited knowledge about critical location is sufficient. Obviously, global methods also have disadvantages, e.g. the wavelength of the naturally excited vibrations is approximately equal to the dimension of the structure or component and so they have relatively low sensitivity to small damages (especially for lower vibration modes).

Although global methods give only a rough estimation of damage location and size they can be successfully used for damage detection. The most commonly used global methods are vibration based methods (Balageas *et al.* 2006). Low frequency vibrations have been applied for diagnostic purposes for many years (Inman *et al.* 2005; Wenzel 2005). The effects of material defects, supporting structures' failures or geometry defects on vibration response of a structure are well known. The relationship between structural vibration and damages of structures is used in their health assessment.

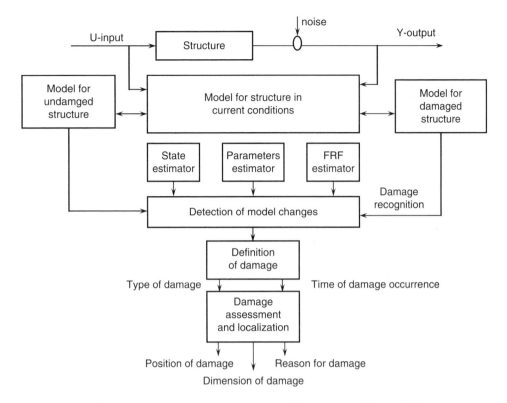

Figure 1.6   Schematic diagram of model based global SHM method

Two types of methods can be distinguished among global methods: signal based (Inman *et al.* 2005) and model based (Natke and Cempel 2000). The signal based methods utilize relations between measured responses of the structure after ambient excitation and possible damages. Signal features in frequency, time and time/frequency domains are the most popular now. The methods are very commonly applied in rotating and reciprocating machinery diagnostics for damage detection, but localization and damage assessment need additional information.

The model based methods employ many different types of models of a monitored structure to detect and localize damage in the structure using relations between the model parameters and particular damages. The idea behind the method is shown in Figure 1.6 (Uhl 2004).

Models of undamaged structure and damaged structure are compared for their parameters or output and differences (residues) are related to given damage and help to localize it. One of the most commonly used models in SHM is a modal model, which can be identified on a real structure with the use of measured external excitation (or vibration excitation caused by operation) and measurements of structural responses at many points.

The modal methods monitor the whole structure by detecting shifts of natural frequencies, increases in damping or changes of vibration modes' shapes. The selected feature should be damage-sensitive. Modal model based techniques can be classified into the following groups (Uhl 2004):

- methods based on perturbation of modal parameters (natural frequency, modal damping);
- methods based on frequency response function (FRF) (stiffness and compliance) variation detection;
- methods based on mode shape analysis;
- methods based on detection of modes' energy;
- methods based on finite element (FE) model updating.

The methods based on modes' shape analysis, such as strain energy analysis or mode shape curvature analysis, are preferred despite the fact that the required SHM system is then more complex than for SHM systems based on natural frequency and modal damping.

The global model based SHM procedures require neither dense sensor network nor sensors located in the vicinity of damage. These methods, however, are less sensitive and have lower spatial resolution compared with the local ones. Their sensitivity and spatial resolution can be improved by a computational model that interprets changes of dynamic properties of a structure.

The global model based methods are employed mainly for the SHM of civil structures. There are several issues that limit the application of these methods, the first is the high cost of a monitoring system caused by very complex cabling. The second is the relatively high influence of environmental conditions on structural dynamic properties, which means that this influence may sometimes dominate in comparison with that caused by serious damage of the structure. The first problem can be solved by a wireless sensor based monitoring system (Uhl *et al.* 2007) and the second one by a special environmental filter which is based on a modal filter [Mendrok and Uhl (2008) and Chapter 8].

Since the global methods are much less sensitive to damage than the local ones, in practical applications, they are used only for damage detection.

1.6   Aspects Related to SHM Systems Design

The entire process of monitoring for possible damage in engineering structures depends on structural design concepts. This can be illustrated using aircraft design. Current design principles of aircraft structures are based on the *safe-life* concept. Load spectra representative of typical operational conditions are first determined. This requires a significant amount of data related to mission profiles, mass distributions and many other parameters. The load spectra and fracture mechanics are then used to evaluate structural components in terms of their service fatigue life. This is followed by a series of fatigue tests of materials, coupons, elements, subcomponents and components, leading finally to the Major Airframe Fatigue Test (MAFT).

In practice, the scatter in design input data (e.g. unknown parameters, change of load conditions, variation of material properties, quality of manufacturing, human errors or structural modifications in service) is quite significant. Therefore various safety factors are imposed on the structure to guarantee the safe fatigue life. The structure is designed for a specific number of flight hours and retired from service afterwards even if no failure occurs.

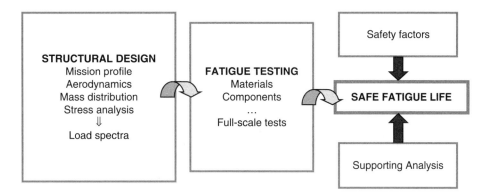

Figure 1.7   Safe-life aircraft design concept

The estimation of operational life of ageing aircraft is even more difficult. The safe-life design concept, illustrated in Figure 1.7, leads in practice to structures that are safe but over-designed. This is not desirable if economy and performance are analysed. Noncritical structural components that are exposed to multiple load paths are often designed using the fail-safe concept. Even if these components develop damage, the structural integrity is not jeopardized due to the assumption that damage can be detected before any catastrophic failure. This requires periodic inspections of components.

Monitoring techniques offering reliable detection, location, estimation of severity and prognosis of damage can lead to the *damage-tolerance* design concept. Detected damage is monitored to maintain the safe life of aircraft in this design concept. Although significant inspection effort is required, this concept can lead to lighter structures and better performance. In fact, the prevention of crack initiation behind the safe-life design concept does not prevent catastrophic failures. Therefore maintenance and inspection of aircraft structures is very important whatever the design concept. It is important to note that fatigue of materials in Aerospace Engineering has significantly contributed to structural design. The *safe-life* and *fail-safe* design concepts, introduced in aerospace, are widely used in many other areas of engineering.

1.6.1   Design Principles

The design of SHM systems requires dedicated procedures and tools because of the costs, the high level of responsibility and the interdisciplinary nature of system design. The procedure for SHM system design consists of the following steps:

1. Assumptions on the type of damage which should be detected.
2. Choice of physical phenomena which is sensitive to damage occurrence.
3. Formulation of monitoring methods – algorithms (for detection, localization and assessment of damage dimension).
4. Simulation study of the method (virtual prototyping).
5. Laboratory validation of the method (physical prototyping).

6. Testing of system performance [probability of detection (POD)].
7. Implementation of SHM system.
8. Operational test of the system.
9. Installation and operation of the system.

A careful description of the structure, environmental conditions and expected load ranges as well as a list of possible damages are necessary to choose appropriate physical phenomena which can be used for structural monitoring. The chosen phenomenon has to be measurable for a particular structure and operation conditions, however, its normal operation should not be disturbed. It is preferred that the phenomenon can be excited by ambient excitation, but in other cases a special actuation system should be designed to excite the structure. The excitation level should be as low as possible because of the limited power supply and high enough to cause measurable responses of the tested structure. While formulating the algorithm for correct damage detection results, the required computational power and volume of data from measurements have to be considered. Minimal computational power and minimization of the amount of required measurement data are important due to the limited availability of power supply and limited memory of the embedded computer power, especially when a wireless embedded SHM system is under design.

The next step in the design procedure is testing the formulated solution by simulation, which requires creating models of the monitored structure, damage scenario, sensors and actuators. The goal of this simulation is to create tools that enable testing of the correctness of the formulated method and choosing a measurement system that will enable measurements of the structural response with sensitivity matched to the monitored damages. Sensitivity analysis of a structure's response to damage accuracy is a basic tool which helps to design a monitoring system. With the use of the model and its simulation the POD can be tested too (Mendrok and Uhl 2008). For a chosen SHM system, the POD should be as high as possible in order to consider the formulated method as useful in the design of SHM for a given structure. But due to undefined localization of the damage and fixed position of the applied sensors, the model based assessment of POD can be a difficult task [Gallina et al. (2011) and Chapter 3].

A much more accurate determination of POD can be done by applying an experimental laboratory test. In such a test damage is defined by a sample with known damage location and dimension and well known properties of a healthy structure. The laboratory test should confirm the sensitivity and probability level of the correct damage detection.

The next step is implementation of the designed SHM system, which requires conformity with related standards that depends on the application area, e.g. aviation standards are completely different from those that have to be obeyed in civil engineering. This is the main reason for completely different designs of SHM systems in both branches.

The modularity of SHM due to the requirement of scalability is an important feature during the implementation phase. In subsequent phases the SHM system is installed on a real structure in the operation environment. The operation test is always required to confirm the correctness of the system design.

A further development of SHM systems requires new automatic algorithms of damage detection, localization and assessment, new state prognosis methods and algorithms, and development of self-diagnosis and self-healing of critical structures.

References

Adams D 2007 *Health Monitoring of Structural Materials and Components*. John Wiley & Sons, Ltd.

Adams D and Farrar C 2002 Identifying linear and nonlinear damage using frequency domain ARX models. *Structural Health Monitoring* **1**, 185–201.

Balageas J, Fritzen C and Guemes A (eds) 2006 *Structural Health Monitoring Systems*. ISTE.

Betz D 2003 Acousto-ultrasonic sensing using fiber Bragg grating. *Smart Materials and Structures* **12**(1), 122–128.

Bhalla S and Soh C 2004 Structural health monitoring by piezo-impedance transducers. *ASCE Journal of Aerospace Engineering* **35**, 154–165.

Cempel C 1991 *Vibroacoustic Condition Monitoring*. Ellis Horwood.

Chung D 2001 Structural health monitoring by electrical resistance measurements. *Smart Materials and Structures* **10**, 624–636.

Dargie W and Poellabauer C (eds) 2010 *Fundamentals of Wireless Sensor Networks: Theory and Practice*. John Wiley & Sons, Ltd.

Doebling S, Farrar C, Prime M and Daniel W 1996 Damage identification and health monitoring of mechanical systems from changes of their vibration characteristics a literature review. Technical report, LA-13070 MS.

Farrar CR, Lieven NAJ and Bement M 2005 An introduction to prognosis. In *Damage Prognosis for Aerospace, Civil and Mechanical Systems*. John Wiley & Sons, Ltd.

Gallina A, Packo, PP, Ambrozinski L, Uhl T and Staszewski WJ 2011 Model assisted probability of detection evaluation of a health monitoring system by using CUDA technology. *Proceedings of IWSHM* 2011 (ed. Chan FK), Stanford University.

Grimberg R, Premel D, Savin A, Le Bihan Y and Placko D 2001 Eddy current holography evaluation of delamination in carbon epoxy composites. *Insight* **34**, 260–264.

Inman D, Farrar C, Lopes V and Steffen J (eds) 2005 *Damage Prognosis for Aerospace, Civil and Mechanical Systems*. John Wiley & Sons.

Lynch J and Loh K 2006 A summary review of wireless sensors and sensor networks for SHM. *The Shock and Vibration Digest* **38**(2), 91–128.

Maldague X 2007 *Nondestructive Testing of Materials Using Infrared Thermography*. Springer.

Mendrok K and Uhl T 2008 Modal filtration for damage detection and localization. *Structural Health Monitoring 2008: Proceedings of the Fourth European Workshop*. Kraków, Poland.

Natke G and Cempel C 2000 *Model Based Diagnostics*. Springer.

Packo P and Uhl T 2011 Multiscale approach to structure damage modelling. *Journal of Theoretical and Applied Mechanics* **49**, 243–264.

Pao Y 1978 Theory of acoustic emission. *Elastic Waves and Non-destructive Testing of Materials* **29**, 107.

Park G, Sohn H, Farrar CR and Inman DJ 2003 Overview of piezoelectric impedance based health monitoring and path forward. *The Shock and Vibration Digest* **35**, 451–463.

Pietrzyk, A and Uhl T 2005 Use of RCM methodology for railway equipment maintenance optimization. *Archives of Transport* **49**(1), 6583.

Priya S and Inman D (eds) 2010 *Energy Harvesting Technologies*. Springer.

Raghavan A and Cesnik CES 2007 Review of guided waves structural health monitoring. *The Shock and Vibration Digest* **39**, 91–114.

Rytter A 1993 *Vibration Based Inspection of Civil Engineering Structures*. PhD thesis, Department of Building Technology and Structural Engineering, Aalborg University, Denmark.

Staszewski WJ, Boller C and Tomlinson GR (eds) 2004 *Structural Health Monitoring Systems of Aerospace Structures*. John Wiley & Sons, Ltd.

Staszewski W and Worden K 2009 Signal processing for damage detection. In *Encyclopedia of Structural Health Monitoring*. John Wiley & Sons, Ltd. pp. 415–421.

Uhl T 2004 The use and challenge of modal analysis in diagnostics. *Diagnostyka* **30**, 151–160.

Uhl T, Hanc A, Tworkowski K and Sekiewicz L 2007 Wireless sensor network based bridge monitoring system. *Key Engineering Materials* **347**, 499–504.

Uhl T, Szwedo M and Bednarz J 2008 Application of active thermography for SHM of mechanical structures. *Structural Health Monitoring 2008: Proceedings of the Fourth European Workshop*. Kraków, Poland.

Wenzel H (ed.) 2005 *Ambient Vibration Monitoring*. John Wiley & Sons, Ltd.

Worden K, Staszewski WJ and Hensman JJ 2011 Natural computing for mechanical systems research: A tutorial overview. *Mechanical Systems and Signal Processing* **25**(1), 4–111.

2

Numerical Simulation of Elastic Wave Propagation

Paweł Paćko
Department of Mechatronics and Robotics, Faculty of Mechanical Engineering and Robotics, AGH University of Science and Technology, Poland

2.1 Introduction

Various physical phenomena are employed for damage detection and localization in health monitoring systems. Recently, elastic waves received a lot of interest due to their inspection potential and nondestructive nature. Particular attention is paid to guided ultrasonic waves, such as Lamb waves (Boller *et al.* 2009; Staszewski *et al.* 2004) that are exploited for damage detection in the maintenance of engineering structures. A broad review on the application of Lamb waves for damage detection in different structural components and materials may be found in Raghavan and Cesnik (2007), Staszewski (2004) and Su and Ye (2009). Despite a considerable amount of research effort, real engineering applications of Lamb waves are still limited. This is due to the complexity associated with Lamb waves propagation, i.e. multimodal nature, dispersion, possible reflections from boundaries and other structural features that produce a wave field that is quite difficult to analyse even without taking into account scattering from defects. Thus, methods that help to understand these phenomena are sought.

Numerical simulations are frequently used to predict the possible wave field, providing an insight into mechanisms that drive the wave interaction with structural features. This enables employment of simulation methods in the design, operation, and improvement of SHM and NDT/NDE systems as a mean of optimizing the system's characteristics, as well as a part of the system to support damage detection and localization. At the design stage of a monitoring system, numerical simulation helps to adjust the settings, choose optimal sensors' positions or

Advanced Structural Damage Detection: From Theory to Engineering Applications, First Edition.
Edited by Tadeusz Stepinski, Tadeusz Uhl and Wieslaw Staszewski.
© 2013 John Wiley & Sons, Ltd. Published 2013 by John Wiley & Sons, Ltd.

examine the system's response for various damage configurations. This may be performed virtually, reducing the number of test samples and physical experiments, and minimizing the time required for the monitoring system design. During the operation phase, the simulation may be employed as a part of the system providing a virtual baseline result or in an inverse analysis for the identification problem. Finally, the conclusions drawn during the virtual testing of a monitoring system lead to an improvement of its performance and inspection capabilities.

Considering the inspection procedure of a structure, the task of simulation is to support an SHM system at as many levels as possible. Starting with the fundamental task of an SHM system, damage detection, some demands on the simulation method for wave propagation may be proposed.

First, once the goal of the system is to detect damage at an early stage, it is necessary to employ high frequency waves. This is because the waveform is significantly affected only by an obstacle with a characteristic dimension that is at least of the order of a wavelength. Despite the increased attenuation, short wavelengths must be employed which implies high spatial resolution of a numerical model. Secondly, due to the high wave speeds for a metallic or composite material, high temporal resolution of the simulation method needs to be applied. These two requirements impose serious limitations on the modelling method, leading directly to large model sizes and long computational times.

The crucial aspect of the simulation of the SHM, as well as NDE/NDT, systems is the modelling of wave interaction with damage. Depending on the material of a structural component, different damage types are considered. Typical damages in metallic structures are fatigue cracks, cavities (e.g. from the casting process), folds and laps introduced in forming operations, highly deformed regions (shear band localizations), degraded regions, corrosion, etc. Cracks may produce a highly nonlinear response to waves due to small scale plasticity effects, interfacial friction at crack faces, thermoelastic effects and others (Buehler 2008; Klepka *et al.* 2011). On the other hand, the most common damages in composite materials are delaminations, debondings, folds, fibre, matrix, and fibre/matrix interface failure, degradation of material properties due to environmental conditions, the temperature dependence of the structure's properties, etc. A reliable damage model is key to the success of the simulation method for SHM.

The wave interaction with damage may be thought of as an interaction with an interface of two media of high impedance mismatch that may additionally be a source of nonlinear effects. This requires special boundary conditions and a very accurate modelling approach, otherwise the structure's response to damage will be of no use and the modelling method discredited.

The possibility to simulate such complex phenomena and the wave interaction with damage allow for full exploitation of the monitoring system's potential. This requires sophisticated modelling techniques which are described in the following section.

2.2   Modelling Methods

This section reviews the selected numerical methods that are employed for the simulation of wave propagation phenomena for health monitoring applications. The key ideas and the necessary mathematical background are provided. Since the main emphasis is on the time evolution of mechanical systems, the time integration techniques are also briefly covered.

2.2.1 Finite Difference Method

The finite difference method (FDM) is one of the most popular methods for wave propagation simulation. The framework for the method was given by Courant, Friedrichs and Lewy in the 1930s (Courant *et al.* 1928). It was further developed during the Second World War and later in the 1950s and 1960s along with the development of computing units. Significant progress for time-dependent problems was made by Crank and Nicolson (1947), Lax and Wendroff (Lax 1961; Lax and Wendroff 1960) and Newmark (1959) by establishing time integration methods and stability criteria. Fundamental contributions to systematization of the theory were made by Collatz (1966, 1986). The method has been widely used in geophysics since then (Abramovici and Alterman 1965; Alford *et al.* 1974; Alterman and Karal 1968; Twizell 1979).

The FDM is an approach to solving differential equations by substituting derivatives by finite difference formulas. Therefore the original equation is transformed into an algebraic equation. There is a wide variety of difference formulas that can be used for spatial and temporal derivatives (Strikwerda 2004). The fundamental relation for deriving difference operators is the Taylor series:

$$f(x + h) = f(x) + h f'(x) + \frac{h^2}{2!} f''(x) + ... + \frac{h^n}{n!} f^{(n)}(x), \qquad (2.1)$$

which relates the unknown value of a function for a point $x+h$ to the known function derivatives at x and the distance at which the value is evaluated, h.

Equation (2.1) may be truncated after a certain number of terms producing a truncation error, and, assuming that the function values are known, used to evaluate the function's derivative. The simplest calculations obtained by truncation of series (2.1) after the second term are the so-called *forward difference estimate* or *backward difference estimate* (Woolfson and Pert 1999):

$$f'(x) = \frac{f(x + h) - f(x)}{h} \quad \text{or} \quad f'(x) = \frac{f(x) - f(x - h)}{h}. \qquad (2.2)$$

Equation (2.1) may be also combined for different points to obtain various finite difference formulas (Strikwerda 2004; Woolfson and Pert 1999).

It is worth noting that formulas obtained by using Equation (2.1) may be used for time or space discretization.

Once a given differential equation is transformed into a finite difference algebraic equation, it is implicitly assumed that both, time and space, have been discretized with certain steps Δt and Δx, respectively. The discretization process leads to discrete steps in time and the domain subdivision into small pieces, called elements or cells. In FDMs it is frequently assumed that the considered domain is composed of cuboidal cells with possibly different edge lengths. Nevertheless, due to the implementation issues it is sometimes convenient to deal with cubic cells. This means that the governing equation of a given problem is evaluated in its discrete form only in nodal points of the grid.

Such an approach to domain discretization leads to a well-known 'staircase' problem of curvature representation. A significant drawback of a regular mesh is the fact that the whole domain of interest is covered by cells of the same size, and cannot be refined where needed, i.e. in corners, fillets, etc., and coarse in other regions of the model, without a special treatment.

On the other hand, the FDM produces a number of relatively simple algebraic equations which can be processed very efficiently by a parallel computing unit. Moreover, due to the same equation structure for each grid point and regular grid, each processor is calculating a similar thread which reduces the computational time significantly.

2.2.2   Finite Element Method

The finite element method (FEM) is a method for solving differential equations by utilization of 'subdomains', called elements, in which it is assumed that the solution type is known. Then a structure is discretized by using elements, composing a mesh, which represents the investigated domain. Although the most intense development of the method is credited to the 1940s and later, the first ideas of FEM date back to the nineteenth century. The first concept of substitution of a general solution to a continuum body by a set of primitive geometrical shapes with assumed solution types was proposed by Kirsch (1868).

There are a few formulations of FEM. The two main groups of approaches are based on: (a) variational method; and (b) weighted residuals methods (Bathe 1996; Segerlind 1984; Zienkiewicz and Taylor 2000). Both provide a means for the solution of mechanical systems.

The variational method makes use of a functional, a kind of integral formulation, that may be used, by invoking the stationary condition, for finding equilibrium. In the case of mechanical systems the mechanical energy of a system may be the functional. Then, the stationary condition corresponds to the minimum of mechanical energy. Therefore it is possible, knowing only the mechanical energy of a system, to find the equilibrium configuration. This provides a versatile and general tool for analysing mechanical systems.

The methods of weighted residuals operate directly on differential formulations for a given problem. Substituting any function that is not the true solution to the differential equation produces an error, called a residual. In such a case, changing the trial solution and evaluating the residual enables finding the closest possible solution to the differential equation. Of course, the smallest possible residual (equal to zero in the limit), provides the best solution. Moreover, the residual is weighted and the product is integrated, to improve the solution process and to impose additional constraints within the domain. Because of different weighting strategies, various types of weighted residual methods may be distinguished (Bathe 1996):

- Collocation method
- Subdomain method
- Galerkin's method
- Least squares method

FEM is well suited for problems involving domains of an arbitrary shape. The 'staircase' problem is eliminated by appropriate mesh construction and refinement, since there are no limitations on the nodes positions (apart from the accuracy considerations). The method is even more attractive due to its ability to simulate not only mechanical systems, but also thermal, electrical and magnetic fields. It is also very popular and available as commercial as well as freeware software.

The drawbacks of FEM are related to the long computational time, especially when high spatial and temporal resolution are required, as is the case for high frequency waves. This is due to the fact that small element sizes and time steps must be used for accuracy.

2.2.3   Spectral Element Method

The name 'spectral element method' (SEM) is used for two known numerical methods, the first after Beskos (Narayanan and Beskos 1978) and the second after Patera (Patera 1984). These approaches are frequently confused. The latter refers to a spectral – spatial approach. In his first work on this subject, Patera (1984) proposed a subparametric approach for standard FEM. This means that the high-order polynomials are used for field variable approximation while the geometry is described by a low order polynomial. He introduced high-order Lagrange interpolants along with the Gauss – Lobatto – Legendre integration rule. This leads to an exactly diagonal mass matrix which naturally allows for efficient explicit time integration. The approach proposed by Patera may be therefore considered as a higher order FEM. It has been originally used in computational fluid dynamics (Korczak and Patera 1985; Patera 1984), but was quickly employed for other areas such as seismic wave propagation (Komatitsch and Tromp 1999; Komatitsch et al. 2005).

The SEM after Beskos (Narayanan and Beskos 1978) is basically the FEM formulated in the frequency domain. It is a combination of some of the features of FEM, the dynamic stiffness method (DSM) and the spectral analysis method (SAM). Since the FEM for wave propagation has been already mentioned, the details will be given for DSM and SAM only.

2.2.3.1   Dynamic Stiffness Method

One of the key concepts is a specific formulation of the stiffness matrix of a system. This may be considered to be a combination of mass, damping, and stiffness matrix for the conventional FEM. The dynamic stiffness matrix is frequency-dependent and it is constructed for each frequency of the analysis separately. The fundamental idea is to construct the matrix by using the exact dynamic shape functions. Since these vary, depending on the frequency, many of the problems encountered in the classical FEM are eliminated, e.g. mesh refinement for high frequency wave components. The shape functions are constructed from the solutions to the governing differential equation, transformed into the frequency domain, and are exact in the sense that they are directly driven by the analytical solution for a given problem. The detailed description of the method may be found in Banerjee (1997), Gopalakrishnan et al. (2008) and Lee (2009).

2.2.3.2   Spectral Analysis Method

There are two major approaches to the time dependent problems. The first is to analyse a system in the time domain by using direct integration methods. The second approach is to decompose a system into a number of frequency components, solve the problem in the frequency domain, and, if necessary, transform the solution back to the time domain. There are different transforms that can be employed for this purpose, e.g. the Laplace transform, the Fourier transform, and the wavelet transform (Gopalakrishnan et al. 2008). The Fourier transform is the most widely used method due to the availability of fast and robust numerical tools, such as fast Fourier transform (FFT). The system is therefore analysed as a set of systems, each for a particular frequency. This type of analysis is widely covered in Gopalakrishnan et al. (2008) and Lee (2009).

2.2.3.3   Spectral Element Approach

The SEM combines the key advantages of the above methods. First, a system is transformed into the frequency domain. Therefore, one has to deal with a number of 'static' systems since the frequency may be considered as a parameter. Then, the dynamic shape functions are built for each frequency. These are derived, directly from the governing equation, which means that the functions are exact and as accurate as the governing equation (Banerjee 1997). The exact solutions have pair components representing incident and reflected waves. This is of particular importance for the artificial boundary conditions that may be formulated at this level by influencing each part independently, e.g. *throw-off* elements. This constitutes a frequency-dependent dynamic stiffness matrix. Then the solution may be obtained in the frequency domain and processed further by the inverse time – frequency transform, e.g. the iFFT, to produce time domain results. The workflow and employed numerical tools are presented in Figure 2.1.

The key advantages of the SEM over the conventional FEM are:

- High accuracy due to the exact form of the shape functions.
- Minimization of the number of degrees of freedom (DOFs) since one element provides a very accurate solution for a regular part of the domain.
- Relatively low computational cost (to the resolution offered).
- Effective for frequency-dependent problems, since it is formulated in the frequency domain.
- Easy formulation of absorbing boundary conditions.
- Locking-free method.
- Explicit availability of the system transfer functions.

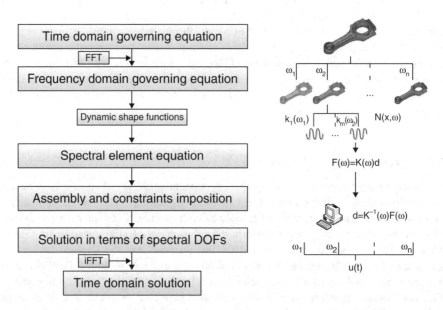

Figure 2.1   The SEM workflow. The main ideas are the Fourier transform and the dynamic shape functions

Despite the aforementioned superiorities, the SEM also has serious drawbacks that limit its application to certain classes of problems. The most important are:

- Exact shape functions are available only for relatively simple systems (this usually poses a problem for multidimensional systems). Approximate methods may be used but the accuracy decreases.
- The method cannot be applied directly to time-variant, nonlinear systems.
- The solution involves inverse time – frequency transformation which may cause numerical problems (e.g. inverse Laplace transform) and deteriorates the solution quality.

2.2.4 Boundary Element Method

The boundary element method (BEM) emerges from the combination of analytical and numerical constituents (Becker 1992). The starting point for the method is a differential equation. Later, the fundamental solution has to be determined. This is done by assuming a concentrated force in the interior of an infinite domain (Kelvin problem) and finding its effect on another point in the domain. Once the solution is found analytically it provides a recipe for the displacement (or traction) in the other point in the domain. The fundamental solution is combined with Betti's theorem, another ingredient of the BEM. Betti's theorem (reciprocal work theorem) states that if there are two stress–strain states in a body in equilibrium, the work done by the first set of stresses on the displacements corresponding to the second state is equal to the work done by the second set of stresses on the displacements corresponding to the first set of strains. The basic form of the theorem is transformed into an equation containing surface terms (surface tractions) and interior terms (volume forces). Subsequently, the fundamental solution is substituted for the known parts of the equation (a selected set of tractions and displacements) resulting in an equation for displacements in the interior, known as the Somigliana identity [after Becker (1992)] or the boundary integral equation (BIE). The first integral equation relating the displacements and tractions at the boundary was proposed by Somigliana in 1886 [after Becker (1992)]. From then up to the 1960s, when the rapid development of numerical methods was observed, there was no progress made regarding the BEM. Then, works by Jaswon (1963) and Symm (1963), and later by Rizzo (1967), and Cruse (1969), provided the fundamentals for the BEM that is known nowadays. In the early works some of the surface integrals were evaluated analytically, thus these are known as semi-direct methods. Later, the theory was extended to involve a fully numerical approach to integration. Summing up, the fundamental solution, defined for the governing differential equation, provides a means for calculating a point force influence on any other point of the domain. The solution is valid for a given differential equation regardless of its geometry. By using Betti's theorem one can relate the known solution (fundamental solution) to the problem at hand.

The BIE enables the calculation of the displacements in the interior of the domain if the displacement and tractions at the boundary and the fundamental solution are known. The discretization of the problem is necessary only for the boundary and is performed by using a standard surface finite element mesh. Since the BIE involves integration over the boundary, standard Gauss integration techniques may be employed.

The BEM provides a powerful tool for solving engineering problems. It may be superior to other numerical approaches for certain classes of problems on account of:

- Ease of model definition – only the surface has to be meshed, stored, and processed.
- High interior field resolution due to the exact, analytical solutions used.
- Lower number of DOFs compared with other methods due to the surface-only description.
- Many numerical approaches developed for other methods, e.g. the FEM, may be directly employed in the BEM.

However, there are also some drawbacks of the method, such as:

- In the case of nonlinear problems the interior must be modelled, which discredits the main advantage of the BEM.
- It is not efficient for plate-like structures due to the low volume-to-surface ratio.
- BEM formulation leads to nonsymmetric and fully populated matrices.

2.2.5   Finite Volume Method

The finite volume method (FVM) utilizes an integrated form of a governing equation. The integral form is obtained directly from the differential equation by integration over a control volume, thus the name finite volume. Since the integral is taken over the volume and there are divergence terms in the governing equation, the Gauss divergence theorem is applied to transform these terms into surface integrals. It is assumed that physical quantities are defined at the grid points, not directly at the control volume surfaces. Thus, the necessary quantities need to be expressed in terms of nodal values. Standard finite difference (FD) approximations are used here. Then the integrated quantities (average quantities) in a control volume are tracked in time. Due to the direct application of the FD formulas, the FVM may be recognized as a FDM applied to a specific form of the differential equation (LeVeque 2002). As mentioned before, the FVM deals with volume averages rather than point values. This enables easy satisfaction of the conservation laws. The details of the derivation and implementation of the FVM may be found in LeVeque (2002) and Versteeg and Malalasekera (2007).

The method has been utilized to develop the elastodynamic finite integration technique (EFIT) (Fellinger et al. 1995; Langenberg and Marklein 2005; Marklein 2002) that has been widely used for elastic wave propagation studies (Rudd et al. 2007; Schubert and Marklein 2002; Schubert et al. 1998).

In the EFIT the Cauchy equation of motion and the deformation rate equation are converted into an integral form and transformed by the Gauss divergence theorem. Then the space is discretized into a grid of cells. The stress and velocity components are evaluated at a particular cell's surfaces and grid points leading to the staggered dual grid model (Fellinger et al. 1995). The equations are then integrated in space and time. The standard central difference operator is applied for time derivatives.

Although the EFIT has been widely used for wave propagation modelling it has certain disadvantages. First, as other typical FDMs it does not allow for explicit treatment of discontinuities. Therefore the defects are represented as grid cells of different materials' properties. Since the spatial discretization is based on the staggered grid concept, the field variables, i.e. stress and velocity components, are split and evaluated at different points in space. This

causes a serious problem when establishing appropriate boundary conditions for a problem
at hand, for instance, the displacement values and stress free boundary conditions cannot be
applied at the same point.

2.2.6   *Other Numerical Methods*

Governing differential equations may be solved by means of many other numerical methods.
Moreover, structural problems may be formulated at the lowest scales by using discrete
rather than continuum approaches. Although not directly applicable to macroscopic problems, these methods are particularly useful for damage modelling, which is an inevitable part
of simulation based testing of the SHM systems.

Other methods than the ones described in detail in the previous sections may be employed
within the continuum mechanics framework. Among the most rapidly growing fields of
numerical methods are the so-called meshless methods (Li and Liu 2007). As the name
implies, no structured mesh describing the domain is analysed. Instead, spatial approximations
are constructed by the use of nodes (particles) only. This eliminates the difficulties encountered in the classical mesh based approaches related to damage initiation and propagation.

Despite the fact that the approximation is based on nodal quantities, it is worth noting
that the underlying governing equation describes a continuum mechanics problem. There is a
number of available meshless methods which may be classified according to the form of the
problem (whether it is defined as a weak or strong form of differential equation):

Strong from based methods

- smoothed particle hydrodynamics (SPH) (Li and Liu 2007)
- vortex method (Bernard 1995)
- meshfree finite difference method (Liszka and Orkisz 1980)

Weak form based methods

- diffuse element method (Breitkopf *et al.* 2000)
- element free Galerkin (Belytschko *et al.* 1996)
- reproducing kernel particle method (Chen *et al.* 1996)
- h-p cloud method (Duarte and Oden 1996)
- partition of unity method (Babuska and Melenk 1996)
- meshless local Petrov – Galerkin (Atluri and Zhu 1998)
- finite point method (Onate *et al.* 1996)
- finite sphere method (De and Bathe 2000)
- free mesh method (Yagawa and Yamada 1996)
- moving particle finite element method (Hao *et al.* 2002)
- natural element method (Sukumar *et al.* 1998)
- reproducing kernel element method (Liu *et al.* 2004)

It is beyond the scope of this book to provide a detailed description of each method. Further
details may be found in Belytschko *et al.* (1996), Fish (2009), Li and Liu (2007), Li *et al.*
(2001) and Lucy (1977). Nevertheless, it is important to mention that the fundamental idea

behind the meshless approximation (it is also claimed to be the fundamental idea for other continuum based approaches) is the so-called partition of unity (Babuska and Melenk 1996).

The partition of unity requires that the interpolating functions possess a property that at any point of the domain their contributions sum to one. This is also a property of the FEM shape functions and other interpolation schemes. Such an approach allows for construction of an interpolation method that decomposes a given function into a set of interpolating parameters. Meshless approximation is constructed from a set of nodes that are not related by any topological relation. Thus, interpolation functions are constructed dynamically and based on the current nodal neighbourhood.

Although meshless methods are particularly profitable in large deformation problems (Chen et al. 1998), they have been successfully employed for wave propagation (Gao et al. 2007; Moosavi et al. 2011; Shobeyri and Afshar 2012).

Another group of particle methods are those which are used for solving discrete problems. It has been already mentioned that discrete models (e.g. atomistic models) are of particular importance when dealing with multiscale modelling. The continuum based approach to materials modelling breaks down when the length scale of the system is decreased. Then, matter cannot be represented as a continuum but rather as a set of atoms (or molecules) that governs its response. In the discrete methods, although they are based on a set of particles, it is assumed that the distribution of parameters is confined to nodal points, i.e. there is no continuum interpolation of quantities in space. Instead, the mutual interactions between particles are described by other forms of direct interactions, e.g. force fields. The discrete methods are widely used for a range of problems starting from the solar system, through granular media, to atomic systems (Allen and Tildesley 1989; Rapaport 2004). The molecular dynamics (MD) method and its modifications play the most prominent role in this field. For a MD based method one has to define particle interactions, which may be two-, three- or many-body interactions. These are frequently called force fields. The system's response is then directly tracked by time integration of the governing equations. Newton's laws are used to establish equilibrium conditions. Considering a set of i particles of mass m_i, and a potential Π (describing the interaction of the ith particle with any other particle), one can write a set of equations for the system:

$$m_i \frac{\delta^2 u_i}{\delta t^2} = -\nabla \Pi. \tag{2.3}$$

Then, by applying a time discretization scheme one may calculate displacements u for a particle i. It is worth noting that the motion of a considered particle is related to (possibly) all other particles through potential Π. This reveals an enormous computational burden associated with MD calculations. Thus, MD simulations have been rediscovered recently due to the availability of powerful parallel processing devices such as GPUs.

Over the last decades various numerical approaches have been combined, taking advantage of the specific features of the methods and enhancing their capabilities. So-called hybrid and multiscale modelling emerged in this way, giving new opportunities for analysing the fundamental phenomena and their influence on the macro-scale response of the material. The aforementioned simulation methods give new possibilities for damage initiation and propagation simulation, extending the potential of application of numerical modelling to the improvement of SHM systems. Both hybrid and multiscale modelling are addressed in Section 2.3.

An interesting approach that should be mentioned as an alternative to conventionally employed methods for elastic waves involves cellular automata (Kluska et al. 2012; Leamy 2008).

The main idea is to establish the state of a cell (or an element) depending on the states of its neighbours. Thus, a set of rules defining possible states, transition conditions, and the neighbourhood are necessary. Two of the most commonly used neighbourhoods are the Moore and von Neumann. Given the states of the neighbours at a certain time instant one may, through a set of transition rules, define the state of the considered cell. Transition conditions may be arbitrary or represent certain mathematical relations.

2.2.7 Time Discretization

Time derivatives are present in any dynamic equation of motion contributing inertia or a velocity based term. During the discretization process such terms are treated as force components and thus a sequence of 'static' solutions for subsequent time steps is sought. Since the procedure is crucial for dynamics, and it is nearly identical for all methods described above, time discretization is treated separately. Specific procedures are suitable for a particular class of problems and possess different advantages and disadvantages in terms of mathematical formulation which is important from the development point of view.

The approaches for time integration of time-dependent problems may be divided into two groups, namely direct integration and mode superposition. The first method evaluates the equations for discrete time steps, according to a particular numerical scheme. The second makes use of a transformation of governing equations of motion, decomposing the system at hand into a set of simple systems which are then integrated in time. The solution in the original domain is obtained by an inverse transform.

For structural problems, based on hyperbolic PDEs, a general form of governing equation may be written as:

$$M\ddot{X} + C\dot{X} + KX = F, \tag{2.4}$$

where M is the mass matrix, C is the damping matrix, K is the stiffness matrix, F is a vector of externally applied loads and X is the displacement vector. In order to follow the response of the system (2.4) in time, the governing equation must be integrated in time.

2.2.7.1 Direct Integration

In the direct integration schemes the time history of a system is tracked by evaluating the system's governing equations at particular time instances. Therefore time derivatives need to be substituted by formal expressions that approximate the time evolution of the system. The finite difference formulas are used for this purpose. Despite a large number of FD formulas that approximate derivatives, only some of them are used in practice. This is due to the fact that a FD operator applied to time derivatives along with additional assumptions for the system provides efficient and robust solution techniques. It is also the time integration scheme that conditions the accuracy of the solution. Two approaches may be distinguished here: explicit and implicit time integration.

Explicit Time Integration
In the explicit time integration approach, the governing equation (2.4) is rewritten for the current time step t, while the solution is sought in the next time step $t + \Delta t$:

$$M^t \ddot{X}^t + C^t \dot{X}^t + K^t X^t = F^t. \qquad (2.5)$$

Then the approximation for time derivatives is introduced. While the most widely applicable approach is the central difference method, it will be presented here:

$$\ddot{X}^t = \frac{1}{\Delta t^2} \left(X^{t-\Delta t} - X^t + X^{t+\Delta t} \right) \quad \text{and} \quad \dot{X}^t = \frac{1}{2\Delta t} \left(X^{t+\Delta t} - X^{t-\Delta t} \right). \qquad (2.6)$$

Substituting (2.6) into (2.5) and solving for X in $t + \Delta t$ gives:

$$\left(\frac{1}{\Delta t^2} M + \frac{1}{2\Delta t} C \right) X^{t+\Delta t} = F^t - \left(K - \frac{2}{\Delta t^2} M \right) X^t - \left(\frac{1}{\Delta t^2} M - \frac{1}{2\Delta t} C \right) X^{t-\Delta t}. \qquad (2.7)$$

The solution of Equation (2.7) is very effective under the following conditions:

- damping matrix is neglected
- lumped mass matrix formulation is used

Then, Equation (2.7) may be written as:

$$X^{t+\Delta t} = 2X^t - X^{t-\Delta t} + \Delta t^2 M^{-1} \left(F^t - K^t X^t \right). \qquad (2.8)$$

It is clearly visible that the most computationally consuming task is the inversion of the matrix M. When the mass matrix M is lumped it is purely diagonal. Therefore its inversion is extremely simple:

$$M^{-1} = diag \left(\frac{1}{m_i} \right), \qquad (2.9)$$

which means that the solution of the inverse is computationally effective. The remaining part of the solution consists of simple matrix manipulations and may be processed very fast.

The main disadvantage of the explicit schemes is the conditional stability, i.e. time step Δt must satisfy:

$$\Delta t \le \Delta t_{cr}, \qquad (2.10)$$

which means that for the method to be stable, the time step must be smaller than the critical value. However, this is the widely used method for wave propagation due to its computational efficiency. Since short time steps are required, Equation (2.8) must be evaluated a number of times. This may be carried out efficiently thanks to the lumped mass inversion that does not require matrix operation and the fact that the stiffness matrix does not need to be inverted.

Implicit Time Integration
Opposite to the explicit approach, the implicit one assumes the equilibrium equation for the next time step $t + \Delta t$:

$$M^{t+\Delta t} \ddot{X}^{t+\Delta t} + C^{t+\Delta t} \dot{X}^{t+\Delta t} + K^{t+\Delta t} X^{t+\Delta t} = F^{t+\Delta t}. \qquad (2.11)$$

Depending on the formulas used to approximate time derivatives in Equation (2.11) the following methods may be mentioned:

- The Houbolt method (Houbolt 1950)
- The Wilson ϕ method (Wilson *et al.* 1973)
- The Newmark method (Newmark 1959)

A common feature of these methods, provided that the parameters are typical is that they are unconditionally stable, thus allowing for time steps that are much larger than for the explicit integration methods. On the other hand, the implicit computations require full matrix inversion operations and are therefore much slower which makes them not preferable for wave propagation applications.

2.2.7.2 Mode Superposition

The direct integration approach, especially for the conditionally stable approach, requires many steps for long analysis. Moreover, for the explicit method, the time step is governed by the smallest element period present in the model, which means that even one small element may increase the computation time enormously. This may be computationally expensive.

The governing equation (2.4) may be transformed in order to minimize the number of operations performed during the solution phase. Following (Bathe 1996) displacements X may be transformed using the formula:

$$X = \Phi D, \tag{2.12}$$

where Φ is the matrix with columns which are mass-normalized eigenvectors calculated for the system (2.4) with damping neglected and D is a vector of modal displacements. After substitution of (2.12) into (2.4) we obtain:

$$\ddot{D} + \Phi^T C \Phi \dot{D} + \Omega^2 D = \Phi^T F, \tag{2.13}$$

where Ω is the diagonal matrix containing eigenvalues corresponding to eigenvector matrix Φ.

If the damping is not included in the analysis, the system (2.13) may be decoupled into a set of equations of the form:

$$\ddot{d}_i + \omega_i^2 d_i = f_i. \tag{2.14}$$

Equation (2.14) shows that the system has been decomposed into a set of independent equations to which any direct integration procedure may be applied. The solution in terms of modal displacements also has a particular feature that makes it attractive. Each equation may be integrated separately with the use of different time step values. For the explicit time integration approach, for which the critical time step depends on the element's period, the time step may be adjusted to a particular modal DOF.

2.3 Hybrid and Multiscale Modelling

In the vast majority of practical applications one is concerned with observable results, i.e. the results which can be interpreted, and that influence the macro-scale response, e.g. wave interaction with microcrack, grain boundary, or inclusion, incorporating nonlinear frictional or microplasticity effects. On the other hand, material behaviour on a global scale is inherently driven by the smallest scale phenomena, i.e. atomic interactions. The polymer composite ageing process or dislocation motion and plasticity may be mentioned by way of example (Buehler 2008; Mlyniec and Uhl 2011).

Although it would be ideal, due to the extreme computational complexity of large molecular systems, it is impossible to investigate any globally meaningful structure or component by means of discrete, atomistic simulations. Thus, the modelling methods based on classical differential formulations, most frequently employed for global scale analysis, assume significant simplifications to the material's behaviour and consider solely its macro-scale response. Differential formulations can be extended by using submodels that predict the system's response, incorporating smaller scale effects, providing e.g. constitutive relations or damage evolution laws (Lemaitre and Chaboche 1994; Li and Liu 2007; Tadmor et al. 1996). Such models, known as multiscale models, have been receiving more and more interest from researchers over the last two decades (Fish 2009; Li and Liu 2007).

Two major types of approach may be distinguished in multiscale modelling, namely, the concurrent and the hierarchical (Packo and Uhl 2011). In the first approach the model is a priori decomposed into different scales which are solved simultaneously. In the second, the information is passed from the highest considered scale to the lowest and the other way around, providing parameters based on calculations at lower scales.

Other model types are hybrid models. These can be classified into the group of concurrent multiscale methods, despite the fact that there is only one scale of interest considered. Again, as in the typical concurrent multiscale methods, one faces a 'domain bridging' problem. As an example, an evolving field of nonlinear acoustics can be mentioned (Klepka et al. 2011), where two time scales, high-and low-frequency excitation, interact with the defects of a structure producing nonlinear effects.

From the technical point of view, hybrid and multiscale models couple various modelling methods, e.g. FEM, SEM, FDM, FVM, MD, meshless, etc., which are used to model material behaviour at different scales. Special transition rules that govern the data flow between the models are then used to build a consistent simulation approach. Therefore, a reliable multiscale model consists of two major, equivalently important parts: simulation methods and a coupling scheme. While it is relatively easy to choose a simulation method for a particular problem, the coupling scheme should be carefully built. This requires a consistent mathematical formulation which correctly mimics the physical phenomena involved.

Apart from the classical continuum models, multiscale modelling usually requires discrete-continuum coupling (Aubertin et al. 2010; Karpov et al. 2006; Park et al. 2005; Xiao and Belytschko 2004). This defines a new class of problems that copes with the transition between a continuum material description and a molecular one. The major difficulties were found to be related to the interpretation of stress–strain measures based on displacements of molecules at any point in space (Admal and Tadmor 2010; Branicio and Srolovitz 2009; Dommelen 2001; Liu and Qiu 2008; Zimmerman et al. 2004).

The problem of the coupling of continuum to continuum as well as continuum to discrete, may be considered from two viewpoints. First, from the static viewpoint, the situation is relatively simple. The scale coupling scheme must satisfy certain conditions, but the solution is established for a given set of boundary conditions at a single time instant. This has been addressed and successfully solved by researchers (Burczynski et al. 2007; Mrozek et al. 2007).

The coupling algorithm becomes much more complicated in the case of dynamics, as it is for wave propagation, due to the stability issues. Depending on the scale of the problem and discretization approach, various problems may occur. For dynamic coupling additional inertia terms must be considered. Moreover, the temporal discretization may differ between submodels, causing additional consequences. Different modelling methods applied for subdomains, as well as scale-related issues, may result in dynamical components propagating in one model that cannot be handled by others. A typical example is that of phonons, which cannot be represented in the coarse scale but are present at the atomic scale (Karpov *et al.* 2006; Li and Liu 2007). Since the waves travelling from the fine towards the coarse scale interface will be reflected producing an artificial model response, it is necessary to construct artificial boundary conditions that absorb these waves (Karpov *et al.* 2006). Such problems are typically solved by employing so-called handshaking domains (Xiao and Belytschko 2004).

Various numerical approaches for multiscale simulations have already been developed. The leaders in the field are Belytshko (Xiao and Belytschko 2004; Zhang *et al.* 2007), Hughes (Garikipati and Hughes 2000), Liu (Karpov *et al.* 2006), Geers (Kouznetsova *et al.* 2004), Yip, Ulm and Pellenq (Pellenq *et al.* 2009; Yip 2005). A considerable amount of work in the field of coupling theories for static problems results in algorithms that enable advantage to be taken of discrete atomic-scale computations and global continuum approaches (Buehler 2008; Fish 2009; Tadmor *et al.* 1996). One of the landmark works is that of macroscopic atomic *ab initio* dynamics (MAAD) (Abraham *et al.* 1998), which concurrently couples two atomic scale discrete methods with the classical finite element approach. Further evolution of the method and accounting for thermal degrees of freedom in the continuum domain led to coarse-grained molecular dynamics (CGMD) (Broughton *et al.* 1999). An excellent example is provided by the Cauchy – Born rule applied for the estimation of elastic constants. The procedure of hierarchical coupling was developed as the quasicontinuum method (Tadmor *et al.* 1996).

The ultimate goal of the multiscale and hybrid methods is to improve the solution without a significant increase in the computational burden. Considering the modelling methods applied to the elastic waves based SHM system, the multiscale methods are an opportunity to improve the damage models by incorporating small scale phenomena. The hybrid models enable efficient interdisciplinary modelling of transducers and temperature-dependent systems, as well as enhancing the capabilities of the modelling methods by taking advantage of the key achievements of each formulation.

The main challenge for wave propagation simulation is the compromise between the calculation speed and accuracy. The models that offer relatively short calculation times, e.g. finite difference methods, fail when dealing with curved geometries, unless the grid size is small enough. On the other hand, small element size increases model size, memory requirements and simulation time. The methods that are devoted to arbitrary shaped domains, such as the FEM, require a considerable computational effort to obtain the solution. Thus, the hybrid approaches, enabling efficient simulation of a domain with curved boundary, have been of particular interest to researchers in recent years (Asensio *et al.* 2007; Chavent and Roberts 1989; Ducellier and Aochi 2009; Eizhong Dai and Nassar 2000; Filoux *et al.* 2009; Galis *et al.* 2008; Karpov *et al.* 2006; Ma *et al.* 2004; Moczo *et al.* 1997; Monorchio *et al.* 2004; Rylander and Bondeson 2000, 2002; Soares 2011; Vachiratienchai *et al.* 2010; Venkatarayalu *et al.* 2004; Wolf *et al.* 2008; Wu and Itoh 1997; Xiao and Belytschko 2004; Zhu *et al.* 2011, 2012).

Considering the coupling schemes, the following cases may be distinguished (where advantages and disadvantages are indicated by plus and minus, respectively):

- Interfacial coupling (coupling through a common surface)
  + Simulation domains are completely decoupled.
  − Boundary conditions (BCs) are hard to establish and implement.
  − BCs must be updated continously.
  − BCs must fit exactly the interface contour.
- Handshake domain (coupling through a partly overlapping region)
  + BCs may be defined independent of the interface contour.
  − Governing equations should account for the overlapping domain (energy).
- Overlaping region (coupling through fully overlapping domain)
  + No problem with BCs directions.
  − Localized model change.

For the interfacial coupling there is no overlapping region. Therefore there are no doubled DOFs in the model. Thus it is the most beneficial approach in terms of model size. However, the boundaries that are coupled should be exactly matched. A coupling scheme should be applied at the boundary and follow its changes. From this point of view this is much more complicated since the shape of the interface may be complex and it should be tracked by both coupled domains. As mentioned before, while considering dynamic coupling, the problem becomes more complicated. Such solutions for static problems may be found in Burczynski *et al.* (2007) and Mrozek *et al.* (2007).

The handshaking zone based models are relatively easy to implement. These models normally smear model properties to match the domains. They may be easily applied to irregular regions, and what is more, the boundaries of the domains may be of different shapes. However, an overlapping region doubles DOFs, which may be highly ineffective. Furthermore, the overlapping zone should be accounted for in the energy considerations of the system to make the whole model consistent. A model based on a handshake approach is presented in Xiao and Belytschko (2004).

A fully overlapping region may be also employed as a coupling scheme. In this approach it is assumed that the solution in a certain region of the model may be decomposed into two components, each associated with a different numerical model. Apart from the models, the additional numerical techniques are used to eliminate phenomena that cannot be handled in both domains. Two of the most prominent works in this field are by Chen *et al.* (1996) and Karpov *et al.* (2006).

A general difficulty encountered in the above coupling schemes is the definition of the constitutive behaviour for the whole model, i.e. the domains and the coupling area. This problem may be considered from two main viewpoints: theory and numerics.

From the theoretical point of view the domain should be recognized as a whole. This means that the equilibrium should be satisfied at every point of the body, regardless of the numerical method applied. Moreover, even for the coupling domain the equilibrium must be satisfied and be consistent with the remaining part of the model, which is sometimes violated in numerical models.

The numerical models employed for the analysis of a given body as well as the coupling strategy should enable the simulation in a way that the domain transition is invisible for the body, as is the case during the standard modelling approaches with the single numerical method. Therefore, the first requirement for the coupling method is that it should enable a consistent connection of different numerical approaches. This very important requirement should also have a strong theoretical background which would ensure that the equilibrium at a particular point of the coupling area is satisfied. Another crucial issue is that the coupling strategy must be stable. It may happen that although the numerical methods are stable when considered separately, their coupling is not. This is especially important in the case of conditionally stable schemes which are sensitive to any introduced disturbance. A less severe consequence of the connection of two explicit numerical approaches is that the overall stability criterion for the model is more restrictive than the criterion governing separate methods.

The requirements for an efficient coupling method are:

- providing an equilibrium that is consistent with the constituent approaches
- stability
- stability criterion for the coupled model does not decrease the efficiency
- minimal number of additional DOFs that are added by coupling
- firm theoretical background
- flexibility

Considering the above requirements for the hybrid method, an efficient and reliable coupling strategy for the FE-FD methods has been developed. However, before proceeding to the method's description, a brief overview of the FD based, efficient method for wave propagation modelling and its implementation will be given.

2.4 The LISA Method

Among various numerical methods presented in this chapter only a few are suitable for efficient simulation of wave propagation considering the demands of high spatio-temporal resolution of the analysis. One of the most widely known and used for this purpose is the FDM. Although it has certain disadvantages, such as regular grids resulting in the 'staircase' problem, it enables very efficient parallelization. Moreover, the algorithm which has been rediscovered, as presented in Packo *et al.* (2012), possesses many other advantages over the standard FDMs. The algorithm and its derivation are presented below.

The Local Interaction Simulation Approach (LISA) is a technique that was proposed in physics (Delsanto *et al.* 1992, 1994, 1997) for wave propagation in complex media with sharp impedance changes. The LISA algorithm was originally designed to be used with a connection machine (a super computer with thousands of parallel processors) and therefore it is well suited for parallel processing. Although the following derivation describes the 2D algorithm, the simulation framework developed for wave propagation in structures allows for fully three-dimensional (3D) simulations.

The LISA can be used for wave propagation in any heterogeneous material of arbitrary shape and complexity. The method discretizes the structure under investigation into a grid of cells. The material properties are assumed to be constant within each cell but may differ between cells. A standard explicit time discretization is employed for time marching. The algorithm can be derived from the elastodynamic wave equation for elastic, isotropic and homogenous media given as (Harris 2004):

$$(\lambda + \mu)\nabla\nabla \cdot W + \mu\nabla^2 W = \rho W_{,tt}, \tag{2.15}$$

where λ and μ are Lame constants, ρ is the material density and W is the vector of particle displacements. For simplicity, the 2D wave propagation case is presented. When the antiplane shear waves are ignored, Equation (2.15) can be simplified as:

$$\rho w_{k,tt} = (\lambda + \mu)w_{l,lk} + \mu w_{k,ll} \text{ and } k = 1, 2; \; l = 1, 2, \tag{2.16}$$

where the comma preceding the subscript denotes differentiation with respect to that variable. The constants λ and μ are the Lame constants for the material given as:

$$\lambda = \frac{Ev}{(1 + v)(1 - 2v)} \text{ and } \mu = \frac{E}{2(1 + v)}, \tag{2.17}$$

where E is the Young's modulus and v is Poisson's ratio.

The above equation can be rewritten in a matrix form as:

$$AW_{,11} + BW_{,22} + CW_{,12} = \rho W_{,tt}, \tag{2.18}$$

where the relevant matrices can be expressed as:

$$A = \begin{bmatrix} \lambda + 2\mu & 0 \\ 0 & \mu \end{bmatrix}; B = \begin{bmatrix} \mu & 0 \\ 0 & \lambda + 2\mu \end{bmatrix}; C = \begin{bmatrix} 0 & \lambda + \mu \\ \lambda + \mu & 0 \end{bmatrix}; W = \begin{bmatrix} w_1 \\ w_2 \end{bmatrix}. \tag{2.19}$$

For the 2D LISA wave propagation simulation, the structure is discretized into cells, as shown in Figure 2.2.

Consider A, B, C, D and F are matrices containing material data, and τ is the 2D stress tensor. The junction of the four cells defines the nodal point P. The second time derivatives across the four cells are required to converge towards a common value Ω at point P. This ensures that if the cell displacements are continuous at P for the two initial times $t = 0$ and $t = 1$, they will remain continuous for all later times. A finite difference scheme is used to calculate spatial derivatives in the four surrounding cells to P. This results in four equations in eight unknown quantities, thus the remaining four equations are needed.

Imposing continuity of the stress tensor τ at the point P and using further finite difference formulae gives four additional equations in the unknown quantities, thus allowing these unknown spatial first derivatives to be solved.

Applying the second-order accurate central difference formula for the time derivative gives the following explicit formula to calculate the displacement components in the next time step:

$$u^{t+\Delta t} = 2u - u^{t-\Delta t} + \frac{\Delta t^2}{\rho \Delta x^2}[\sigma_5 u_5 + \sigma_7 u_7 + \mu_6 u_6 + \mu_8 u_8 - 2(\sigma + \mu)u - \frac{1}{4}\sum_{k=1}^{4}(-1)^k v_k v$$

$$- \frac{1}{4}\sum_{k=1}^{4}(-1)^k v_k v_k + \frac{1}{2}[(g_4 - g_1)v_5 + (g_1 - g_2)v_6 + (g_2 - g_3)v_7 + (g_3 - g_4)v_8]]$$

$$\tag{2.20}$$

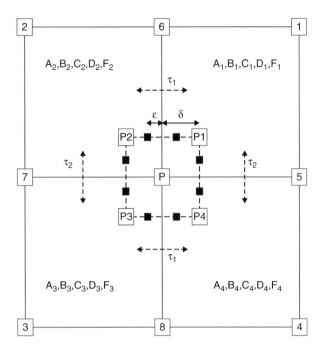

Figure 2.2 Intersection of four LISA cells for the 2D case. Each cell contributes to the common node P, thus the equilibrium conditions are considered separately and matched by using stress continuity relations

$$v^{t+\Delta t} = 2v - v^{t-\Delta t} + \frac{\Delta t^2}{\rho \Delta x^2}[\mu_5 v_5 + \sigma_6 v_6 + \mu_7 v_7 + \sigma_8 v_8 - 2(\sigma + \mu)v - \frac{1}{4}\sum_{k=1}^{4}(-1)^k v_k u$$

$$-\frac{1}{4}\sum_{k=1}^{4}(-1)^k v_k u_k - \frac{1}{2}[(g_4 - g_1)u_5 + (g_1 - g_2)u_6 + (g_2 - g_3)u_7 + (g_3 - g_4)u_8]],$$

$$(2.21)$$

where $u = w_1$, $v = w_2$, Δx is the grid spacing (assumed equal for x and y directions) and Δt is the time step.

Equations (2.20) and (2.21) are the principal displacement equations of LISA in two dimensions. A more detailed derivation can be found in Delsanto *et al.* (1994).

It is important to note that the local interaction nature of boundaries in the model is one of the major advantages of the LISA algorithm when used for wave propagation. The sharp interface model (SIM) is used to average physical properties at the interface grid points which represent intersections of the four elementary cells. In other words, cells are treated as discontinuous and displacements and stresses are matched at interface grid points. The SIM allows for a more physical and unambiguous treatment of interface discontinuities for different layers of material than the typical FD schemes. The classical FD algorithms require parameters smoothing across material interfaces, which depends on the applied scheme, so therefore are

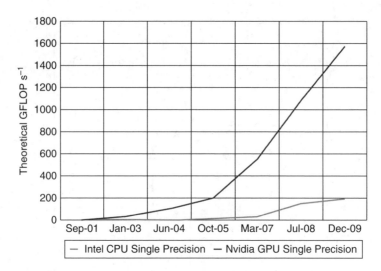

Figure 2.3 GPU versus CPU performance. A comparison of computing capabilities of CPUs and GPUs in terms of number of floating point operations per second

not very accurate for sharp interfaces of high impedance mismatch. The SIM leads to more accurate results when wave propagation problems in complex media with complex boundaries are studied. The algorithm has been successfully used for the Lamb wave propagation investigations for damage detection (Ambrozinski *et al.* 2010; Delsanto *et al.* 1994, 1997; Lee and Staszewski 2003a,b, 2007a,b,c; Packo *et al.* 2011).

2.4.1 GPU Implementation

Very fast evolution and recent developments in GPUs have demonstrated that new architectures available in low-cost graphics cards, called CUDA (Computer Unified Device Architecture), can be efficiently used for numerical simulations. The CUDA technology has been used to develop extremely efficient and accurate modelling tools in medicine, finance, seismology and computer aided design, as shown in Krawezik and Poole (2009), Roberts *et al.* (2011), Seitchik *et al.* (2009) and Stone *et al.* (2007).

The CUDA technology was introduced a few years ago by Nvidia in GPUs for graphics acceleration. The current Nvidia GPUs – widely available in graphics cards of high-end desktop PCs and laptops – have several computing multiprocessors, consisting of hundreds of cores, and capable of running thousands of light-weight threads in parallel. This technology, when combined with an on-board high-bandwidth memory, exceeds the performance and computing power of desktop CPUs. Following the data published by Nvidia (Nvidia Corporation, 2010), Figure 2.3 illustrates the computational power of the CUDA. The data show that the computational capability of modern graphics cards is much larger than the similar power offered by CPUs; recent developments of GPUs make them at least 8–10 times faster than CPUs.

It was realized very quickly that the computational power of such units can be employed very efficiently for numerical simulation of difficult scientific problems. When numerical tasks can be decomposed into many relatively simple sub-tasks, the parallel computing architecture developed by Nvidia can be used to speed up the computations. The CUDA has been used for example in numerical simulation applications related to medicine, where it was used to accelerate the postprocessing of ultrasonographic images (Techniscan), biology for the first simulations of virus behaviour in water at the University of Illinois in Urbana-Champaign (Nanoscale Molecular Dynamics), finance (risk analysis), seismology (exploration of mineral reserves; SeismicCity) and engineering, where it is used for graphics acceleration in CAD applications and speeding up numerical simulations, e.g. finite element codes (ANSYS), matrix solvers (Accelerware), fluid dynamics (Autodesk Moldflow) and others (Krawezik and Poole 2009; Roberts *et al.* 2011; Seitchik *et al.* 2009; Stone *et al.* 2007).

2.4.2 Developed GPU-Based LISA Software Package

The LISA algorithm has been implemented using GPUs. As already mentioned, GPUs offer much more computing power than the CPUs for a lower price. Moreover, a standard GPU is equipped with more than 400 cores and an on-board fast memory that make it a suitable solution for parallel processing tasks. Considering the LISA algorithm, and the requirements that have been set during derivation of the method it is better suited for GPU calculations than CPU for the following reasons:

- Iteration equations may be qualified as so-called 'light weight threads' that do not require a single processor to be very powerful.
- Direct integration of computing units and fast memory enable extremely efficient calculations.
- Equations have been optimized in terms of minimization of variables, and thus memory consumption.
- Variables that can be calculated from other basic quantities are calculated directly in the equation. This is because Nvidia graphics cards are optimized to process 'multiply and add' operations.

The remaining part of this section provides an overview of the framework that has been developed and implemented. This encompasses analysis definition strategy, analysis management and postprocessing capabilities.

The analysis is defined through a text file containing the model geometry, material properties, actuators, sensors, excitation signals and basic simulation parameters. Thus, the model may be easily modified and managed either by a direct input file modification or by an external program. Auxiliary tools have been developed to support the analysis definition and results postprocessing. First, the model geometry may be read from a standard FE mesh. Thus the model used for FE calculations can be imported and the results compared. Furthermore, the developed coupling procedures enable the performance of accurate and efficient simulations of piezoelectric transducers, thermomechanical problems and the improvement of the solution for curved boundaries. Postprocessing tools allow for a convenient visualization of the results.

The analysis process may be managed either interactively through a set of MATLAB® functions, or as a standard solver that reads the input file and produces the results. Thus the simulation framework provides a flexible solution that is ready-to-use or may be employed

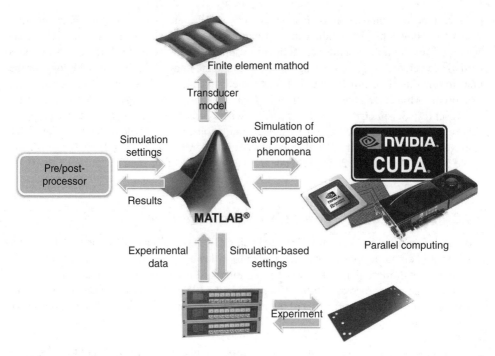

Figure 2.4   The general scheme of the simulation framework based on the LISA method. The LISA is supplemented by the coupled FE solver enabling the transducer modelling. The system is integrated with measurement hardware enabling a comparison of experimental and simulation results as well as defining simulation based settings in experiments

as a core solver for various applications. A dedicated solution for phased arrays design has been developed (Packo *et al.* 2010). The software has been also integrated with measurement hardware allowing for direct comparison of the results between physical and numerical experiments, as well as application of simulation based settings in measurements. The structure of the simulation framework is presented in Figure 2.4.

2.4.3   *cuLISA3D Solver's Performance*

The performance of the developed solver is superior to other implementations of the same algorithm as well as other numerical methods applied to a wave propagation problem. The results obtained for different hardware platforms are presented in this section. The code optimization was performed twofold. In the first step the prototypical MATLAB® code was created. Then it was optimized in terms of memory consumption and preliminary profiled for calculation speed. Then, based on the MATLAB® code, a C++ implementation on GPU with CUDA was developed. A few variants have been investigated, and the best in terms of memory consumption/calculation speed has been chosen. The results presented in this section were obtained for the final version of the code that minimizes the number of variables involved, maximizing the number of 'multiply-and-add' operations. In this way some parameters in the governing equations are calculated by the GPU internally.

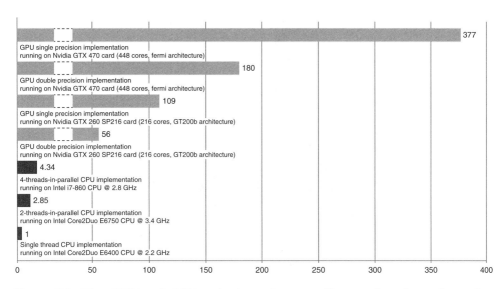

Figure 2.5 The GPU-based LISA solver's performance. The speed-up factor for various implementations (Bielak *et al.* 2010)

In order to test the performance of the solver a simple model has been calculated using various graphics cards and standard CPUs. Moreover, for the GPU cases the double and single precision models have been used. In all the CPU implementations the results are presented for double precision. Figure 2.5 summarizes the speed-up factors obtained. As a reference, a single thread-CPU result is used.

It is clearly visible that even for the fastest CPU and the slowest GPU the difference in computational time is large.

Subsequently, various model sizes were simulated and the average number of time steps per second was measured. The results are presented in Figure 2.6. Again, the GPU based calculations are superior to the CPU ones. Apart from the calculation speed it is worth noting that the GPUs are much cheaper than the CPUs, thus the simulations may be performed on a standard PC. The speed-up factors for currently available GPUs are of the order of a few hundreds. This will further increase due to the development of the hardware architecture. The software architecture is fully flexible and adjusts to the hardware detected. Thus, multiple-GPU calculations may be performed with a linear speed-up.

2.5 Coupling Scheme

The GPU based LISA solver developed within the MONIT project enables extremely efficient wave propagation modelling. Since the simulations may be performed using multiple-GPU workstations, the models consisting of nearly 2 billion DOFs are simulated on a standard desktop PC in minutes. However, the general disadvantage of the regular grid based methods, the 'staircase' problem, deteriorates the solution for curve-shaped domains. Thus, a combination of the efficiency of the developed method and the flexibility and versatility of the FEM in terms of modelling of arbitrary geometries, enhances the capabilities of the simulation

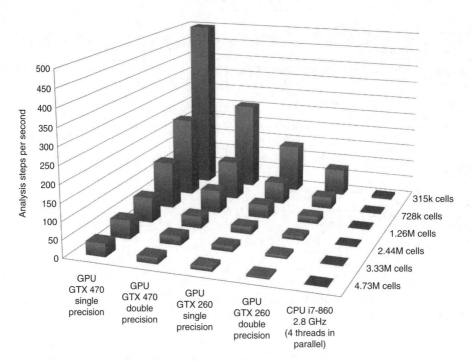

Figure 2.6 The GPU-based LISA solver's performance. The number of analysis steps per second for various model sizes and implementations (Bielak *et al.* 2010)

framework for virtual testing of the SHM systems. Such a hybrid approach enables an efficient and accurate simulation of wave propagation in structures of arbitrary shape and its interaction with defects. Since the nonlinear effects may be easily treated in the FEM, the hybrid approach is capable of simulating the phenomena accompanying the defect's response to the disturbance, e.g. microplasticity, friction, thermoelastic effects, etc.

The most obvious approach to coupling different numerical methods is by direct coupling. This requires formulation of both methods in the same, consistent mathematical form and establishing coupling conditions by direct matrix aggregation. However, numerical methods are frequently based on different theoretical concepts which makes it difficult, or even impossible, to arrive at a consistent formulation for both. Furthermore, such direct coupling is usually provided with no physical justification. It also implicitly assumes direct access to all numerical codes employed for the solution, which is not always the case. This is of huge practical importance since there is a wide variety of commercially available CAE programs that are speed-optimized and have a broad range of functionalities. Direct equations based coupling usually does not allow for utilization of these programs and is therefore highly ineffective.

The coupling method proposed in this section makes use of the fundamental equilibrium conditions for a body. The body is in equilibrium at a particular time instant (i.e. dynamic equilibrium) when at every point of the body the sum of forces is zero. According to the general governing equilibrium equation these forces may be due to internal stresses, thermal

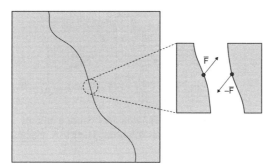

Figure 2.7 Equilibrium conditions for a body. The imaginary interface within a body and a close-up of the force equilibrium conditions for each part of the domain

expansion, external forces or tractions or inertial forces. The equivalent equilibrium condition for a body may be approached from the energy point of view by stating that the body is in equilibrium when the gradient of the potential energy with respect to the generalized coordinates is equal to zero. Figure 2.7 presents the equilibrium conditions for a body.

However, from the theoretical point of view the force equilibrium for a body must be amended by compatibility conditions. Therefore the two conditions that should be enforced at every point on the interface are:

- displacement compatibility
- stress vector compatibility

To enforce the above compatibility conditions during a hybrid analysis with different numerical methods is not a trivial task. Therefore the following discussion is limited to the case of the FE-FD coupling, for the models with consistent time stepping and compatible meshes. However, it may be easily extended to other numerical methods, meshes and time marching schemes.

There are many ways to implement the above conditions. However, a few restrictions should be considered in order for the coupling scheme to be reliable, flexible and, what is the most severe limitation, stable. Reliability of the scheme is ensured by the theoretical framework. What is understood by flexibility is the ability to employ the scheme for commercially available CAE programs without interrupting the code of the solvers. The coupling procedure is therefore implemented as a flexible tool that may operate externally to the core solvers and uses standard inputs and outputs that are available for the CAE program. One of the most severe limitations to the coupling method is the stability of the whole model. Since the developed coupling method concerns coupling of the explicit time integration methods, it is considered to be the most difficult case. This is because the explicit schemes are only conditionally stable. The last important point is the calculation speed. If the method is efficient, it will not limit the performance of the methods employed for the analysis. Otherwise the whole hybrid model will not be superior to the solutions obtained individually by the methods. Thus, the amount of data passed between the models should be minimized, and the coupling routine should be implemented efficiently.

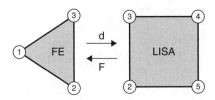

Figure 2.8 Data flow between the hybrid model domains. Since explicitly available at the boundary, the FE displacements (*d*) are treated as the master solutions. The LISA response on the FE domain is the reaction force (*F*) to the displacements

First, the displacement matching at the boundary must be carried out. This may be realized simply by imposing the calculated displacements for a boundary from one domain to the other. However, this may be done twofold, depending on whether the data are passed from the FE to FD or the other way around. Since for the FD the quantities that are to be calculated are available for internal grid points only, i.e. in order to calculate the displacement for a particular grid point one needs to use the data from at least adjacent grid points, or apply special boundary conditions to mimic the material behaviour at the boundary, the natural choice is to use the FE as a 'master' solution. The displacement values for boundary nodes are naturally calculated without a special boundary treatment.

There is another consequence to construction of artificial boundary conditions for boundary nodes in FD. It leads to serious changes in the structure of the equation that is to be solved, in comparison with an ordinary grid point. This is highly inefficient in terms of parallel processing because of different threads being processed. As already mentioned, the GPU processing is suited to operate on threads that have identical structure. Otherwise a significant efficiency drop is observed. Thus, such an approach should be avoided to maintain high efficiency of the method.

The displacement matching by direct constraint imposition has certain consequences. Since a boundary point, that is common for both domains, has certain FE and FD related mass properties, additional inertial force must be considered. What is more, the elastic response of the FD domain must be taken into account in the FE solution, Since the excitation of the FD domain is by FE displacements, additional self-coupling terms occur in the FE governing equation. This is of particular importance for the stability of the coupling scheme.

The elastic response of the FD domain is the stress-related coupling condition, defined above. Stress coupling for a compatible mesh and consistent time stepping may be recognized as an elastic force matching at a boundary node. The reaction force calculation for the FD model due to the FE excitation is calculated using a FD stiffness matrix. The general idea of the data flow between the domains of the hybrid model is presented in Figure 2.8.

Up to this point, the general coupling conditions above and the background for their calculation have been provided. The displacements are explicitly available in both considered methods, while the stresses (or forces) are calculated for the boundary of the domain. These quantities will be used to enforce the equilibrium for the coupled hybrid model. The detailed data flow for the hybrid model is inferred from the direct coupling procedure, assuming that both numerical approaches use the explicit time integration technique. Considering the

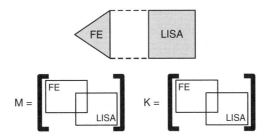

Figure 2.9 Matrix based coupling approach. The matrices are assembled in a standard manner, as for mixed-mesh FE models

elastic wave propagation, this is the most common integration technique. It is also the most challenging in terms of stability requirements.

An immediate result of direct coupling of a set of equations are additional components in stiffness and mass matrices. In the case of compatible meshes, as in the above discussion, the assembly procedure may be carried out in exactly the same manner as for the standard FE mesh. The idea of FE-FD coupling, considered as a mixed-mesh FE approach, is presented in Figure 2.9.

It is clearly visible how the two systems are coupled. If there are no external forces applied the system is in equilibrium when the forces for a given nodal point are zero. The above conditions must hold for any time step. While analysing the set of equations with coupled matrices presented in Figure 2.9, three distinct cases are distinguished: the LISA domain, the FE domain and the coupled FE–LISA domain. For the first two domains, the standard equations are recovered. However, for the coupled model additional coupling terms are introduced. Thus, instead of direct matrix coupling an alternative procedure is proposed that may be employed for commercially available solvers.

Considering the above restrictions, the following coupling methodology is proposed:

- Displacements are calculated from the FE model. This means that in terms of displacements the FE is the 'master' model. The displacements calculated from the FE domain are naturally calculated, thus there is no need for artificial boundary conditions.
- Displacements from FE provide the excitation for the LISA model. A set of transducers at the FD domain boundary is applied. Displacements are calculated by the FE model. Since the LISA governing equation cannot be applied to the boundary nodes without any modifications, this approach enables utilization of a standard FD code. This ensures the displacement compatibility.
- Having the displacements exerted by the FE domain onto the LISA part of the model, and knowing the consistent stiffness and mass matrices, the reaction force is calculated.

Since the equilibrium conditions must be established for both models in order to calculate the solution in the further time steps, attention must be paid to the proper force – displacement equilibrium calculation in time. In other words, in order to calculate the next time step by the FEM, the model must already know that it is coupled to the FD domain instead of the traction-free boundary condition. The case of the FD response to the excitation is similar. Otherwise the excitations between the models would cause nonphysical reflections and lead to instabilities, as the explicit scheme is very sensitive to any disturbances.

Once the coupling forces have been established, the data flow in time should be investigated. It is assumed that both equations, i.e. FE and LISA, are solved in time by the explicit procedure. This is a more demanding task since the explicit approach is only conditionally stable and sensitive to any parameter changes.

The explicit time integration is described in detail in Section 2.2.7. However, an additional difficulty is to couple two different methods in time, retaining the consistent time marching. In other words, the goal is to couple two explicit approaches by solving them subsequently and obtaining the solution for both coupled methods at the same time (to avoid the leap-frog scheme). Since the explicit procedure is based on the governing equation written for the current time step, only current and previous time steps are required to calculate the response for the next time step. Thus it may be assumed that the solutions at present and previous time instances are known, and satisfy the boundary conditions. It is also inferred from the governing equation at time t that the boundary conditions are prescribed for the current time step.

To arrange the data flow in time the direct coupling analogy is utilized. The starting point is the general governing equation for dynamics without damping:

$$M\ddot{X} + KX = F, \tag{2.22}$$

where M is the global (hybrid) mass matrix, K is the global stiffness matrix, X is the displacement vector for all DOFs in the model, and F is the external force vector. The FE DOFs are all DOFs including the coupled boundary nodes. In the explicit time integration approach the governing equation (2.22) is written for the present time step t:

$$M^t\ddot{X}^t + K^tX^t = F^t. \tag{2.23}$$

Decomposing global hybrid matrices and substituting into (2.23), and subdividing external forces into those related to the LISA and FE parts, the following equation for the hybrid model is obtained:

$$(M^{FE,t} + M^{LISA,t})\ddot{X}^t + (K^{FE,t} + K^{LISA,t})X^t = F^{FE,t} + F^{LISA,t}, \tag{2.24}$$

where the summation indicates the assembly process. Three distinct model domains are distinguished:

- the FE domain, where the standard FEM is applied;
- the LISA domain, where the standard LISA method is applied;
- the coupled FE – LISA domain, where the coupling procedure is employed.

The first two cases were treated in detail and the calculations follow the standard procedure. For a fully coupled region the mass and stiffness matrices may be decomposed as presented in Figure 2.10.

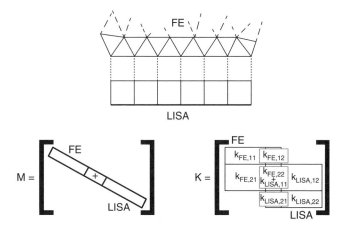

Figure 2.10 Hybrid FE – LISA domain coupling. For the explicit scheme the mass matrices are assumed to be lumped, thus the coupling introduces a single DOF-dependent force. Coupling forces due to the elastic effects involve three force components

Assuming that the boundary nodes are FE-driven, and the external force vector is zero, it follows from the Equation (2.24) and Figure (2.10):

$$M^{LISA,t}\ddot{X}^{FE,t} + M^{FE,t}\ddot{X}^{FE,t} + K_{21}^{FE,t}X^{FE,t} + K_{22}^{FE,t}X^{FE,t}$$
$$+ K_{11}^{LISA,t}X^{FE,t} + K_{12}^{LISA,t}X^{LISA,t} = 0, \tag{2.25}$$

where

$$M^{LISA,t}\ddot{X}^{FE,t} = F_{inertial}^{LISA,t} \tag{2.26}$$

$$K_{11}^{LISA,t}X^{FE,t} + K_{12}^{LISA,t}X^{LISA,t} = F_{elastic}^{LISA,t}. \tag{2.27}$$

Thus Equation (2.25) may be rewritten as:

$$M^{FE,t}\ddot{X}^{FE,t} + K_{21}^{FE,t}X^{FE,t} + K_{22}^{FE,t}X^{FE,t} + F_{inertial}^{LISA,t} + F_{elastic}^{LISA,t} = 0. \tag{2.28}$$

The first three terms in Equation (2.28) recover the standard FE approach. The only difference is the additional coupling forces $F_{inertial}^{LISA,t}$ and $F_{elastic}^{LISA,t}$ that represent the response of the LISA domain to the applied displacement excitation.

Several remarks should be made here:

- Due to the coupling equation's structure in (2.28), a standard FE based software package may be used. The only changes to be made are the coupling forces $F_{inertial}^{LISA,t}$ and $F_{elastic}^{LISA,t}$.
- It is clearly seen that the $F_{elastic}^{LISA,t}$ term consists of two components, i.e. $K_{11}^{LISA,t}X^{FE,t}$ and $K_{12}^{LISA,t}X^{LISA,t}$. For a standard formulation this would be a standard force response of the

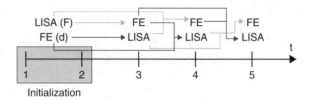

Initialization

Figure 2.11 Time marching scheme for the explicit time integration. After the initialization phase the models are calculated simultaneously on the basis of the solutions from FE and LISA domains in two previous time steps

LISA cells due to prescribed displacement. However, it is worth noting that the term $K_{11}^{LISA,t}X^{FE,t}$ used to calculate the LISA response is in fact driven by the FE displacement. Thus, a kind of self-response of the FE part of the hybrid model is introduced.

- The above point has severe consequences for the stability of the FE model, since additional LISA-related components increase the stiffness of the whole model.

The procedure proposed above defines the coupling force components for the model. It also provides a time schedule for force application. At the beginning of the analysis it is assumed that no special starting procedure is employed. This means that the two time step solutions are already known for the calculation of the system's response at time $t + \Delta t$. In the first analysis step the FE model is solved. At this stage the knowledge of the reaction forces from the LISA domain from the previous time step is required. If not specified, zero initial reaction forces are applied. After calculation of the FE displacements, the LISA iteration is performed. For the first iteration the displacements from the previous FE time step are needed. Again, zero displacements are assumed as an initial condition. After performing the FE and LISA iterations the solutions for $t + \Delta t$ are known. These solutions are utilized in the next time step as an input. This enables advancing the solution by the same time step and obtaining the displacements for the same time instance, ensuring the coupling conditions at all times. The time marching scheme is presented in Figure 2.11.

The method has been fully implemented for the LISA and commercial FE solver. In terms of efficiency, the proposed approach minimizes the amount of data that is transferred between the models (i.e. proportional to the number of nodes at the interface times the dimensionality of the problem). Force calculation for the LISA domain may be carried out using a GPU since it involves a number of simple operations that are repeated for each cell.

Possible applications of the proposed method cover a broad range of problems encountered in classical modelling approaches, such as efficient simulation of arbitrary shaped domains. Due to the FE solver's capabilities, it may be used to solve multiphysics problems such as piezo-electro-mechanical coupling, wave interaction with defects incorporating frictional, thermoelastic effects and microplasticity. Examples of multiphysical applications are presented in Figure 2.12.

As a summary, the main features of the proposed method are listed below:

- firm theoretical background
- flexibility in terms of methods that can be coupled

Figure 2.12 Challenges in numerical simulation of elastic wave propagation – complex geometries, multiphysical phenomena, e.g. friction, microplasticity, thermoelasticity, piezo-electo-mechanical coupling, etc.

- generality
- efficiency
- consistent time marching for both methods
- flexibility in terms of commercial programs that can be used

2.6 Damage Modelling

Reliable damage modelling is a key issue when considering numerical simulations as a means to predict structures' performance and lifetime, and provide a way to simulate and improve the SHM and NDT/NDE systems. According to the level of consideration, damage modelling methods can be classified analogously to multiscale modelling methods (Packo and Uhl 2011) as single and multiscale models (meso, micro and nano).

In the first group two main approaches are generally considered, i.e. theoretical and phenomenological. With the theoretical approach one considers macroscopic crack behaviour in terms of a fracture mechanics framework, considering stress intensity factors (Leski 2007; Wnuk 1990) or energy release rate (Wnuk 1990; Xie and Biggers Jr 2007), which can be used as damage indicators. Structural damage can also be caused by materials' degradation or other phenomena causing material to lose its load-carrying capabilities. Such deterioration is usually related to microscale phenomena and it is generally covered by phenomenological models on a macro scale (Just and Behrens 2004). A more precise treatment of this type of damage is through multiscale modelling. In the macro models group a cohesive zones approach (Ural *et al.* 2009; Xie and Biggers Jr 2006) may be also mentioned. Its constitutive behaviour is described by a traction versus separation relation. Damage is therefore modelled as an evolution of interface bonding parameters finally leading to surface separation.

Multiscale damage modelling concerns the evaluation of microstructural changes and upscaling their influence to the macro scale (Gurson 1977; Krajcinovic 1996; Needleman and Tvergaard 1984). The implementation of models at lower scales (meso, micro and nano scales) and data flow between subsequent scales during analysis, allows for more precise results.

While there is quite a lot of phenomenological models operating at a macro scale, these rarely have any strong theoretical background. There are a few models proposed for micro scale and upscaled (Lemaitre 1996). Modelling of structures' damage at the lowest scales (molecules and atoms) is a new field of interest and a reliable and useful scheme has not been fully discovered yet.

Prediction of damage initiation and propagation is inevitable for the simulation of health monitoring systems. Possible damage-prone regions of a structure as well as damage types, and possible configurations may be inferred from a numerical analysis. This provides valuable information for a system's improvement and optimization in terms of likely damage configurations and maximizing the possibility of its detection. Subsequently, the damaged structure is subjected to a virtual health monitoring examination. Then, various effects influencing the wave response must be taken into account, depending on the damage type, size and localization. First, the wave scattering from defect must be considered. If the damage is a crack, delamination, void, inclusion or other type of the material's discontinuity then it may be modelled by direct, local change of material properties. However, since a sharp discontinuity is considered, the accuracy of the damage model relies on the numerical model's ability to simulate the wave interaction with the interface of high impedance mismatch. Since the vast majority of elastic waves based monitoring strategies analyse the reflection from defects, this is the most important phenomenon that must be simulated accurately. What is more, nonlinear effects may arise during wave propagation and interaction with damage, such as friction on the crack faces, temperature-related wavefield changes, e.g. inhomogeneous temperature field or thermoelasticity, plasticity at the crack tip, etc. These may be simulated by using contact algorithms (Martowicz et al. 2012), coupled thermomechanical analysis considering the influence of temperature field on the material properties, thermoelastic coupling based on strain calculations (Packo et al. 2010), nonlinear material models, etc. The broad range of phenomena that should be taken into account, especially nonlinear effects, requires significantly larger computational resources. However, these complex structure's responses are frequently confined to the damage region that is relatively small compared with the structure dimensions and thus may be efficiently analysed by hybrid and multiscale approaches.

2.7 Absorbing Boundary Conditions for Wave Propagation

The importance of absorbing boundary conditions (ABCs) may be considered from at least two viewpoints, i.e. as a way of model size reduction and as a part of the hybrid and multiscale schemes.

Numerical modelling of high frequency wave propagation in structures is an extremely time consuming task. Since the waves, especially GWs, propagate over long distances, a large part of the component or even the whole component must be included in the analysis. It may be proved mathematically that waves decay with distance r from the source as $(r^{\frac{d-1}{2}})^{-1}$, where d is the number of dimensions. This means that the waves may be acquired relatively far from the

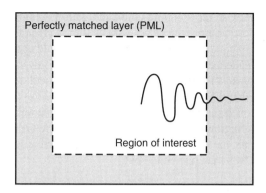

Figure 2.13   The perfectly matched layer. An object is surrounded by a material layer that absorbs waves

source. Considering small element size, discretization of the whole structure produces models of very large sizes. It is therefore of particular interest to provide a means for model size reduction. This may be achieved by truncating the domain to a particular part that is important from the application point of view. This cannot be realized by simply removing parts of the model, since this causes wave reflections from new boundaries and artificial distortions of the response. Thus the absorbing boundaries allow for the model extension to an infinite domain. A similar problem frequently occurs in hybrid and multiscale modelling for the transition region. It often happens that the ranges of wavelengths that may be represented in the domains are different, causing reflections and deteriorating the solution accuracy.

ABCs are realized in many different ways by establishing a formula that is applied directly at the interface and mimics the behaviour of the removed part of the domain, or by means of an additional boundary layer. First attempts at ABCs were based on viscous damping layers (Lysmer and Kuhlemeyer 1969). An alternative approach has been proposed in (Clayton and Engquist 1977; Reynolds 1978), and is based on application of special boundary conditions derived from one way wave equation. Recently, utilizing the periodic properties of the structure has been used to construct absorbing conditions for multiscale models (Karpov *et al.* 2006). In the approach called perfectly matched layer (PML) proposed by Berenger (1994) the object is surrounded by an absorbing layer that eliminates the waves reflection from the boundary, which is similar to the viscous ABC. However, the superiority of the PML lies in the more effective damping properties of the damping layer. The general idea of PML is presented in Figure 2.13.

The robustness and efficiency of the ABCs determine their applicability to wave propagation problems.

A boundary condition based on a mathematical formulation, although more efficient in terms of additional degrees of freedom added to the model, relies fully on the accuracy of the formula used to damp out the waves. Thus the damping properties are often angle and frequency dependent and significant reflections occur. Moreover, a special boundary treatment is needed which decreases the solution efficiency when using parallel processing units.

For an ABC utilizing a damping layer the truncated model is surrounded by an artificial layer of material that absorbs any disturbance that enters it. The additional material layer should not be large, otherwise the solution, compared with the complete model, will not be profitable. Thus the goal is to absorb the waves with as short a layer as possible. It should also be noted that the boundary condition provided by the layer should absorb *any* disturbance, i.e. wave, regardless of its incidence angle and frequency. Once the governing equation within the absorbing layer is the same as for the remaining part of the model, there is no difference for the processing unit, whether the ABC region or the actual model is calculated.

Due to the aforementioned superiorities of absorbing layers over other boundary conditions, the PML based damping regions have been implemented in the GPU based LISA package. Among the damping layers, the PML approach is the most effective and robust. The following advantages over other ABCs may be pointed out:

- the solution in the undamped region is *not* changed;
- the material is reflectionless (since it is the same material) (only numerical reflections due to discretization may occur);
- since it is reflectionless it does not affect the solution for 'not damped' domain (no reflections are generated in the backward direction).

The main difficulty in obtaining the solution to the problem is in establishing a reliable and stable numerical solution to the modified governing equation. The solution constructed for the LISA method allowed high efficiency to be maintained, minimal possible increase in memory consumption, and highly efficient damping properties to be obtained. The ratio of incident to reflected wave for a model/ABC interface is well below 1%, even for high damping properties mismatch.

2.8   Conclusions

Various numerical methods, presented in the previous sections, may be applied to simulate a system's dynamics. Their usefulness relies on the compromise between computational effort and accuracy offered. Thus, to assess the potential for wave propagation simulation for each method, their advantages and disadvantages have been provided.

Elastic waves that propagate in structures possess many features that define particular requirements for a simulation method. At first, elastic waves propagate in solids in two modes, namely longitudinal and transversal, with drastically different speeds. For a structural material like steel or aluminium the ratio of longitudinal to transversal wave speed is approximately 2. Thus, a numerical method should enable simulation of the two components and their interaction, with possibly the same accuracy. Secondly, waves interact with any interface of different materials, which may be understood as an interface of material layers, e.g. in composite materials, or structure/air boundary, e.g. wave reflection from an external boundary of an internal defect such as a crack or delamination, etc. In the case of external boundaries and defects, the interface separates two media of drastically different impedances which is an additional challenge for the simulation method.

Subsequently, waves propagating in structural materials are very fast. Depending on the particular material properties wave may propagate with speed of 6000 m s^{-1} and more. Therefore high temporal resolution, i.e. small time stepping, is required. Finally, since excitation frequencies are of the order of tens or hundreds of kilohertz, the generated elastic waves are relatively short. Dealing with wavelengths of the order of millimetres requires high spatial discretization of a domain, thus small element sizes. Rough estimates, as well as practical considerations suggest 8–20 elements per wavelength.

The aforementioned requirements for the simulation method define rather strict conditions that imply serious numerical problems. However, by combining different modelling approaches it was possible to build an extremely efficient simulation framework for the simulation of elastic waves propagation. Calculations are performed using GPUs enabling the simulations to be speeded up a few hundred times. Modelling capabilities are enhanced by coupling of the LISA based GPU solver to the commercially available FE package, maintaining high efficiency.

Most of the results for wave propagation simulations presented in this book have been obtained using the developed GPU-based software package.

References

Abraham F, Brodbeck D, Rudge WE, Broughton JQ, Schneider D, Land B, Lifka D, Gerner J, Rosenkrantz M, Skovira J and Gao H 1998 Ab initio dynamics of rapid fracture. *Modelling and Simulation in Materials Science and Engineering* **6**, 639–670.

Abramovici F and Alterman Z 1965 Computations pertaining to the problem of propagation of a seismic pulse in a layered solid. *Methods of Computational Physics Academic Press* **4**, 349–379.

Admal NC and Tadmor EB 2010 A unified interpretation of stress in molecular systems. *Journal of Elasticity* **100** (1–2), 86.

Alford RM, Kelly KR and Boore DM 1974 Accuracy of finite-difference modeling of the acoustic wave equation. *Geophysics* **39**(6), 834–842.

Allen P and Tildesley DJ 1989 *Computer Simulation of Liquids*. Oxford Science Publications. Clarendon Press.

Alterman Z and Karal FC 1968 Propagation of elastic waves in layered media by finite difference methods. *Bulletin of the Seismological Society of America* **58**, 367–398.

Ambrozinski L, Packo P, Stepinski T and Uhl T 2010 Ultrasonic guided waves based method for SHM simulations and an experimental test. *Fifth World Conference on Structural Control and Monitoring*. Tokyo, Japan. http://www.bridge.t.u-tokyo.ac.jp/WCSCM5/ (last accessed 7 January 2013).

Asensio M, Ayuso B and Sangalli G 2007 Coupling stabilized finite element methods with finite difference time integration for advection – diffusion – reaction problems. *Computer Methods in Applied Mechanics and Engineering* **196**(3536), 3475–3491.

Atluri SN and Zhu TL 1998 A new meshless local Petrov-Galerkin (MLPG) approach to nonlinear problems in computer modeling and simulation. *Computer Modeling and Simulation in Engineering* **3**, 187–196.

Aubertin P, Rethore J and de Borst R 2010 A coupled molecular dynamics and extended finite element method for dynamic crack propagation. *International Journal for Numerical Methods in Engineering* **81**(1), 72–88.

Babuska I and Melenk JM 1996 The partition of unity method. *International Journal of Numerical Methods in Engineering* **40**, 727–758.

Banerjee JR 1997 Dynamic stiffness formulation for structural elements: A general approach. *Computers and Structures* **63**(1), 101–103.

Bathe KJ 1996 *Finite Element Procedures*. Prentice Hall.

Becker AA 1992 *The Boundary Element Method in Engineering: A Complete Course*. McGraw-Hill.

Belytschko T, Krongauz Y, Organ D, Fleming M and Krysl P 1996 Meshless methods: An overview and recent developments. *Computer Methods in Applied Mechanics and Engineering* **139**(14), 3–47.

Berenger JP 1994 A perfectly matched layer for the absorption of electromagnetic waves. *Journal of Computational Physics* **114**, 185–200.

Bernard PS 1995 A deterministic vortex sheet method for boundary layer flow. *Journal of Computational Physics* **117**(1), 132–145.

Bielak T, Packo P, Uhl T and Barszcz T 2010 Performance analysis of the cuLISA3d solver. Technical report, AGH University of Science and Technology, EC Systems.

Boller C, Chang F and Fujino Y 2009 *Encyclopedia of Structural Health Monitoring*. John Wiley & Sons, Ltd.

Branicio PS and Srolovitz DJ 2009 Local stress calculation in simulations of multicomponent systems. *Journal of Computational Physics* **228**(22), 8467–8479.

Breitkopf P, Touzot G and Villon P 2000 Double grid diffuse collocation method. *Computational Mechanics* **25**, 199–206.

Broughton JQ, Abraham FF, Bernstein N and Kaxiras E 1999 Concurrent coupling of length scales: Methodology and application. *Physical Review B* **60**(4), 2391–2403.

Buehler MJ 2008 *Atomistic Modeling of Materials Failure*. Springer.

Burczynski T, Mrozek A and Kus W 2007 A computational continuum-discrete model of materials. *Bulletin of the Polish Academy of Sciences* **55**, 85–89.

Chavent G and Roberts J 1989 A unified physical presentation of mixed, mixed-hybrid finite elements and usual finite differences for the determination of velocities in waterflow problems. Research report RR-1107, INRIA.

Chen SJ, Oliveira Lima Roque CM, Pan C and Button ST 1998 Analysis of metal forming process based on meshless method. *Journal of Materials Processing Technology* **8081**, 642–646.

Chen SJ, Pan C, Wu CT and Liu WK 1996 Reproducing kernel particle methods for large deformation analysis of nonlinear structures. *Computer Methods in Applied Mechanics and Engineering* **139**(14), 195–227.

Clayton R and Engquist B 1977 Absorbing boundary conditions for acoustic and elastic wave equations. *Bulletin of the Seismological Society of America* **67**, 1529–1540.

Collatz L 1966 *The Numerical Treatment of Differential Equations*. Springer-Verlag.

Collatz L 1986 *Differential Equations: An Introduction with Applications*. John Wiley & Sons, Ltd.

Courant R, Friedrichs K and Lewy H 1928 Uber die partiellen differenzengleichungen der mathematischen physik. *Mathematische Annalen* **100**, 32–74.

Crank J and Nicolson P 1947 A practical method for numerical integration of solution of partial differential equations of heat-conduction type. *Mathematical Proceedings of the Cambridge Philosophical Society* **43**, 5067.

Cruse TA 1969 Numerical solutions in three dimensional elastostatics. *International Journal of Solids and Structures* **5**, 1259–1274.

De S and Bathe KJ 2000 The method of finite spheres. *Computational Mechanics* **25**, 329–345.

Delsanto P, Schechter R and Mignogna R 1997 Connection machine simulation of ultrasonic wave propagation in materials. III: The three-dimensional case. *Wave Motion* **26**(4), 329–339.

Delsanto P, Schechter RS, Chaskelis HH, Mignogna RB and Kline R 1994 Connection machine simulation of ultrasonic wave propagation in materials. II: The two-dimensional case. *Wave Motion* **20**(4), 295–314.

Delsanto P, Whitcombe T, Chaskelis H and Mignogna R 1992 Connection machine simulation of ultrasonic wave propagation in materials. I: The one-dimensional case. *Wave Motion* **16**(1), 65–80.

Dommelen LV 2001 Physical interpretation of the virial stress. *The Royal Society* **317**(1988), 1–7.

Duarte CA and Oden JT 1996 An *h-p* adaptive method using clouds. *Computer Methods in Applied Mechanics and Engineering* **139**, 237–262.

Ducellier A and Aochi H 2009 A coupled 2D finite-elements-finite-difference method for seismic wave propagation: Case of a semi-circular canyon. *9th International Conference on Mathematical and Numerical Aspects of Waves Propagation*. Paris, France. http://hal.archives-ouvertes.fr/docs/00/57/42/38/PDF/preprint.pdf.

Eizhong Dai W and Nassar R 2000 A hybrid finite element-finite difference method for solving three-dimensional heat transport equations in a double-layered thin film with microscale thickness. *Numerical Heat Transfer, Part A: Applications* **38**(6), 573–588.

Fellinger P, Marklein R, Langenberg KJ and Klaholz S 1995 Numerical modeling of elastic wave propagation and scattering with EFIT-elastodynamic finite integration technique. *Wave Motion* **21**(1), 47–66.

Filoux E, Levassort F, Calle S, Certon D and Lethiecq M 2009 Single-element ultrasonic transducer modeling using a hybrid FD-PTSD method. *Ultrasonics* **49**(8), 611–614.

Fish J (ed.) 2009 *Multiscale Methods: Bridging the Scales in Science and Engineering*. Oxford University Press.

Galis M, Moczo P and Kristek J 2008 A 3-D hybrid finite-difference-finite-element viscoelastic modelling of seismic wave motion. *Geophysical Journal International* **175**(1), 153–184.

Gao L, Liu K and Liu Y 2007 A meshless method for stress-wave propagation in anisotropic and cracked media. *International Journal of Engineering Science* **45**(28), 601–616.

Garikipati K and Hughes TJR 2000 A variational multiscale approach to strain localization formulation for multidimensional problems. *Computer Methods in Applied Mechanics and Engineering* **188**(13), 39–60.

Gopalakrishnan S, Chakraborty A and Mahapatra DR 2008 *Spectral Finite Element Method*. Springer.

Gurson AL 1977 Continuum theory of ductile rupture by void nucleation and growth, 1. Yield criteria and flow rules for porous ductile media. *Journal of Engineering Materials and Technology-Transactions of the ASME* **99**(1), 2–15.

Hao S, Park H and Liu W 2002 Moving particle finite element method. *International Journal for Numerical Methods in Engineering* **53**, 1937–1958.

Harris JG 2004 *Linear Elastic Waves*. Cambridge Texts in Applied Mathematics. Cambridge University Press.

Houbolt JC 1950 A recurrence matrix solution for the dynamic response of elastic aircraft. *Journal of the Aeronautical Sciences* **17**, 540–550.

Jaswon MA 1963 Integral equation methods in potential theory. I. *Proceedings of the Royal Society of London. Series A. Mathematical and Physical Sciences* **275**(1360), 23–32.

Just H and Behrens A 2004 Investigations on the use of damage indicating criteria in FE simulations of cold and semi-hot bulk forging operations. *ASME Conference Proceedings* **2004**(47020), 453–461.

Karpov EG, Yu H, Park HS, Liu WK, Wang QJ and Qian D 2006 Multiscale boundary conditions in crystalline solids: Theory and application to nanoindentation. *International Journal of Solids and Structures* **43**(21), 6359–6379.

Kirsch EG 1868 Die fundamentalgleichungen der theorie der elasticitt fester krper, hergeleitet aus der betrachtung eines systems von punkten, welche durch elastische streben verbunden sind. *Zeitschrift des Vereines deutscher Ingenieure* **12**, 481–487, 553–570, 631–638.

Klepka A, Jenal RB, Staszewski WJ and Uhl T 2011 Fatigue crack detection using nonlinear acoustics and laser vibrometry. *ICSV 18: 18th International Congress on Sound and Vibration*, p. 235.

Kluska P, Staszewski WJ, Leamy MJ and Uhl T 2012 Lamb wave propagation modeling using cellular automata. *Structural Health Monitoring 2012: Proceedings of the Sixth European Workshop* **2**, 1446–1454.

Komatitsch D and Tromp J 1999 Introduction to the spectral element method for three-dimensional seismic wave propagation. *Geophysical Journal International* **139**(3), 806–822.

Komatitsch D, Tsuboi S and Tromp J 2005 The spectral-element method in seismology. In *Seismic Earth: Array Analysis of Broadband Seismograms*. (eds Levander A and Nolet G), vol. 157 of *Geophysical Monograph*. American Geophysical Union, pp. 205–228.

Korczak K and Patera A 1985 A spectral element method applied to unsteady flows at moderate Reynolds number. In *Ninth International Conference on Numerical Methods in Fluid Dynamics* (eds Soubbaramayer and Boujot J), vol. 218 of *Lecture Notes in Physics*. Springer, pp. 314–319.

Kouznetsova V, Geers MGD and Brekelmans WAM 2004 Multi-scale second-order computational homogenization of multi-phase materials: a nested finite element solution strategy. *Computer Methods in Applied Mechanics and Engineering* **193**(4851), 5525–5550.

Krajcinovic D 1996 *Damage Mechanics*. North-Holland Series in Applied Mathematics and Mechanics. Elsevier.

Krawezik G and Poole G 2009 Accelerating the ANSYS direct sparse solver with GPUs. *Symposium on Application Accelerators in High Performance Computing, SAAHPC*.

Langenberg KJ and Marklein R 2005 Transient elastic waves applied to nondestructive testing of transversely isotropic lossless materials: a coordinate-free approach. *Wave Motion* **41**(3), 247–261.

Lax PD 1961 On the stability of difference approximations to solutions of hyperbolic equations with variable coefficients. *Communications on Pure and Applied Mathematics* **14**(3), 497–520.

Lax PD and Wendroff B 1960 Systems of conservation laws. *Communications on Pure and Applied Mathematics* **13**, 217–237.

Leamy MJ 2008 Application of cellular automata modeling to seismic elastodynamics. *International Journal of Solids and Structures* **45**(17), 4835–4849.

Lee BC and Staszewski WJ 2003a Modelling of Lamb waves for damage detection in metallic structures: Part I. Wave propagation. *Smart Materials and Structures* **12**(5), 804.

Lee BC and Staszewski WJ 2003b Modelling of Lamb waves for damage detection in metallic structures: Part II. Wave interactions with damage. *Smart Materials and Structures* **12**(5), 815.

Lee BC and Staszewski WJ 2007a Lamb wave propagation modelling for damage detection: I. Two-dimensional analysis. *Smart Materials and Structures* **16**(2), 249.

Lee BC and Staszewski WJ 2007b Lamb wave propagation modelling for damage detection: II. Damage monitoring strategy. *Smart Materials and Structures* **16**(2), 260.

Lee BC and Staszewski WJ 2007c Sensor location studies for damage detection with Lamb waves. *Smart Materials and Structures* **16**(2), 399.

Lee U 2009 *Spectral Element Method in Structural Dynamics*. John Wiley & Sons, Ltd.

Lemaitre J 1996 *A Course on Damage Mechanics*. Springer.

Lemaitre J and Chaboche J 1994 *Mechanics of Solid Materials*. Cambridge University Press.

Leski A 2007 Implementation of the virtual crack closure technique in engineering FE calculations. *Finite Elements in Analysis and Design* **43**(3), 261–268.

LeVeque R 2002 *Finite Volume Methods for Hyperbolic Problems*. Cambridge Texts in Applied Mathematics. Cambridge University Press.

Li S and Liu WK 2007 *Meshfree Particle Methods*. Springer.

Li S, Qian D, Liu W and Belytschko T 2001 A meshfree contact-detection algorithm. *Computer Methods in Applied Mechanics and Engineering* **190**, 3271–3292.

Liszka T and Orkisz J 1980 The finite difference method at arbitrary irregular grids and its application in applied mechanics. *Computers & Structures* **11**(12), 83–95.

Liu B and Qiu X 2008 How to compute the atomic stress objectively? *Journal of Computational and Theoretical Nanoscience* **6**(5), 1081–1089.

Liu WK, Han W, Lu H, Li S and Cao J 2004 Reproducing kernel element method. Part I: Theoretical formulation. *Computer Methods in Applied Mechanics and Engineering* **193**, 933–951.

Lucy LB 1977 A numerical approach to the testing of the fission hypothesis. *Astronomical Journal*, **82**, 1013–1024.

Lysmer J and Kuhlemeyer RL 1969 Finite dynamic model for infinite media. *Journal of the Engineering Mechanics Division* **95**, 859–878.

Ma S, Archuleta RJ and Liu P 2004 Hybrid modeling of elastic P-SV wave motion: A combined finite-element and staggered-grid finite-difference approach. *Bulletin of the Seismological Society of America* **94**, 1557–1563.

Marklein R 2002 Ultrasonic wave and transducer modeling with the finite integration technique (FIT). *Ultrasonics, IEEE Symposium* **1**, 563–566.

Martowicz A, Packo P, Staszewski WJ and Uhl T 2012 Modelling of nonlinear vibro-acoustic wave interaction in cracked aluminium plates using local interaction simulation approach. *Proceedings of 6th European Congress on Computational Methods in Applied Sciences and Engineering (ECCOMAS 2012)*. University of Vienna, Austria.

Mlyniec A and Uhl T 2011 Modelling and testing of ageing of short fibre reinforced polymer composites. *Proceedings of the Institution of Mechanical Engineers* **226**, 16–31.

Moczo P, Bystricky E, Kristek J, Carcione JM and Bouchon M 1997 Hybrid modeling of P-SV seismic motion at inhomogeneous viscoelastic topographic structures. *Bulletin of the Seismological Society of America* **87**(5), 1305–1323.

Monorchio A, Rubio Bretones A, Mittra R, Manara G and Gmez Martn R 2004 A hybrid time-domain technique that combines the finite element, finite difference and method of moment techniques to solve complex electromagnetic problems. *IEEE Transactions on Antennas and Propagation* **52**, 2666–2674.

Moosavi MR, Delfanian F, Khelil A and Rabczuk T 2011 Orthogonal meshless finite volume method in elastodynamics. *Thin-Walled Structures* **49**(9), 1171–1177.

Mrozek A, Kus W and Burczynski T 2007 Application of the coupled boundary element method with atomic model in the static analysis. *Computer Methods in Materials Science* **7**, 284–288.

Narayanan GV and Beskos DE 1978 Use of dynamic influence coefficients in forced vibration problems with the aid of fast Fourier transform. *Computers and Structures* **9**(2), 145–150.

Needleman A and Tvergaard V 1984 An analysis of ductile rupture in notched bars. *Journal of the Mechanics and Physics of Solids* **32**(6), 461–490.

Newmark NM 1959 A method of computation for structural dynamics. *Journal of Engineering Mechanics* **85**, 67–94.

Nvidia Corporation 2010 *NVIDIA CUDA C Programming Guide*.

Onate E, Idelsohn S, Zienkiewicz OC, Taylor RL and Sacco C 1996 A stabilized finite point method for analysis of fluid mechanics problems. *Computer Methods in Applied Mechanics and Engineering* **139**(14), 315–346.

Packo P, Ambrozinski L and Uhl T 2011 Structure damage modelling for guided waves-based SHM systems testing *Modeling, Simulation and Applied Optimization (ICMSAO)*. Kuala Lumpur, pp. 1–6. http://dx.doi.org/10.1109/ICMSAO.2011.5775618.

Packo P, Bielak T, Spencer AB, Staszewski WJ, Uhl T and Worden K 2012 Lamb wave propagation modelling and simulation using parallel processing architecture and graphical cards. *Smart Materials and Structures* **21**, 075001–1075001–13.

Packo P and Uhl T 2011 Multiscale approach to structure damage modelling. *Journal of Theoretical and Applied Mechanics* **49**(1), 243–264.

Packo P, Uhl T and Staszewski WJ 2010 Coupled thermo-mechanical simulations of Lamb waves propagation in structures with damage. *18th International Congress on Sound and Vibration*. Rio de Janeiro.

Park HS, Karpov EG, Klein PA and Liu WK 2005 Three-dimensional bridging scale analysis of dynamic fracture. *Journal of Computational Physics* **207**(2), 588–609.

Patera AT 1984 A spectral element method for fluid dynamics: Laminar flow in a channel expansion. *Journal of Computational Physics* **54**(3), 468–488.

Pellenq RJM, Kushima A, Shahsavari R, van Vliet KJ, Buehler MJ, Yip S and Ulm FJ 2009 A realistic molecular model of cement hydrates. *Proceedings of the National Academy of Sciences* **106**, 16102–16107.

Raghavan A and Cesnik CES 2007 Review of guided-wave structural health monitoring. *The Shock and Vibration Digest* **39**(2), 91–114.

Rapaport D 2004 *The Art of Molecular Dynamics Simulation*. Cambridge University Press.

Reynolds AC 1978 Boundary conditions for the numerical solution of wave propagation problems. *Geophysics* **43**(6), 1099–1110.

Rizzo FJ 1967 An integral equation approach to boundary value problems of classical elastostatics. *Quarterly of Applied Mathematics* **25**, 83–95.

Roberts M, Jeong WK, Vazquez-Reina A, Unger M, Bischof H, Lichtman J and Pfister H 2011 Neural process reconstruction from sparse user scribbles. *Medical Image Computing and Computer Assisted Intervention (MICCAI '11)*. Springer, pp. 621–628.

Rudd KE, Leonard KR, Bingham JP and Hinders MK 2007 Simulation of guided waves in complex piping geometries using the elastodynamic finite integration technique. *The Journal of the Acoustical Society of America* **121**(3), 1449–1458.

Rylander T and Bondeson A 2000 Stable FEM-FDTD hybrid method for Maxwells equations. *Computer Physics Communications* **125**, 7582.

Rylander T and Bondeson A 2002 Application of stable FEM-FDTD hybrid to scattering problems. *IEEE Transactions on Antennas and Propagation* **50**, 141–144.

Schubert F and Marklein R 2002 Numerical computation of ultrasonic wave propagation in concrete using the elastodynamic finite integration technique (EFIT). *Proceedings of the 2002 IEEE Ultrasonics Symposium*, pp. 799–804.

Schubert F, Peiffer A, Kohler B and Sanderson T 1998 The elastodynamic finite integration technique for waves in cylindrical geometries. *The Journal of the Acoustical Society of America* **104**(5), 2604–2614.

Segerlind LJ 1984 *Applied Finite Element Analysis*. John Wiley & Sons, Ltd.

Seitchik A, Jurick D, Bridge A, Brietzke R, Beeney K, Codd J, Hoxha F, Pignol C and Kessler D 2009 The Tempest Project-addressing challenges in deepwater Gulf of Mexico depth imaging through geologic models and numerical simulation. *The Leading Edge* **28**(5), 546–553.

Shobeyri G and Afshar MH 2012 Corrected discrete least-squares meshless method for simulating free surface flows. *Engineering Analysis with Boundary Elements* **36**(11), 1581–1594.

Soares D 2011 Coupled numerical methods to analyze interacting acoustic-dynamic models by multidomain decomposition techniques. *Mathematical Problems in Engineering* **2011**, 1–28.

Staszewski WJ 2004 Health monitoring using guided ultrasonic waves. In *Advances in Smart Technologies in Structural Engineering*. Computational Methods in Applied Sciences (eds Holnicki-Szulc J and Mota Soares CA). Springer, pp. 117–162.

Staszewski WJ, Boller C and Tomlinson GR (eds) 2004 *Health Monitoring of Aerospace Structures: Smart Sensor Technologies and Signal Processing*. John Wiley & Sons, Ltd.

Stone JE, Phillips JC, Freddolino PL, Hardy DJ, Trabuco LG and Schulten K 2007 Accelerating molecular modeling applications with graphics processors. *Journal of Computational Chemistry* **28**(16), 2618–2640.

Strikwerda J 2004 *Finite Difference Schemes and Partial Differential Equations*, 2nd edn. Society for Industrial and Applied Mathematics.

Su Z and Ye L 2009 *Identification of Damage Using Lamb Waves: From Fundamentals to Applications*. Lecture Notes in Applied and Computational Mechanics. Springer.

Sukumar N, Moran B and Belytschko T 1998 The natural element method in solid mechanics. *International Journal for Numerical Methods in Engineering* **43**, 839–887.

Symm GT 1963 Integral equation methods in potential theory. II. *Proceedings of the Royal Society of London* **A275**, 3346.

Tadmor EB, Ortiz M and Phillips R 1996 Quasicontinuum analysis of defects in solids. *Philosophical Magazine A* **73**(6), 1529–1563.

Twizell EH 1979 An explicit difference method for the wave equation with extended stability range. *BIT Numerical Mathematics* **19**, 378–383.

Ural A, Krishnan VR and Papoulia KD 2009 A cohesive zone model for fatigue crack growth allowing for crack retardation. *International Journal of Solids and Structures* **46**(1112), 2453–2462.

Vachiratienchai C, Boonchaisuk S and Siripunvaraporn W 2010 A hybrid finite differencefinite element method to incorporate topography for 2D direct current (DC) resistivity modeling. *Physics of the Earth and Planetary Interiors* **183**(34), 426–434.

Venkatarayalu NV, Beng GY and Li LW 2004 On the numerical errors in the 2-D FE/FDTD algorithm for different hybridization schemes. *IEEE Microwave and Wireless Components Letters* **4**, 168–170.

Versteeg H and Malalasekera W 2007 *An Introduction to Computational Fluid Dynamics: The Finite Volume Method*. Pearson Education Limited.

Wilson EL, Farhoomand I and Bathe KJ 1973 Nonlinear dynamic analysis of complex structures. *International Journal of Earthquake Engineering and Structural Dynamics* **1**, 241–252.

Wnuk M 1990 *Nonlinear Fracture Mechanics*. Courses and lectures, International Centre for Mechanical Sciences. Springer-Verlag.

Wolf S, Favreau P and Ionescu IR 2008 Hybrid unstructured FEM-FDM modeling of seismic wave propagation. application to dynamic faulting. http://sylvie.wolf.free.fr/WFI_preprint.pdf (last accessed 7 January 2013).

Woolfson MM and Pert GJ 1999 *An Introduction to Computer Simulation*. Oxford University Press.

Wu RB and Itoh T 1997 Hybrid finite-difference time-domain modeling of curved surfaces using tetrahedral edge elements. *IEEE Transactions on Antennas and Propagation* **45**, 1302–1309.

Xiao S and Belytschko T 2004 A bridging domain method for coupling continua with molecular dynamics. *Computer Methods in Applied Mechanics and Engineering* **193**(1720), 1645–1669.

Xie D and Biggers Jr SB 2006 Progressive crack growth analysis using interface element based on the virtual crack closure technique. *Finite Elements in Analysis and Design* **42**(11), 977–984.

Xie D and Biggers Jr SB 2007 Calculation of transient strain energy release rates under impact loading based on the virtual crack closure technique. *International Journal of Impact Engineering* **34**(6), 1047–1060.

Yagawa G and Yamada T 1996 Free mesh method: A new meshless finite element method. *Computational Mechanics* **18**, 383–386.

Yip S 2005 *Handbook of Materials Modeling*. Springer.

Zhang S, Zhu T and Belytschko T 2007 Atomistic and multiscale analyses of brittle fracture in crystal lattices. *Physical Review B* **76**, 094114.

Zhu B, Chen J and Zhong W 2012 A hybrid finite-element/finite-difference method with implicit-explicit time stepping scheme for Maxwell's equations. *International Journal of Numerical Modelling: Electronic Networks, Devices and Fields* **25**(5–6), 607–620.

Zhu B, Chen J, Zhong W and Liu QH 2011 A hybrid FETD-FDTD method with nonconforming meshes. *Communications in Computational Physics* **9**, 828–842.

Zienkiewicz O and Taylor R 2000 *The Finite Element Method: Solid Mechanics*. Butterworth-Heinemann.

Zimmerman JA, Webb III EB, Hoyt JJ, Jones RE, Klein PA and Bammann DJ 2004 Calculation of stress in atomistic simulation. *Modelling and Simulation in Materials Science and Engineering* **12**(4), S319–S332.

3

Model Assisted Probability of Detection in Structural Health Monitoring

Alberto Gallina, Paweł Paćko and Łukasz Ambroziński
Department of Mechatronics and Robotics, Faculty of Mechanical Engineering and Robotics, AGH University of Science and Technology, Poland

3.1 Introduction

Structural Health Monitoring (SHM) systems are generally able to report the presence of a crack (detection problem), the location, extent of the damage (estimation problem) and, in the fully developed cases, the remaining life of the structure (prognosis problem). Statistical tools and machine learning algorithms are widely used to face the detection and estimation problem (Worden and Manson 2007). With regard to the detection of structural damage, and focusing on the issue of the system reliability assessment, the engineer is concerned mainly with four different probabilities (Kabban and Derriso 2011): (i) the probability that the system detects damage when it is not, commonly referred to as probability of false alarm (PFA), (ii) the probability that the system detects damage when it is present, commonly referred to as probability of detection (POD), (iii) the probability that the system fails to detect damage when it is present and (iv) the probability that the system detects no damage when there is none. Since (i) and (iv) sum up to one, as well as (ii) and (iii), only two probabilities need to be assessed to give an insight into the system detection reliability. Typically, POD and PFA are evaluated. Furthermore, PFA and POD are not independent of each other but related through the definition of the system threshold that indicates the system response level above which the structure is considered damaged. In addition to POD and PFA, their inverse conditional probability can also be computed. That is the positive predictive probability (PPP), that quantifies the probability that the damage does exist when the system indicates no damage,

Advanced Structural Damage Detection: From Theory to Engineering Applications, First Edition.
Edited by Tadeusz Stepinski, Tadeusz Uhl and Wieslaw Staszewski.
© 2013 John Wiley & Sons, Ltd. Published 2013 by John Wiley & Sons, Ltd.

and the negative predictive probability (NPP) that estimates the probability that damage does not exist when the system indicates no damage. Using Bayes' rule these probabilities can be estimated from POD and PFA once the likelihood of damage is given. In practice, the likelihood of damage is generally unknown and difficult to calculate with the desired accuracy. Thus, the evaluation of PPP and NPP is usually skipped and only POD and PFA are computed. Nevertheless, the assessment of POD and PFA is not a trivial issue. The POD technique was introduced to study the reliability of systems for nondestructive evaluation (NDE). All statistical procedures behind the POD methods were meant for analysing data from experiments on real structures. However, in the case of SHM systems empirical studies do not suffice to yield reliable POD measurements. Indeed, since the sensors are permanently mounted on the hosting structure, the whole SHM-structure system is unique and cannot be reproduced within adequate precision, due to inherent uncertainties related to the installation of the SHM system on the hosting structure. As a consequence, the amount of experimental tests required for a reliability assessment problem based only on empirical data would be extraordinarily large and, in practice, infeasible. To support POD analyses, computer models have been introduced. Computer models, simulating complex physical phenomena, can replace costly experimental tests with faster numerical calculations. This approach, in the literature referred to as model assisted probability of detection (MAPOD), proved to be promising from the view point of a standard procedure for SHM reliability assessment.

In this context, this chapter intends to highlight the advantages of an enhanced computer modelling technique which makes use of massively parallel processing. Modern computing units are equipped with a large number of processors that may be used to process the data extremely efficiently, enabling the acceleration of numerical analyses. Thus, a number of iterations, required to produce the data for POD analysis, may be performed in a short time using a standard desktop PC. The application of this approach to an ultrasound SHM is presented utilizing a developed simulation framework for wave propagation. A detailed discussion of modelling techniques for elastic wave propagation in structures and the developed solver details are given in Chapter 2.

This chapter is structured in the following way. In Section 3.2 a short review of POD methods is given. Section 3.3 focuses on theoretical aspects of POD and issues related to the implementation of the method. Section 3.4 introduces the MAPOD technique, while the extension of MAPOD to SHM systems is presented in Section 3.5. Then an enhanced MAPOD technique based on GPU parallel processing is presented in Section 3.6.

3.2   Probability of Detection

The advent of damage tolerance decision rules as a way to determine the integrity of a component led to the development of statistical tools for assessing the POD for NDE inspection techniques and systems. The first work started with the National Aeronautics and Space Administration (NASA) space shuttle program in 1969 to determine the largest flaw that could be missed for various NDT methods (Georgiou 2006). The methodology introduced by NASA was soon adopted by the United States Air Force (USAF) that in the early 1970s published a simple approach (Pettit and Hoeppner 1972) where the POD was calculated by averaging the number of detected cracks of specific size over the total number of cracks of the same size in the trial. In the mid 1970s the binomial distribution method was proposed by (Yee *et al.* 1976) for imposing confidence bounds to the POD estimates. These methods were straightforward with no sophisticated statistics and entailed a very large number of tests when determining POD values for different crack sizes. Since these pioneering works, extensive research activity and experimental programs have followed over the course of the years. The analysis of experimental data allowed researchers to reveal common patterns between damage levels of a structure and NDE system responses in many test cases. On the basis of these observations Berens and Hovey (1981) introduced probabilistic models for POD as functions of the crack size only. The proposed models were:

- Lockheed
- Probit
- Log Probit
- Log Odds (linear scale)
- Log Odds (log scale)
- Arcsine

Fitting the models to a set of available data, they pointed out better performances for the Log Odds and Log Probit (also commonly referred to as Log Normal) models. Currently, these two probabilistic models are also the most widely used in POD evaluation. Experimental studies (Berens and Hovey 1983) evidenced higher accuracy of probabilistic models than of the former binomial approaches. However, all studies presented so far were hit/miss analyses. This means that the NDE system response was a discrete signal indicating only whether the crack was present or not. The next evolution of the POD evaluation was given by the so-called signal response analysis where a continuous NDE system response was elaborated. Signal response analysis represents an improvement of the hit/miss analysis in the sense that more information is conveyed from the data signal. Therefore, a smaller number of experiments is expected to achieve a reliable POD assessment. More recently, an enhanced nonparametric statistical technique based on the original binomial approach has been presented in Generazio (2011). This approach tries to overcome inherent pitfalls of the parametric methods that impose a functional form of the POD function. References giving an exhaustive literature review of historical developments and existing applications of POD to NDT can be found e.g. in Wall *et al.* (2009) and Matzkanin and Yolken (2001).

3.3   Theoretical Aspects of POD

As already introduced in Section 3.2, POD analysis may be classified into hit/miss or signal response analysis. The basic theoretical aspects and analytical issues appearing when implementing both approaches are described in the next two subsections.

3.3.1   Hit/Miss Analysis

In a hit/miss approach the outcome signal from the inspection is a binary value 0/1 indicating the fact that the system failed to detect (0) or succeeded in detecting (1) the crack in the test structure. The passage from this discrete outcome to a continuous POD curve is given by assuming a probabilistic model relating the POD to the crack size, commonly denoted by a.

The usual choices in this case are a Log Odds or Log Normal model. In the Log Odds the probability of detection $P(a)$ is modelled as

$$P(a) = \frac{\exp(\alpha + \beta a)}{1 + \exp(\alpha + \beta a)} \tag{3.1}$$

with α and β parameters to estimate controlling the POD curve. Taking the natural logarithm of both sides of Equation (3.1) the following linear relationship is obtained

$$Y(a) = \alpha + \beta X(a) \tag{3.2}$$

with

$$Y(a) = \frac{P(a)}{1 - P(a)} \quad \text{and} \quad X(a) = \ln(a). \tag{3.3}$$

The POD curve is built by calculating the α and β coefficients from experimental data. For this purpose, two approaches are commonly used, the range interval method (RIM) and maximum likelihood estimation (MLE). A description of RIM and MLE for Log Odds models in hit/miss analysis is given in Forsyth and Fahr (1998).

When using the Log Normal probabilistic model the POD is described by

$$P(a) = 1 - Q(z) \tag{3.4}$$

with $Q(z)$ the standard normal survivor function

$$Q(z) = \frac{1}{\sqrt{2\pi}} \int_{-\infty}^{z} e^{-0.5z^2} dz \tag{3.5}$$

and z the standard normal variate related to the crack size a through

$$z = \frac{\ln(a) - \mu}{\sigma}. \tag{3.6}$$

The parameters μ and σ in Equation (3.6) are called *location* and *scale* parameters and play a similar role to α and β in the Log Odds model. Location and scale parameters are calculated via MLE. Also in this case a description of the method is beyond the scope of this book and can be found in Forsyth and Fahr (1998). Here, it seems important to stress that, in general, RIM is a robust algorithm but with poor accuracy, while MLE is more accurate but entails an optimization step. Moreover, the optimization performed in MLE is commonly a local optimization. This requires a starting point that is not too far away from the global optimum. Otherwise, the procedure gets trapped into a local minimum. Thus, the features of RIM and MLE may be combined leading to a robust and accurate procedure for parameter estimates. The procedure is as follows:

- α and β are calculated by RIM.
- The RIM estimates are the starting values of the local optimization for MLE. A simplex algorithm (Nelder and Mead 1965) can be employed for this purpose.

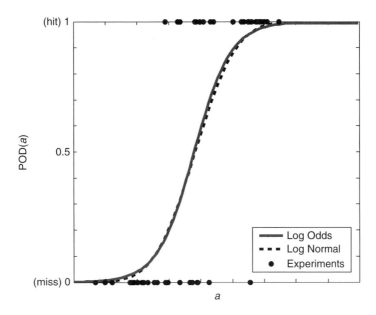

Figure 3.1   Example of hit/miss analysis for Log Odds and Log Normal probabilistic models

- An initial estimate of μ and σ can be obtained from α and β by generating a Log Normal POD curve similar a to Log Odds POD curve by the optimization process.
- The obtained μ and σ values are the initial values of the local optimization for MLE.

An example of the proposed sequential approach applied to a set of randomly generated data is given in Figure 3.1.

3.3.2   Signal Response Analysis

In the signal response approach the outcome from the system is a continuous signal denoted by $\hat{a}$. $\hat{a}$ is usually strongly correlated with the crack size a. The experimental evidence shows that the $\log(a)$–$\log(\hat{a})$ correlation is generally linear. This allows one to write the following linear relationship

$$y = \beta_0 + \beta_1 x + \epsilon, \tag{3.7}$$

where $y = \log(\hat{a})$, $x = \log(a)$ and ϵ is a random error. Equation (3.7) represents a linear function whose coefficients β_0 and β_1 can be evaluated via the ordinary least squares (OLS) method. However, OLS necessitates the random error to be independent and identically distributed (i.i.d.). Furthermore, for POD calculation purposes it is also required that ϵ is normally distributed. Often, these properties are verified, at least partially, and in any case can be always checked. If they are completely wrong, a change of base can be attempted using, for instance, a different probabilistic model, such as one of those listed in Section 3.2.

Assuming here that ϵ is i.i.d. with zero mean and standard deviation δ, from the properties of linear regression $y(x)$ is a normal random variable with mean $\bar{y}(x) = \beta_0 + \beta_1 x$ and standard deviation also δ.

Since due to the noise, the system possibly gives a signal different from zero even in the case of an undamaged structure, there is a need to fix a threshold value $\hat{a}_{th}$ that establishes the boundary between damage and no damage. Thus, the crack is considered to be present in the structure as long as $\hat{a} > \hat{a}_{th}$, that is $y > y_{th}$. The choice of $\hat{a}_{th}$ is a crucial point of the signal response approach and it is usually made on the basis of PFA. This issue will be better explained in Section 3.3.4. Once the threshold is assigned, the POD of a crack of size a can be written in terms of x and y as follows:

$$POD(x) = \int_{y_{th}}^{\infty} \frac{1}{\sqrt{2\pi}\delta} e^{-\frac{[y-(\beta_0+\beta_1 x)]^2}{2\delta^2}} \, dy. \tag{3.8}$$

Introducing the standard normal variate

$$z = \frac{y - (\beta_0 + \beta_1 x)}{\delta} \tag{3.9}$$

and plugging it into Equation (3.8), it follows that

$$POD(x) = \int_{z_{th}}^{\infty} \frac{1}{\sqrt{2\pi}} e^{-\frac{z^2}{2}} \, dz \tag{3.10}$$

with

$$z_{th} = \frac{y_{th} - (\beta_0 + \beta_1 x)}{\delta}. \tag{3.11}$$

After a few simple manipulations (Petrin $et\ al.$ 1993), Equation (3.10) can be written as

$$POD(x) = 1 - Q\left(\frac{x - \mu}{\sigma}\right) \tag{3.12}$$

with Q the standard normal survivor function

$$Q(u) = \frac{1}{\sqrt{2\pi}} \int_{u}^{\infty} e^{-0.5v^2} \, dv \tag{3.13}$$

and μ and σ defined as

$$\mu = \frac{y_{th} - \beta_0}{\beta_1} \quad \text{and} \quad \sigma = \frac{\delta}{\beta_1}. \tag{3.14}$$

Hence, in a response signal analysis the POD curve is controlled by the two parameters μ and σ or, equivalently, by the three inter dependent parameters β_1, β_2 and δ. The analytical representation of POD in the case of signal response analysis is straightforward. Referring to Figure 3.2 the POD at a generic crack size x' is represented by the integral of the Gaussian

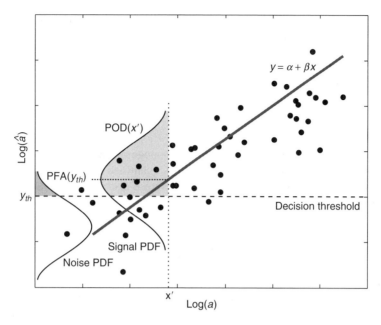

Figure 3.2 Graphical representation of POD for signal response analysis

curve with mean $y(x')$ and standard deviation δ and with limits y_{th} and $+\infty$. The most straightforward method for POD evaluation is OLS. However, the use of OLS becomes problematic in the presence of censored data. As in the hit/miss analysis, also in this case MLE provides a robust approach for parameter estimation. Nevertheless, as already explained, MLE needs a 'good' starting point for its optimization. In this case it is convenient to solve first a pilot OLS analysis without censored data and use the calculated μ and σ as initial values for the MLE optimization. For a detailed description on how to apply MLE in response signal analysis see Petrin *et al.* (1993).

3.3.3 Confidence Bounds

Several approaches have been developed in the literature for placing confidence bounds on the POD: the Wald method, the likelihood ratio method (LRM) and a fast algebraic method proposed by Harding and Hugo (2003) for signal response POD only. Both, the Wald method and LRM make use of the asymptotically normal behaviour of the maximum likelihood estimators Cheng and Iles (1983, 1988) and give similar results for large sampling. However, the Wald method is claimed to underestimate the confidence bounds when the number of samples is limited. A detailed explanation of how to build POD confidence bounds is provided in Petrin *et al.* (1993) and Meeker and Escobar (1995). Comparative analyses among the methods are given, for instance, in Harding and Hugo (2003) and Annis and Knopp (2007). In this context it seems important to point out that confidence bounds depend on the procedure used to calculate them and are not a property of the available experimental data (Generazio 2011). Confidence bounds are usually plotted along with POD curves with the scope to highlight a particular flaw size $a_{u/v}$. The value $a_{u/v}$ represents the minimum crack size that the system is able to detect with probability u and confidence v. For instance, for aerospace applications $a_{90/95}$ is of interest. Confidence bounds and in particular the $a_{u/v}$ statistic, are considered by the NDT community as a valid summary metric for system capability and, therefore, are useful for certification purposes.

3.3.4 Probability of False Alarm

The probability of false alarm represents the percentage of undamaged specimens that the system classifies as damaged. Analytically, PFA is quantified from the integral of the measurement noise probability density function. If the noise distribution is Gaussian, then PFA can be given by

$$\text{PFA} = \int_{y_{th}}^{\infty} \frac{1}{\sqrt{2\pi}\delta_n} e^{-\frac{(y-\bar{y}_n)^2}{2\delta_n^2}} \, dy, \qquad (3.15)$$

where $\bar{y}_n$ is the noise mean and δ_n the noise standard deviation. Following the PFA definition, $\bar{y}_n$ and δ_n could be calculated by testing undamaged structures or, alternatively, by extracting information from the available measurements of damaged items (MIL-HDBK-1823A 2009). From Equation (3.15) it is evident that choosing a large y_{th} would advantageously decrease PFA. On the downside, recalling Equation (3.8), this choice would also move the POD curve toward higher a and $a_{u/v}$ (for fixed u and v) values, which is an undesired effect. To make a wise decision on y_{th} it is useful to plot on the same diagram $a_{u/v}(y_{th})$ and $\text{PFA}(y_{th})$. This allows one to pick the threshold value that results in the best compromise between PFA and $a_{u/v}$. A graphical description of the PFA in the case of signal response analysis is given in Figure 3.2.

3.4 From POD to MAPOD

It is well-known to the POD practitioner that POD evaluations call for large experimental programmes. It is stated in the literature that a reliable hit/miss analysis should collect measurements from at least 60 different crack sizes. Taking advantage of the extended information, for a signal response analysis the recommended minimum number of crack lengths would be about 30. Moreover, repetitive tests should be also performed for the same crack size in order to account for existing uncertainty related to the human factor, operational conditions and other parameters that may change from test to test. Therefore, in many situations the required experimental effort is excessive with respect to the available time and budget. In order to circumvent this problem, MAPOD approaches have been developed. In a MAPOD approach, computer models are built to replace real experiments, with the aim to speed up the POD analysis by alleviating the experimental burden. The first MAPOD applications were proposed by Martinez and Bahr (1984) where numerical models were used to replace real NDT inspections and experimental noise was added to the numerical signal. The potential of this methodology was soon evident and gave rise to various research groups (e.g. Iowa State University, U.S. Air Force Laboratory, NASA, UK National NDT Centre, and Structures and Composite Laboratory, Stanford University). Special attention should be paid to the MAPOD working group that promotes 'the increased understanding, development and implementation of model assisted POD methodologies'.

Nowadays, two different MAPOD approaches are of main interest Thompson *et al.* (2009), the transfer function approach and the full model assisted approach. In the former, proposed in Harding *et al.* (2007), physics based methods are used to transfer the POD curve calculated for a certain inspection process to another instance with different inspection parameters (e.g. different material of the tested structure, different flaw type). The latter approach employs physics based models to directly propagate the uncertainty of a given set of inspection parameters. Eventually, the numerical signal may be further combined with experimental noise accounting for the variability of the remaining parameters not considered in the physics based model. Applications of MAPOD to NDE can be found, for instance, in Knopp *et al.* (2007) for eddy current measurements and in Harding *et al.* (2009) for ultrasound measurements.

3.5 POD for SHM

With the advent of SHM techniques researchers wondered how POD methods, initially developed for NDE, could be employed for SHM systems too. Indeed, SHM systems show differences with respect to classical NDT systems. First of all, as opposed to NDE, SHM systems are permanently mounted on the structure. This leads to very reproducible repeated measurements. Conversely, the reproducibility of repeated NDE is generally poor mainly due to the human factor, i.e. the influence of the operator that performs the inspection. On the other hand, for NDE the sensors and hardware of the system may be calibrated before any usage. This is not possible for SHM systems whose sensors are subject to in situ and ageing effects that make SHM performances, and consequently POD measures, time dependent. Thus, the main types of uncertainty that should be captured during the reliability assessment for NDE are the human factor, variations in the transducer/structure contact interface, crack morphology and local structure properties; while for SHM those are environmental conditions, ageing effects in the structure and sensors and damage morphology. Importantly, since the process of mounting the SHM system transducers and geometric/mechanical properties of the hosting structure are not perfectly reproducible, the pair hosting structure/SHM systems cannot be doubled. This means that measured SHM performances are strictly connected to the tested systems and cannot be applied to another hosting structure/SHM system pair since the aforementioned uncertainties may significantly affect the system capabilities. All these factors make the reliability assessment of SHM systems more involved than for NDE. At present, there still exists no standard procedure for this issue, although it is widely accepted that POD evaluations based only on experimental tests are not feasible. Recently, protocols for SHM reliability assessment have been proposed by Mueller *et al.* (2011) and Aldrin *et al.* (2010) and both make use of computer models. Therefore, the development of computer models that are able to accurately simulate important physical phenomena involved in SHM measurements becomes a fundamental aspect. Unfortunately, accurate computer models in general resort to a large amount of computer power. This fact can limit the application of MAPOD approaches. Aldrin *et al.* (2010) pointed out two different ways to mitigate the sampling effort. The first way is to create enhanced statistical models that include the dependency on several others factors in addition to the crack length, as proposed originally by Hoppe (2010). This gives a better insight into the inspection parameter/SHM system performance relationship and, therefore, helps to reduce the experimental/numerical burden. The second way employs efficient uncertainty quantification procedures, such as polynomial chaos methods (Xiu 2010) that significantly reduce the amount of required samples with respect to traditional sampling approaches. Application of the protocol is illustrated in Aldrin *et al.* (2011, 2012).

A third direction that may be followed to considerably speed up the MAPOD calculations makes use of parallel computing techniques. Parallel computing may tremendously cut down on simulation time and still be integrated with enhanced statistical models and sampling techniques that lead to a practical MAPOD approach. However, access to powerful computing units, such as clusters, is still limited due to the costs associated with the hardware and space requirements. Recently, graphics cards have been successfully employed for scientific computations, see e.g., Packo and Uhl (2011) and Packo *et al.* (2012). Due to the integration of a large number of computing units with a high-speed on-board memory, these units offer much more computational power than standard CPUs, for a much lower price. It is therefore possible to carry out a number of simulations in a short time, using a standard PC, which is the crucial issue to evaluate the POD characteristics.

In the next section a full-model MAPOD approach for an ultrasonic based SHM system is presented.

3.6   MAPOD of an SHM System Considering Flaw Geometry Uncertainty

It is widely acknowledged that for a physical setup POD evaluation a large number of specimens has to be prepared and equipped with the SHM system, which makes it very time consuming and expensive. These factors may be reduced by a model based approach. The MAPOD technique has been employed to study the reliability of an SHM system. The performance of the proposed GPU based approach has been examined in terms of calculation speed and quality of results. As the most influencing factor, the variability of the crack configuration has been investigated. The virtual framework consists of a thin rectangular plate with an array of transducers attached on one of its edges. The transducers are part of the SHM system based on the phased array technique. Since in a realistic scenario the cracks do not always have the same orientation and position with respect to the transducers' array, the reliability of the SHM system to this variability is examined.

3.6.1   SHM System

The MAPOD approach was used to analyse the SHM system for plate-like structures monitoring. The system was based on beamforming of Lamb waves with the use of a standard delay and sum (DAS) beamformer. An extensive study of the technique and 2D arrays design can be found in Chapter 7.

However, in the presented approach a simple case of a uniform linear array (ULA) of transducers in single transmitter, multiple receivers (STMR) setup was examined. In order to avoid equivocal damage localization, the array placed on the edge of the evaluated plate was considered. One of the transducers was used as both emitter and receiver, while the rest were used only as sensors. The wave generated by the emitter propagates in the structure and is reflected

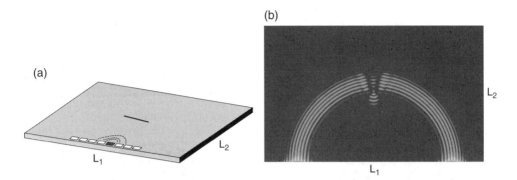

Figure 3.3 Schematic diagram of the simulated model (a) and the displacement field generated by the cuLISA3D solver (b)

back when it encounters a crack. The returning wave is captured by the array of sensors. Then, the signals recorded by the sensors are elaborated. First, a linear mapping algorithm, proposed by Liu and Yuan (2010), is applied to remove the effect of dispersion. Secondly, the signals are delayed and summed. When the assumed incident angle is equal to the direction of arrival (DOA) of the damage reflected echo, signals from all sensors are coherently summed, and the resultant noise summed from the successive sensors is reduced. Damage detection and localization is performed with the imaging technique described in Chapter 7 for the STMR case. The output of the system is defined as the value of the most intense pixel presented in the examined area.

3.6.2   Simulation Framework

For the considered SHM system it was necessary to investigate the propagation of high frequency waves, i.e. Lamb waves in plate-like structures; therefore high spatial and temporal resolution was needed. The cuLISA3D software, described in detail in Chapter 2, used in this research fulfils these requirements. Compared with other simulation methods, the simulation time is reduced by a factor of thousands offering higher spatial and temporal resolution. These features are particularly useful in MAPOD evaluations. A cell size of 0.5 mm and a time step of 0.05 μs have been used in calculations. For an analysis consisting of 10000 time steps that correspond to 500 μs, the calculation time was approximately 4 min. An exemplary snapshot of the displacement field generated by the piezo-actuator, calculated by the cuLISA3D solver, is shown in Figure 3.3(b).

3.6.3   Reliability Assessment

The modelled SHM system consisted of a ULA of 8 transducers spaced at a distance of 5 mm. The array was localized on the edge of a rectangular aluminium plate with dimensions of $500 \times 300 \times 2$ mm. Figure 3.3(a) shows a schematic representation of the virtual system. The receiving transducers are shown as small white rectangles while the emitter is a small shaded rectangle. The emitter was used to excite 3 cycles of sine at a frequency of

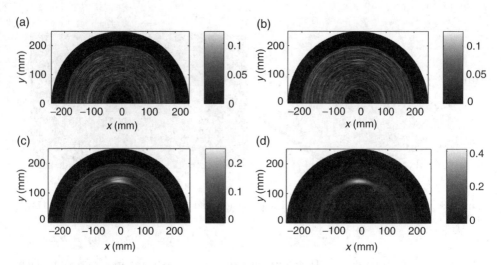

Figure 3.4 Imaging results obtained for the crack lengths: 3.01 mm (a), 5.71 mm (b), 8.66 mm (c) and 14.62 mm (d)

100 kHz, modulated with a Hanning window. The wavelength of the enhanced A_0 mode at the excitation frequency was 12 mm.

The numerical experiment was repeated for 120 different crack configurations. The flaws were all straight but differed in length, position and orientation. The considered crack sizes were within the 3–15 mm interval. The choice of the investigated cracks size range is due to the expected smallest defect that may be detected for a given wavelength. The position of the crack mid point was randomly chosen within a 10×6 mm rectangle located in the middle of the plate in order to model the crack position variability. Since the basic information about the structure is usually assumed to be known, the orientation of the crack with respect to the transducer line is also known and subjected to random angle within a $\pm 4°$ interval.

The wave propagation was simulated in the cuLISA3D software. The output of the numerical simulations were the time signals of the out-of-plane displacement components for each crack configuration. Then, an empirical white Gaussian noise of 15 dB was added to the recorded sensor signals, in order to consider *in situ* factors that may produce measurement to measurement variability. The damage imaging algorithm presented in Section 3.6.1 was applied to the signals obtained for all of the 120 crack configurations. The exemplary damage images obtained for crack lengths of 3.01, 5.71, 8.66 and 14.62 mm can be seen in Figure 3.4 (a–d), respectively. In Figure 3.4 (a) and (b) no damage symptoms can be observed, which is not surprising, since the crack sizes were smaller than half of the wavelength. However, areas with a higher contrast can be observed for damage images presented in Figure 3.4 (c) and (d). Moreover, it can be seen that the bigger the damage, the higher the value of the pixels at the damage-related areas.

The analysis of all 120 flaws produces the data to be used in the statistical analysis for constructing the POD curves. Since the SHM system provides a continuous signal, a signal response analysis is performed, as described in Section 3.3. Figure 3.5 (a) shows the resulting $a−\hat{a}$ scatter plot. Here, it is possible to see that, apart from a cluster of points with low crack

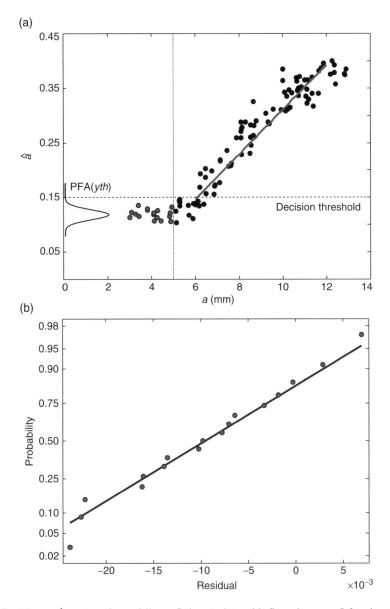

Figure 3.5 The $a - \hat{a}$ scatter plot and linear fitting to data with flaw size $a > 5$ for signal response analysis (a) and normal plot of residuals of data with flaw size $a < b$ (b)

size, a is linearly correlated with $\hat{a}$. Hence, a linear model $y = \beta_0 + \beta_1 x$ is reasonable with $x = a$ and $y = \hat{a}$. In this case a good choice for the decision threshold appears evident already from the scatter plot. Indeed, assuming $a_{th} = y_{th} = 0.14$, censors all data that do not show linear correlation. Nonetheless, PFA for $y_{th} = 0.14$ was calculated. Thus, the POD analysis provides an additional guide on the threshold levels to optimize the system's performance.

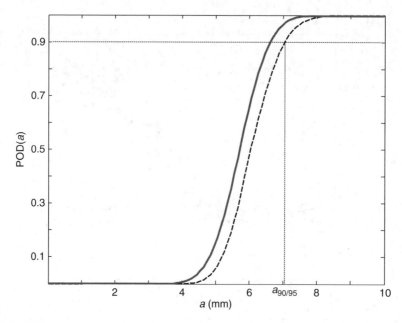

Figure 3.6 POD curve obtained from the signal response analysis of simulated data. The dashed line denotes the 95% lower confidence bound

Since no experiment simulating an undamaged plate was carried out, noise probability density and PFA were evaluated from the data according to the procedure given in MIL-HDBK-1823A (2009). Considering only signal values with $a < 5$, that show no $a - \hat{a}$ correlation, a Gaussian noise distribution was built and tested. The resultant normal plot, presented in Figure 3.6, proves the adequacy of the assumed probabilistic model for the noise. With the assumed decision threshold PFA$(y_{th} = 0.14) = 1\%$. The POD curve obtained from the signal response analysis is shown in Figure 3.6 along with the 95% lower confidence bound that has been calculated using the likelihood ratio method. The resultant $a_{90/95}$ is 7.07 mm which is in agreement with theoretical considerations of the wave interaction with defects that limits the detection capability of the system to crack sizes larger than half-wavelength. Thus, for the wavelength of 12 mm used in this analysis, the result obtained is in very good agreement.

3.7 Conclusions

The calculation of the POD characteristics for an SHM system is of particular importance for the reliability assessment and serves as a first step in certification. It has been proven helpful in system's performance optimization by guiding the choice of the damage detection thresholds. However, the number of experiments required to obtain reliable results is significant. Since the specimens must be damaged in order to be used in evaluations, a large number of samples needs to be produced. Moreover, the experimental campaign for such a setup is very time-consuming and should be repeated for many environmental and human factors. This makes the traditional approach to POD evaluation costly. Taking advantage of reliable numerical models and massively parallel processing techniques, the physical experiments may be conducted virtually. A wide variety of environmental and human factors may be taken into account in numerical simulations. Due to the utilization of GPUs, the simulations are speeded up significantly, allowing for the increase in the number of virtual experiments. Moreover, the GPU based approach may be employed with no special hardware requirements, so is available for a wide engineering audience. For the elastic waves based SHM systems the developed cuLISA3D simulation framework has been proven reliable and extremely efficient.

References

Aldrin JC, Medina EA, Lindgren EA, Buynak C, Steffes G and Derriso M 2010 Model-assisted probabilistic reliability assessment for structural health monitoring systems. *AIP Conference Proceedings* **1211**(1), 1965–1972.

Aldrin JC, Medina EA, Lindgren EA, Buynak CF and Knopp JS 2011 Case studies for model-assisted probabilistic reliability assessment for structural health monitoring systems. *AIP Conference Proceedings* **1335**(1), 1589–1596.

Aldrin JC, Sabbagh HA, Murphy RK, Sabbagh EH, Knopp JS, Lindgren EA and Cherry MR 2012 Demonstration of model-assisted probability of detection evaluation methodology for eddy current nondestructive evaluation. *AIP Conference Proceedings* **1430**(1), 1733–1740.

Annis C and Knopp J 2007 Comparing the effectiveness of $a_{90/95}$ calculations. *AIP Conference Proceedings* **894**(1), 1767–1774.

Berens A and Hovey P 1981 Evaluation of nde reliability characterization. Technical report, University of Dayton Research Institute.

Berens AP and Hovey PW 1983 *Statistical Methods for Estimating Crack Detection Probabilities*, Probabilistic Fracture Mechanics and Fatigue Methods: Applications for Structural Design and Maintenance, ASTM STP 798 (eds Bloom JM and Ekvall JC). American Society for Testing and Materials, pp. 79–94.

Cheng RCH and Iles TC 1983 Confidence bands for cumulative distribution functions of continuous random variables. *Technometrics* **25**(1), 77–86.

Cheng RCH and Iles TC 1988 One-sided confidence bands for cumulative distribution functions. *Technometrics* **30**(2), 155–159.

Forsyth S and Fahr A 1998 *An Evaluation of Probability of Detection Statistics*, RTO AVT Workshop on 'Airframe Inspection Reliability under Field/Depot Conditions'. Brussels, Belgium.

Generazio E 2011 Binomial test method for determining probabiliy of detection apability of fracture critical applications. Technical report, NASA.

Georgiou G 2006 Probability of detection (POD) curves. Technical report, Jacobi Consulting Limited.

Harding C, Hugo G and Bowles S 2007 Model-assisted probability of detection validation of automated ultrasonic scanning for crack detection at fastener holes. *Proceedings of the 10th Joint FAA/DoD/NASA Conference on Aging Aircraft*. Palm Springs, California, USA.

Harding CA and Hugo GR 2003 Statistical analysis of probability of detection hit/miss data for small data sets. *AIP Conference Proceedings* **657**(1), 1838–1845.

Harding CA, Hugo GR and Bowles SJ 2009 Application of model-assisted POD using a transfer function approach. *AIP Conference Proceedings* **1096**(1), 1792–1799.

Hoppe WC 2010 A parametric study of eddy current response for probability of detection estimation. *AIP Conference Proceedings* **1211**(1), 1895–1902.

Kabban C and Derriso M 2011 Certification in stuctural health monitoring systems. *Proceedings of the Eighth International Workshop on Structural Health Monitoring*. Stanford, USA, pp. 2429–2436.

Knopp JS, Aldrin JC, Lindgren E and Annis C 2007 Investigation of a model-assisted approach to probability of detection evaluation. *AIP Conference Proceedings* **894**(1), 1775–1782.

Liu L and Yuan FG 2010 A linear mapping technique for dispersion removal of Lamb waves. *Structural Health Monitoring* **9**, 75–86.

Martinez J and Bahr A 1984 Statistical detection model for eddy current systems. *Review of Progress in QNDE* **3A**, 499–510.

Matzkanin G and Yolken H 2001 *Probability of Detection (POD) for Nondestructive Evaluation (NDE)*. Texas Research Institute/Austin, Incorporated.

Meeker WQ and Escobar LA 1995 Teaching about approximate confidence regions based on maximum likelihood estimation. *The American Statistician* **49**(1), 48–53.

MIL-HDBK-1823A 2009 Nondestructive evaluation system reliability assessment. Technical report, US Department of Defense.

Mueller I, Janapati V, Banerjee S, Lonkar K, Roy S and Chang F 2011 On the performance quantification of active sensing SHM systems using model assisted POD methods. *Proceedings of the Eighth International Workshop on Structural Health Monitoring*. Stanford, USA, pp. 2417–2428.

Nelder JA and Mead R 1965 A simplex method for function minimization. *The Computer Journal* **7**(4), 308–313.

Packo P, Bielak T, Spencer AB, Staszewski WJ, Uhl T and Worden K 2012 Lamb wave propagation modelling and simulation using parallel processing architecture and graphical cards. *Smart Materials and Structures* **21**, 075001–1075001–13.

Packo P and Uhl T 2011 Elastic wave propagation simulation system based on parallel processing architecture *19th international conference on Computer Methods in Mechanics*. Warsaw, Poland.

Petrin C, Annis C and Vukelich S 1993 A recommended methodology for quantifying NDE/NDI based on aircraft engine experience. Agard lecture series 190. National Atlantic Treaty Organization.

Pettit D and Hoeppner D 1972 Fatigue flaw growth and NDT evaluation for preventing through cracks in spacecraft tankage structures. Technical report CR NAS 9-11722 LR 25387, NASA.

Thompson RB, Lindgren E, Swindell P, Brasche L, W. W and Forsyth D 2009 Recent advances in model-assisted probability of detection *4th European-Americal Workshop on Reliability of NDE*. Berlin, Germany.

Wall M, Burch SF and Lilley J 2009 Review of models and simulators for NDT reliability (POD). *Insight – Non-Destructive Testing and Condition Monitoring* **51**(11), 612–619.

Worden K and Manson G 2007 The application fo machine learning to structural healt monitoring. *Philosophical Transactions of The Royal Society A* **365**, 515–537.

Xiu 2010 *Numerical Methods for Stochastic Computations. A Spectral Method Approach*. Princeton University Press.

Yee B, Chang F, Couchman J, Lemon G and Packman P 1976 Assessment of NDE reliability data. Technical report, NASA.

4

Nonlinear Acoustics

Andrzej Klepka

Department of Mechatronics and Robotics, Faculty of Mechanical Engineering and Robotics, AGH University of Science and Technology, Poland

4.1 Introduction

Ultrasonic methods used in the SHM and NDT techniques can be divided into two main groups. The first group includes methods based on the principles of linear acoustics. As a result of acoustic wave interaction with damage, phenomena such as reflection, transmission, absorption of acoustic energy or mode conversion are generated (Kessler *et al.* 2002; Kundu 2003; Raghvan 2007; Rizzo *et al.* 2009; Staszewski 2004; Wilcox *et al.* 2007). The second group includes methods based on nonlinear acoustical phenomena. Nonlinear acoustical phenomena can be related to various imperfections of atomic lattices (intrinsic or material nonlinearity) and/or nonsymmetric thermoelastic behaviour of interfaces (e.g. cracks, contacts, rubbing surfaces). The former group, with well-known intrinsic or material global nonlinearity, has been investigated since the early 1960s (Bateman *et al.* 1961; Breazeale and Ford 1965) and has been used to detect material imperfections (e.g. microcracks in materials; Jhang and Kim 1999; Nagy 1994). The latter is a local nonlinearity and has received a lot of interest in theoretical and applied research in the last 15 years. The above-mentioned phenomena can produce nonlinear effects in the form of higher, sub- and super-harmonics generation (Buck and Morris 1978; Morris *et al.* 1979; Solodov *et al.* 2004), frequency mixing (Bruneau and Potel 2009) and shifting (Haller and Hedberg 2006), modulation of acoustic waves by low frequency vibration (Aymerich *et al.* 2010; Klepka *et al.* 2012a; Meo and Zumpano 2005; Staszewski and Parsons 2006; Van Den Abeele *et al.* 2000b), slow dynamics effects (Bentahar *et al.* 2009; Guyer *et al.* 1998) or reverberation (Van Den Abeele *et al.* 2002).

Advanced Structural Damage Detection: From Theory to Engineering Applications, First Edition.
Edited by Tadeusz Stepinski, Tadeusz Uhl and Wieslaw Staszewski.
© 2013 John Wiley & Sons, Ltd. Published 2013 by John Wiley & Sons, Ltd.

Table 4.1 Crack-induced nonlinear mechanisms in solids

Atomic scale $10^{-10} - 10^{-6}$	• Intrinsic elastic nonlinearity due to anharmonicity of interatomic potential (Nazarov 1988) • Nonfriction and nonhysteretic dissipation locally enhanced by thermoelastic coupling (local concentration of stress and increased temperature gradient) (Landau and Lifshitz 1986; Zaitsev *et al.* 2002a,b) • Local variation in the elasticity of the defect produced by either hysteretic or purely elastic (nondissipative and nonhysteretic) nonlinearity of the defect (cascade mechanism) (Nazarov *et al.* 2002) • Drastic increase in nonlinear elasticity at defects (microcracks and microcontacts); strong strain concentration due to breaking interatomic bonds (Van Den Abeele *et al.* 2000b) • Hysteresis in stress–strain; amplitude dependent dissipation (Meo *et al.* 2008) • Friction, adhesion hysteresis (Pecorari 2004) • Stick-slip friction between crack surfaces, adhesion hysteresis (Woolfries 1998)
Mesoscopic scale 10^{-4}	• Crack induced nonlinearity (variation in elastic moduli), Hertzian type nonlinearity (contact between crack surfaces) (Kim *et al.* 2003; Ostrovsky and Johnson 2001; Solodov 1998) • High compliant elastic inclusion with an arbitrary weak deviation from the linear model (rheological model) (Zaitsev and Sas 2000) • Local stiffness reduction leading to natural frequency shift (Muller *et al.* 2005; Van Den Abeele *et al.* 2000a) • Bilinear stiffness (closing – opening crack) (Friswell and Penny 2002; Zaitsev and Sas 2000)
Macroscopic scale 10^{-2}	• Clapping mechanism (Belyaeva *et al.* 1997; Moussatov *et al.* 2003) • Wave modulation due to impedance mismatch (discontinuity due to closing – opening crack) (Duffour *et al.* 2006; Kim and Rokhlin 2002)

Despite much research effort, there is still very little understanding of what is the physical mechanism related to these nonlinearities. Table 4.1. summarizes the major developments related to previous work in this area and the major theoretical models. The question still remains why material sensitivity of nonlinear effects is so strong compared with linear responses.

There are two major difficulties associated with these problems. The first one is the diversity of the proposed nonlinear mechanisms. Similar nonlinear effects can be manifested by different mechanisms and vice versa. For example, energy dissipation can be modelled using frictional, hysteretic or thermoelastic mechanisms. Hysteresis, in turn, involves both elasticity and dissipation, and could be linear or nonlinear. The second difficulty is that various experimental evidence – related to these nonlinear mechanisms – has been observed. It is often very difficult, if not impossible, to separate all the mechanisms involved. It is also important to note that nonlinearities may result not only from cracks but also from other non-damage related effects, such as friction between elements at structural joints or boundaries, overloads, material connections between transducers and monitored surfaces, electronics and instrumentation measurement chain. Nevertheless, nonlinear acoustics remains very attractive for fatigue crack detection. Physical understanding of the nonlinear mechanisms involved is thus very important for implementation and real engineering applications.

4.2 Theoretical Background

The basic equation describing nonlinear acoustics is a nonlinear equation of motion, written in the form

$$\rho \frac{\partial^2 u}{\partial t^2} = \frac{\partial \sigma}{\partial x}. \tag{4.1}$$

where ρ is the density of the medium at equilibrium, t is the time, u is the displacement, σ is the stress and x is distance of propagation. Nonlinearity of this equation results from two facts:

1. The relationship between stress σ and strain ε is nonlinear (physical nonlinearity). It can be expressed in a general form as

$$\sigma = \int K(\varepsilon, \dot{\varepsilon}) d\varepsilon, \tag{4.2}$$

where parameter K can contains terms describing both classical and nonclassical nonlinearity. For example, a form of K for a medium with classical first- and second-order nonlinearity and nonclassical hysteresis nonlinearity can be written as

$$K(\varepsilon, \dot{\varepsilon}) = K_0\{1 - \beta\varepsilon - \delta\varepsilon^2 - \alpha[\Delta\varepsilon + \varepsilon(t)\text{sign}(\dot{\varepsilon})]\}, \tag{4.3}$$

where K_0 is linear modulus, $\Delta\varepsilon$ is local strain amplitude, $\dot{\varepsilon} = d\varepsilon/dt$ is strain rate and β and δ are classical nonlinear perturbation parameters. For the hysteretic term of Equation (4.3)

$$\text{sign}(\dot{\varepsilon}) = 1 \quad \text{if} \quad \dot{\varepsilon} > 0 \quad \text{sign}(\dot{\varepsilon}) = -1 \quad \text{if} \quad \dot{\varepsilon} < 0 \tag{4.4}$$

and α is the hysteresis coefficient.

2. The relationship between the strain and the displacement gradient for longitudinal waves is given by

$$\varepsilon = \frac{\partial u}{\partial x} + \frac{1}{2}\left(\frac{\partial u}{\partial x}\right)^2. \tag{4.5}$$

This nonlinearity, called kinematic or geometric nonlinearity, does not appear for shear plane waves.

It follows that in the material, the strain σ can depend on both stress σ and the sign of the strain rate. The quoted example, of stress–strain characteristic, including hysteresis behaviour, is of course ideal. This means that in each loading – unloading cycle, material returns to the same state at the same point. In fact this characteristics can be modified by phenomena like cyclic creeping and cyclic hardening. The hysteresis model can also consist of many elements, including parts dependent on strain rate sign and independent of it. The part depending on strain rate sign ($\dot{\varepsilon}$) describes a loop on the stress–strain characteristic plane. This term is responsible for hysteretic nonlinear dissipation. The energy of dissipation of the acoustics field is positive and equal to the surface area of the hysteresis loop. The part independent of the strain rate sign is responsible for orientation of the hysteresis loop. Many studies reported that for micro inhomogeneous media different types of nonlinear dissipation mechanism can be observed (in addition to dissipation mechanism related to hysteretic nonlinearity).

It is generally known that acoustic wave propagating in a structure can induce an additional absorption or transparency for another wave (Nazarov 1988; Zaitsev *et al.* 2002a). Although there is no clear explanation for this phenomenon, it can be ascribed to the processes of relaxation in media. It is generally known that physical parameters of the system are related to external excitation (reach given equilibrium state related to excitation with some delay). It causes memory effect in the structures and means that parameters of the system depend on both values of external excitation and its history. This implies a stress σ dependence on strain ε and additional parameters related to internal degrees of freedom of the structure. In the presence of an acoustics wave, relaxation time increases, which causes the resonant curve to be shifted (Bruneau and Potel 2009).

In granular media a different type of nonlinearity has the major contribution to nonlinear acoustics. This nonlinearity comes from the Hertz contacts (Landau and Lifshitz 1986; Meo *et al.* 2008; Nazarov *et al.* 2002; Zaitsev *et al.* 2002b). This nonlinearity arises from the contact between surfaces. During loading action the rigidity of the contacts is changed (changes of the contacts' surfaces). It should be noted that there are other reasons that may modify the stress–strain characteristics including amplitude level or velocity of loading – unloading, which causes thermal fluctuations.

A nonlinear form of Hooke's law (4.3) causes distortion in the signal response of structure excited by purely harmonic excitation. As a result, additional components in signal response can appear. The number of these components and their characters depends on the type of nonlinearity. Figure 4.1 presents examples of the stress–strain relationship and spectra of strain for a structure excited by a harmonic wave in the case of a linear system, and systems with quadratic and cubic nonlinearities and hysteresis. The above-mentioned phenomena can be generated in structures due to material imperfections and damage; in the case of damaged structures the scale of nonlinearities is significantly higher. Nonlinear phenomena caused by damage can be divided into several groups. However, all of them can make a contribution to the total nonlinear characteristics of the system simultaneously. The basic nonlinear effect caused by damage will be described below.

4.2.1 Contact Acoustics Nonlinearity

Contact acoustics nonlinearity (CAN) is a nonlinear effect resulting from mechanical interaction between damage interfaces. Two mechanisms of nonlinear vibrations caused by damage

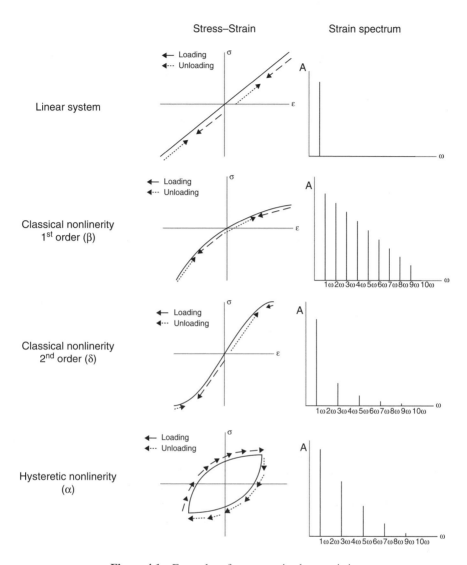

Figure 4.1   Examples of stress–strain characteristics

are considered. The first is *clapping*, which is manifested as a mechanical diode effect. It results from asymmetry in stress–strain characteristics for a damaged structure. This causes the stiffness of the material to be higher for the compression than for tensile phase. The clapping mechanism is approximated by (Pecorari and Solodov 2006; Woolfries 1998)

$$\sigma = K\left[1 - H\left(\varepsilon\right)\frac{\Delta K}{K}\right]\varepsilon, \tag{4.6}$$

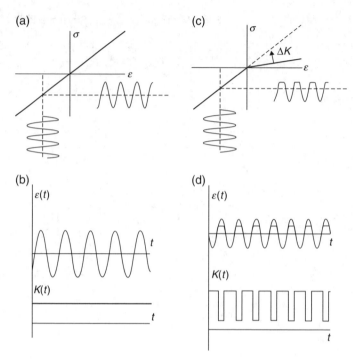

Figure 4.2   Stiffness modulation and wave deformation for an intact (a,b) and damaged structure (c,d)

where $H(\varepsilon)$ is the Heaviside function and $\Delta K = K - (d\sigma/d\varepsilon)$ for $\varepsilon > 0$. The bimodular contact caused by a harmonic acoustic wave results in pulse-type modulation of material stiffness and wave deformation [Figure 4.2(d)]. The modulating function is directly connected to the frequency of the longitudinal acoustic wave. Therefore, the spectrum of the response function will contain the harmonics of the fundamental frequency of the acoustics wave (both even and odd). The amplitudes of these harmonics are modulated by the *sinc* function. Figure 4.2 presents an example of stress–strain characteristics and wave deformation for a linear (intact) [Figure 4.2(a)] and nonlinear (damaged) [Figure 4.2(c)] structure.

The second nonlinear mechanism which causes changes in the stiffness characteristic is connected to the interaction of a share acoustic wave and damage. In this case a mechanical connection of the damage interface is made by friction forces. If the amplitude of the share wave is low, the damage interfaces are displaced in micro-slip mode between the neighbouring roughness areas. This effect is independent of the direction of motion and additionally changes the stiffness characteristics twice (symmetrical nonlinearity). In this case only odd harmonics are generated in spectra of the system response. When the amplitude of the acoustic wave increases enough, the contact static friction forces become broken. Then the surfaces of the crack start sliding relative to each other in stick and slip mode. This means that first, the asperities coupled by adhesion force deform elastically and then plastically (slip)(Solodov 1998). It causes a cyclic change between the static and kinematic friction phase which turns the strain–stress characteristics into hysteretic (Awrejcewicz and Olejnik 2005; Pecorari and Solodov

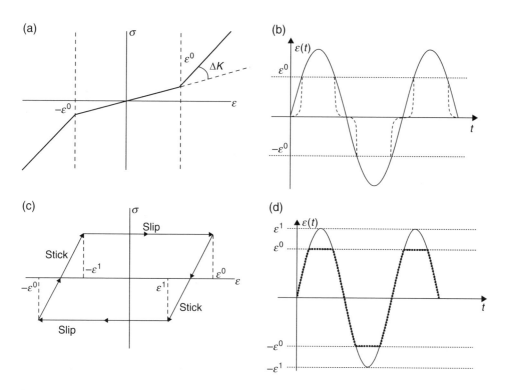

Figure 4.3 Stress–strain characteristics and wave deformation for micro-slip mode (a,b) and stick and slip mode (c,d)

2006; Woolfries 1998). As in the case of the micro-slip mode, changes are independent of the direction of displacement and change stiffness twice per cycle of loading, causing generation of only odd harmonics. Figure 4.3 presents stress–strain characteristics for micro-slip as well as stick and slip mode. More details on different types of friction forces in dynamical systems can be found in Ostrovsky and Johnson (2001).

4.2.2   Nonlinear Resonance

Apart from the above-mentioned phenomena there are other mechanisms that cause nonlinear behaviour of the damaged structure. One of them is nonlinear resonance, which appears due to closing and opening action of the crack. To understand this phenomenon let us assume that the crack consists of a series of oscillators with mass m_i and stiffness k_i (not ideal surfaces of the crack interfaces). The natural frequency of a given oscillator ω_0 is determined by mass (the mass of material inside the damage area) and stiffness. Simplifying the problem, the equation of motion for a given oscillator can be written as

$$\ddot{x} + \omega_0 x = f(t) + F(x),\tag{4.7}$$

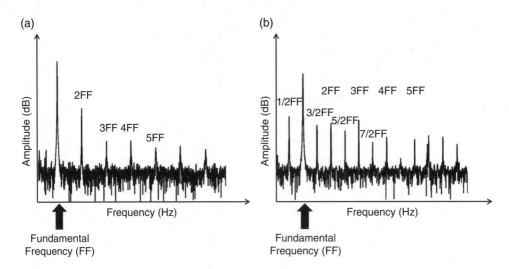

Figure 4.4 Example of spectra: CAN effect – asymmetry of stiffness characteristic (a) and imperfect interface of damage (b)

where x is displacement, $f(t) = f_0 \cos(\omega_0 t)$ is excitation with frequency ω_0 and $F(x) \approx \cos(\omega_1 t - \omega_0 t)$ is the nonlinear interaction force. Assuming that $\omega_1 - \omega_0 \approx 0$, resonance at frequency $\omega_0 = \omega_1/2$ is expected (generation of first-order subharmonic). The higher order subharmonics appear when a multiple of ω_0 is used. The subharmonics generation problem becomes more complicated when the damage is assumed as a series of resonators with different natural frequencies. In such a case a pair of resonant frequencies centred around subharmonics appear (Woolfries 1998). Figure 4.4 shows an example of system response spectra that exhibit higher harmonics due to the CAN effect, asymmetry of stress–strain characteristics and imperfect interface of the damage. A similar effect has been observed as a result of adhesion forces between crack tips (Kim *et al.* 2003).

Another phenomenon which has a contribution to nonlinear resonance of damaged structure is hysteresis in the stress–strain relationship (already discussed above). In addition to the above-mentioned effects it also causes a change (decrease) in the velocity of the acoustic wave propagating in the damaged structure (Bruneau and Potel 2009). Decreasing wave velocity results in a shift of the resonant frequency proportional to the amplitude of the acoustic wave.

4.2.3 Frequency Mixing

The process of harmonics generation is a special case of frequency mixing which assumes that a monofrequency wave is introduced to the damaged structure. In the case of two waves with different amplitude and frequency, the interaction between these two waves and damage results in different effects. Let us assume that the excitation signal introduced to the damaged structure is a linear combination of two components and is described by

$$u(t) = A_1 \sin(\omega_1 t) + A_2 \sin(\omega_2 t), \tag{4.8}$$

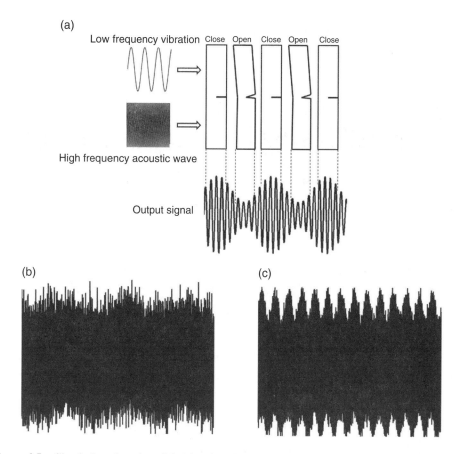

Figure 4.5   Classical explanation of the signal modulation process in a cracked structure (a). Example of signal response for an intact (b) and damaged (c) structure

where A_1 and A_2 are amplitudes of the particular components and ω_1 and ω_2 are their frequencies, but $\omega_1 \gg \omega_2$. In this case, the solution of Equation (4.1) consists of terms corresponding to higher harmonics of ω_1 and ω_2 and mixed terms responsible for the generation of sidebands around frequency ω_1. The order of sidebands is dependent on the type of nonlinearity. For example in the case of 'classical' quadratic nonlinearity, the first-order sidebands $\omega_1 \pm \omega_2$ with amplitude proportional to $\beta A_1 A_2$ appear. In the case of cubic nonlinearity the second-order sidebands $\omega_1 \pm 2\omega_2$ with amplitude proportional to $C (A_1)^2 A_2$ appear, where C is a constant dependent on β and δ. Classically, the modulation of an acoustic wave is explained by the closing and opening action of the crack during vibrations. This process is schematically shown in Figure 4.5(a). Figure 4.5(b,c) presents an example of a signal's response acquired during a nonlinear acoustic test for an undamaged (b) and impacted (c) composite plate. As a result of signal modulation the spectrum of the structure's response contains additional components (sidebands). Figure 4.6. presents an example of signal response spectra acquired during a vibro-acoustic modulation test. The amplitude of the sidebands

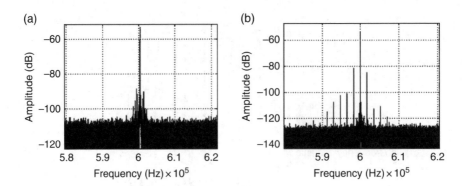

Figure 4.6   Examples of ultrasonic response power spectra for nonlinear acoustics test: (a) undamaged plate; and (b) damaged plate

depends mainly on the modulation intensity, which in turn depends on stiffness changes. Therefore, the modulation intensity factor R is defined as

$$R = \frac{(A_{L1} + A_{R1}) + (A_{L2} + A_{R2}) + \ldots + (A_{Ln} + A_{Rn})}{A_0}, \tag{4.9}$$

where $A_{L1\ldots Ln}$ and $A_{R1\ldots Rn}$ are amplitudes of the subsequent left and right sidebands, respectively, and A_0 is the amplitude of the high frequency acoustic wave. The intensity of the modulation ratio can be correlated with damage size. In many studies parameter R is used as the damage index (Aymerich *et al.* 2010; Friswell and Penny 2002; Muller *et al.* 2005; Van Den Abeele *et al.* 2000b).

Apart from the 'classical' mechanisms leading to modulation, there are others which take into account different phenomena, for example the *Luxembourg – Gorky* (L-G) effect. Originally, this phenomenon was observed for radio waves in 1933 in the Netherlands. The modulation transfer made the radio signal sent by a strong Luxembourg radio station audible during a Dutch radio broadcast. A similar effect was observed in Gorky, where a strong signal sent by a Moscow radio station was heard. This effect was also discovered in elastic media (Zaitsev *et al.* 2002a,b).

The mechanism of modulation transfer is completely different from the classical nonlinear effects. The classical vibro-acoustic modulation is related to the elastic part of the stress–strain characteristics while the L-G effect is related to the inelastic part. The principle of L-G cross modulation is shown in Figure 4.7. A high frequency acoustic wave with frequency ω_1 (carrier) is modulated by a low frequency signal with frequency ω_M ($\omega_1 \gg \omega_M$). As a result, the 'pumping' acoustic wave is produced. This wave is sent into the structure together with a probing wave with frequency ω_2 simultaneously. If the structure is cracked, the new signal components will appear in the signal response spectrum with frequency $\omega_2 \pm \omega_M$ and $\omega_2 \pm 2\omega_M$. This means that modulation transfer occurs from the pumping acoustic wave to the probing wave.

Modulation transfer phenomena can be explained by nonlinear coupling between the strain field and an auxiliary field for smaller strain levels that are not sufficient to open the crack but sufficient enough to perturb crack surfaces. A mechanism based on nonlinear coupling

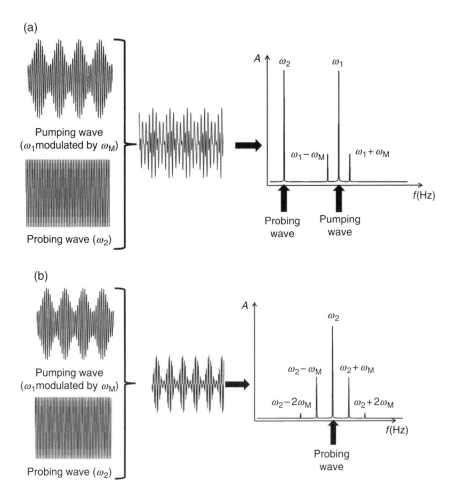

Figure 4.7 Principle of Luxembourg – Gorky effect: (a) damaged structure; and (b) undamaged structure

between strain and temperature fields was proposed in Moussatov *et al.* (2003) and suggested as the analogue of the L-G effect (Duffour *et al.* 2006).

Two theoretical explanations related to wave scattering and temperature were proposed to explain this nonlinear coupling mechanism (Kim and Rokhlin 2002): (1) an elastic wave creates a temperature disturbance from which there is a further elastic-wave scattering; and (2) the temperature carries aspects of the elastic waves that created it and transfers these to the scattered wave. This results in both a cross-modulation effect and amplitude (strain level) dependent acoustic wave attenuation.

In Klepka *et al.* (2012a) it has been proven in an experimental way that a crack does not have to be open to modulate the signal. However, when the crack is open, modulation intensity is enhanced. It should be noted that these phenomena occur for specific values of low-frequency excitation, corresponding to the mode shape of the structure. Figure 4.8 and Figure 4.9(a)

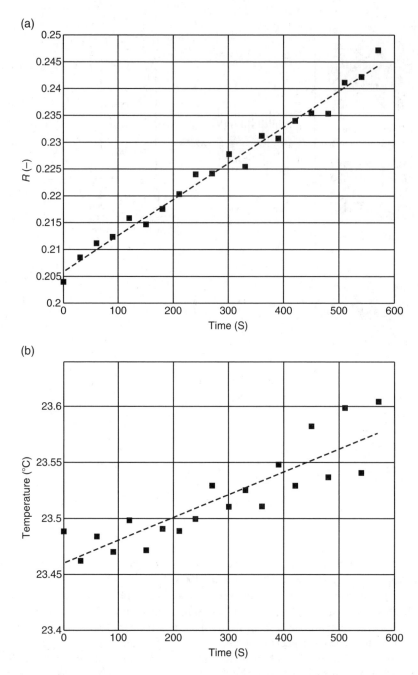

Figure 4.8 Experimental evidence of vibro-acoustic modulation dependence on coupled strain – temperature field: (a) modulation intensity versus time; and (b) temperature versus time. Reproduced from Klepka A., Staszewski WJ., Jenal R., Szwedo M., Uhl T., and Iwaniec J. 2012a. Nonlinear acoustics for fatigue crack detection experimental investigations of vibro-acoustic wave modulations. *Structural Health Monitoring* **25**, 197–211, © 2012 Sage Publications

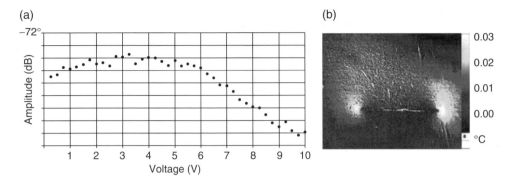

Figure 4.9 Experimental evidence of the elastic wave L-G effect demonstrating high frequency (HF) amplitude attenuation (a). Infrared thermographic image of the temperature field during a vibro-acoustic modulation test (b). Reproduced from Klepka A., Staszewski WJ., Jenal R., Szwedo M., Uhl T., and Iwaniec J. 2012a. Nonlinear acoustics for fatigue crack detection experimental investigations of vibro-acoustic wave modulations. *Structural Health Monitoring* **25**, 197–211, © 2012 Sage Publications

present experimental evidence of vibro-acoustic modulation dependence on coupled strain – temperature field and L-G effect. Figure 4.9(b) presents the temperature field around the damage area during a vibro-acoustic modulation test.

4.3 Damage Detection Methods and Applications

Various methods have been developed in the last decades to detect fatigue cracks. This includes classical, well-established NDT/E approaches such as ultrasonic testing, acoustic emission, eddy current, radiography, magnetic particle and liquid penetrant techniques, together with new approaches based on guided ultrasonic waves, comparative vacuum monitoring or microwave techniques, as discussed in Boller *et al.* (2009) and Staszewski *et al.* (2004). The classical methods based on ultrasonic and guided ultrasonic waves utilize linear effects associated with wave propagation. Wave reflection and scattering is analysed to locate damage and assess its severity.

Recent years have shown a considerable growth of interest in nonlinear damage detection methods based on ultrasonic waves. These methods are more sensitive to structural damage at a very early stage. Nonlinear acoustics is particularly attractive for contact-type damage detection, such as fatigue cracking. An extremely important advantage of methods based on nonlinear acoustics is the fact that these phenomena do not occur (or are poorly visible) in intact structures. The damage causes an instantaneous activation of mechanisms that generate nonlinear behaviour of the structure. Another advantage is the sensitivity of the methods based on nonlinear acoustics. It enables the detection of defects at a very early stage. In contrast to linear methods, there are no limitations related to wavelength of an acoustic wave and damage size. In Table 4.2 the applicability of nonlinear acoustics methods is presented.

Existing damage detection methods based on nonlinear acoustics can be divided into two main groups. The first group includes classical methods based on acousto-elastic phenomena, harmonic generation and frequency mixing. These methods have been known for many

Table 4.2 Applicability of nonlinear acoustics methods

Crystal lattice defects	Microstructures	Microdefects	Macrodefects
Vacancies	Voids	Microcracks	Cracks
Dislocations	Grain boundaries defects	Micropores	Inclusion
Guinier – Preston zones	Corrosion pits	Delamination	
Interstitial atoms			

Table 4.3 Examples of the use of different nonlinear mechanisms for damage detection

Technique	Application	References
Second-harmonic analysis, second- third-order acousto-elastic coefficient	Steel, titanium, aluminium, etc.	Cantrell and Yost (1994), Staszewski and Parsons (2006), Boller et al. (2009) and Klepka et al. (2010)
Nonlinear resonance modes	Polystyrene, glass fibre reinforced polymer, glass, steel pipes	Nagy (1998), Buck (1990) and Ostrovsky and Johnson (2001) and Solodov (2010)
Vibro-acoustic modulation	Composite structure, glass, chiral structures, aluminium, steel	Van Den Abeele et al. (2000a,b), Duffour et al. (2006), Aymerich et al. (2010) and Klepka et al. (2012a,b)
Luxembourg – Gorky effect	Granular materials, glass, aluminium	Zaitsev and Sas (2000), Zaitsev et al. (2002a,b) and Aymerich and Staszewski (2010a)
Subharmonics and selfmodulation	Polystyrene, composites	Solodov et al. (2004), Pecorari (2004) and Awrejcewicz and Olejnik (2005)
Instantaneous amplitude and frequency	Aluminium	Donskoy et al. (2001) and Hu et al. (2010)

years and nonlinear mechanisms which generate nonlinear behaviour of a structure are clearly defined. However, there are studies that reveal nonclassical nonlinear effects in the classical methods of nonlinear acoustics (Klepka et al. 2012a). The second group includes 'modern' (nonclassical) methods which are based on quite different phenomena than the classical methods. They include methods based on cross-modulation or dissipation. Examples of the use of different nonlinear mechanisms for the detection of defects in various types of materials are presented in Table 4.3.

Previous studies in nonlinear acoustics demonstrate that nonlinear phenomena observed in practice can be attributed not only to structural damage but also to material behaviour, boundary conditions and measurement instrumentation chain. It is very important to control all these undesired effects when the nonlinear acoustics tests are performed. The methods of excitation, transducers and their placing as well as excitation frequency have to be chosen carefully to obtain good damage detection sensitivity. Various methods can be used to excite the monitored structures, including speakers (Ballada *et al.* 2004), shakers (Aymerich *et al.* 2010; Klepka *et al.* 2012a), hammers (Ballada *et al.* 2004), lasers (Kolomenskii *et al.* 2003) and piezoceramics transducers (Staszewski and Parsons 2006). The measuring techniques are based on devices such as accelerometers and lasers.

Methods based on noncontact optical/laser measuring techniques are particularly attractive for nonlinear acoustic applications. Low-frequency laser scanning vibrometers have been used for many years to perform vibration and modal analysis. More recent applications include high-frequency ultrasound measurements. 3D scanning laser vibrometers provide combined out-of-plane and in-plane measurements. A number of 3D configurations have been reported, implemented commercially and used for elastic wave propagation measurements.

The problem of excitation method of the structure is also essential in a nonlinear acoustic test. Selecting an appropriate excitation source and its location, as well as providing an appropriate strain level, allows the nonlinear mechanism to be revealed, which is essential for a nonlinear acoustic experiment.

This chapter provides a brief overview of the applications of nonlinear acoustics to damage detection. Some of the work presented was performed within the research project N N501158640 sponsored by the Polish National Science Center.

4.3.1   Nonlinear Acoustics for Damage Detection

This section describes the application of methods making use of nonlinear effects for damage detection. By using various nonlinear acoustics techniques it is possible to detect defects in metallic structures, composites, rocks and concrete, glass, Plexiglas, and many others.

4.3.1.1   Metallic Structures

Metallic structures are widely used in many branches of engineering including civil infrastructures, the aerospace industry and the automotive industry. Failures of metallic structures are mainly attributed to fatigue damage caused by cyclic loading. Fatigue in materials is a local damage, which is progressive and permanent. This type of damage can occur even when the loads are smaller than the static yield stress of the material. This types of damage is very dangerous and can lead to catastrophic consequences.

One of the methods of nonlinear acoustics used for damage detection in metallic structures is a method based on frequency mixing phenomenon (see Section 1.5). One of its varieties is based on vibro-acoustic modulation. The method is based on simultaneous excitation of the structure by a high by frequency acoustic wave and low-frequency (modal) excitation. This method was used by Klepka *et al.* (2012a) to detect fatigue damage in an aluminium plate. In the first phase of the experiment a numerical model was created for the purpose of crack edge divergence (CED) analysis. Then the natural frequency (associated with the different dynamic behaviour of damage) was selected for the nonlinear acoustics test. The selected mode shapes, and fatigue crack specimen are presented in Figure 4.10.

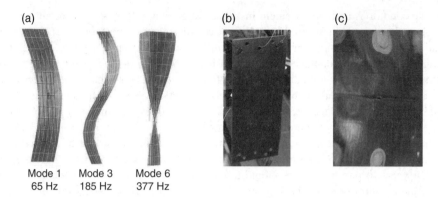

Figure 4.10 Selected mode shapes (a), specimen under investigation (b) and fatigue crack (c). Reproduced from Klepka A., Staszewski WJ., Jenal R., Szwedo M., Uhl T., and Iwaniec J. 2012a. Nonlinear acoustics for fatigue crack detection experimental investigations of vibro-acoustic wave modulations. *Structural Health Monitoring* **25**, 197–211, © 2012 Sage Publications

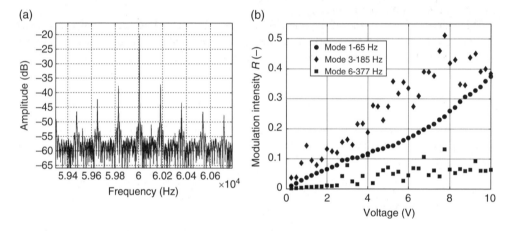

Figure 4.11 Power spectrum illustrating vibro-acoustic modulations (a). Modulation intensity for different levels of low-frequency excitation (b). Reproduced from Klepka A., Staszewski WJ., Jenal R., Szwedo M., Uhl T., and Iwaniec J. 2012a. Nonlinear acoustics for fatigue crack detection experimental investigations of vibro-acoustic wave modulations. *Structural Health Monitoring* **25**, 197–211, © 2012 Sage Publications

The plate was simultaneously modally excited (low-frequency excitation) with the use of various amplitude levels. Modulated responses were acquired and the Fourier analysis was applied to obtain power spectra. Figure 4.11(a) gives an example of the power spectrum for the 6 V low-frequency excitation amplitude.

Figure 4.11(b) gives the intensity of the modulation parameter for three different vibration modes and various amplitude levels of low-frequency excitation. The values of R for the 3rd (crack mode III) and 6th (crack mode II) vibration mode excitation are scattered compared with the results for the 1st (crack mode I) vibration mode excitation. The results show

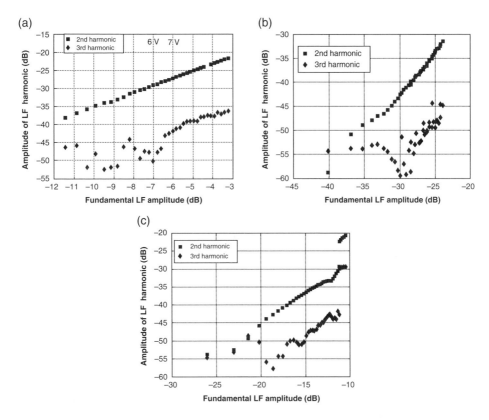

Figure 4.12 Analysis of modulation sidebands amplitude: (a) 1st vibration mode; (b) 3rd vibration mode; and (c) 6th vibration mode. Vertical dashed lines indicate the area of crack opening. LF, low-frequency. Reproduced from Klepka A., Staszewski WJ., Jenal R., Szwedo M., Uhl T., and Iwaniec J. 2012a. Nonlinear acoustics for fatigue crack detection experimental investigations of vibro-acoustic wave modulations. *Structural Health Monitoring* **25**, 197–211, © 2012 Sage Publications

that modulation intensity generally increases with excitation level, as expected. However, the largest values of R were obtained for the 3rd vibration mode excitation whereas the largest strain level was induced to the plate by the 1st vibration mode excitation. It is important to note that opening/closing action was observed only for the latter excitation. This suggests that the opening/closing action of the crack is not the major factor contributing to vibro-acoustic modulations. Figure 4.12 gives the amplitude of the second and third harmonics versus the amplitude of the fundamental low-frequency harmonic for the vibro-acoustic nonlinear test performed using three different vibration mode excitations.

The analysis of harmonics and sidebands presented in this chapter demonstrates a possible twofold nonlinear mechanism for the analysed large crack and low-frequency excitation levels. Some evidence of the classical nonlinear elasticity (first-order perturbation due to quadratic second harmonic, cubic third harmonic, quadratic second-order sideband dependence and linear first-order sideband dependence) can be observed for all three vibration

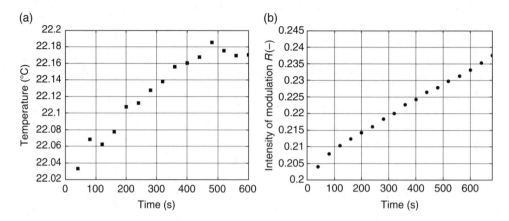

Figure 4.13 Experimental evidence of vibro-acoustic modulation dependence on coupled strain – temperature field: temperature versus time (a); and modulation intensity versus time (b) (Klepka *et al.* 2012b)

modes analysed. Some evidence of hysteretic behaviour can be observed in the pattern of generated harmonics (quadratic third harmonic) and sidebands (linear second-order sideband dependence) when the plate is excited by the 3rd vibration mode.

The pattern of harmonics and sidebands is clearly affected by the opening crack and the classical perturbation nonlinearity is probably related to the elastic behaviour associated with the crack. However, the analysis clearly shows that this mechanism is not dominant when nonlinear modulations are analysed. First, a strong pattern of modulation sidebands was observed for the 3rd and 6th vibration mode excitation when the crack was always closed. Secondly, the largest intensity of modulations was achieved for the 3rd vibration mode despite the fact that the strain level associated with this excitation was much smaller than the strain level for the 1st vibration mode. The 3rd vibration mode exhibited some evidence of hysteretic elasticity.

One more experiment was performed to reveal the possible link between temperature change and vibro-acoustic modulations. The cracked aluminium plate was excited with constant-amplitude low-frequency vibration and high-frequency ultrasonic signals for 600 s. It is important to note that the crack did not open for the excitations levels used. The temperature was then measured in the vicinity of the crack. Vibro-acoustic responses were captured simultaneously with temperature measurements to estimate the intensity of modulation (R parameter). The results are presented in Figure 4.13. Again, the temperature did not change when the plate was excited with the 3rd mode excitations and increased by approximately 0.12 (4.13a) and 0.05 °C when the plate was excited by the 1st and 6th vibration modes, respectively. At the same time the intensity of modulation [Figure 4.13(b)] increased by 20% but only for the 1st vibration mode excitation.

There are three main conclusions from the temperature analysis. First, the link between temperature and vibro-acoustic modulations has been found only for the 1st vibration mode (crack mode I – opening – closing) for strain levels that did not open the crack but were sufficient to perturb crack surfaces. Interestingly, this mode exhibited some evidence of the

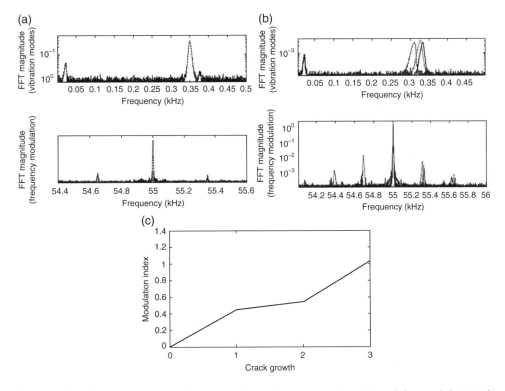

Figure 4.14   The spectra of acoustic response for undamaged specimen (a) and damaged (b) sample with different crack size. (c) Modulation index versus crack size. Reproduced by permission from Haron M. and Adams DE. 2008. Implementation of nonlinear acoustic techniques for crack detection in a slender beam specimen. *Health Monitoring of Structural and Biological Systems* 2008. Edited by Kundu T, 69350P–69350P–12, © 2008 SPIE

nonclassical amplitude attenuation (the L-G effect). Secondly, no heat generation (temperature change) has been observed for the 3rd vibration mode excitation (crack mode III – tearing) that produced the largest intensity modulation and strong evidence of hysteretic behaviour. Thirdly, some heat generation (probably due to friction) has been observed for the 6th vibration mode excitation that produced the smallest level of modulation intensity. However, the temperature change did not produce any changes in vibro-acoustic modulations.

Intensity of the modulation parameter is often used as a damage index. In Haron and Adams (2008) the slender beam specimen is investigated in order to detect the fatigue crack. A modal impact hammer and a piezoelectric transducer are used to excite the structure by broad-band impulse and high frequency acoustic wave, respectively. Experiments were performed for different damage size and for each of them the intensity modulation factor was calculated. Figure 4.14 presents results of the performed tests.

A very important issue raised in the article is the problem of carrying out the nonlinear acoustics experiment. The authors analyse in detail the influence of the measuring apparatus, boundary conditions and variability of environmental conditions. Nonlinear acoustics is also

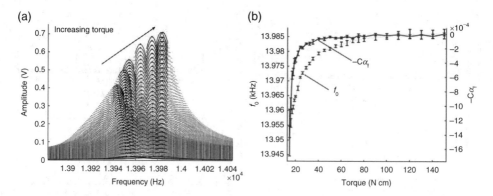

Figure 4.15 Spectra of the system response for different excitation amplitudes and different torques (a). Evolution of linear and nonlinear elastic parameters versus torque (b). Reproduced with permission from Riviere J., Renaud G, Haupert S., Talmant M., Laugier P. and Johnson P. Nonlinear acoustic resonances to probe a threaded interface. *Journal of Applied Physics* **107**, © 2010 American Institute of Physics

used for SHM. In Riviere *et al.* (2010) a nonlinear acoustic resonance technique was used for the detection of torque changes in a threaded interface. The method is based on the identification of nonlinear parameters associated with nonlinear hysteretic behaviour. These parameters are responsible for the shift of resonance frequencies (nonlinear elastic parameter) and damping variation (nonlinear dissipative parameter). The results are presented in Figure 4.15(a).

The analyses performed [Figure 4.15(b)] clearly show that the nonlinear elastic parameter is more sensitive for the detection of torque changes than its linear counterpart. Further analysis shows the simultaneous presence of different types of nonlinearity associated with classical second-order nonlinearity (cubic) and nonclassical hysteretic nonlinearity.

4.3.1.2   Composites

A very important area of SHM is the monitoring of composite structures. Various methods have been developed in the last few decades to detect different types of damages in composites structures. These include classical, well-established NDT/E approaches such as vibration based analysis (Gherlone *et al.* 2005), acoustic emission (AE) (Scholey *et al.* 2006), X-ray radiography (Staszewski *et al.* 2004), thermo- and shearography (Staszewski *et al.* 2004), eddy current (Gros (1995)), vibrothemography (Pieczonka *et al.* 2011), GW techniques (Ambrozinski *et al.* 2012) and A- and C-scanning ultrasonic inspection techniques (Cantwell and Morton 1992; Pearson *et al.* 2011). The classical methods based on ultrasonic and guided ultrasonic waves utilize linear effects associated with wave propagation. Those methods work properly for relatively large damages, comparable with the wavelength where phenomena like reflections and scattering can be used for damage detection. In recent years there has been an increasing interest in nonlinear methods of damage detection (Farrar *et al.* 2007; Iwaniec 2011). The studies conducted so far clearly indicate that such methods are more sensitive and can detect damage at a very early stage.

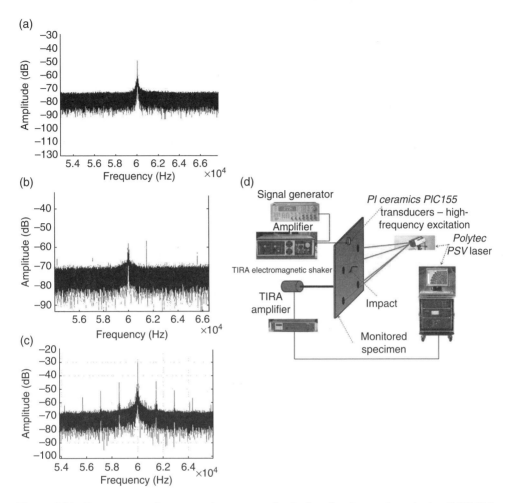

Figure 4.16 Power spectra from acoustic responses for the 1st vibration mode excitation (1430 Hz):
(a) undamaged plate, (b) damaged plate after the first 9 J impact, (c) damaged plate after the second 30
J impact. (d) Experimental setup for nonlinear acoustic tests

Among the nonlinear methods used for damage detection in composite materials the previously described method of vibro-acoustic modulation is applied. In Klepka *et al.* (2012b) the method was used to detect defects in a composite chiral panel induced by low velocity impact.

The study was conducted for a number of known levels of impact energy. After each impact test, the composite panel was subjected to a vibro-acoustic modulation experiment for three different low frequency excitations corresponding to the selected mode shapes. The intensity of modulation parameter was used as a damage index. The response signal was acquired by a Polytec scanning vibrometer. The experimental arrangement and an example of response signal spectra for different impact energy levels are shown in Figure 4.16.

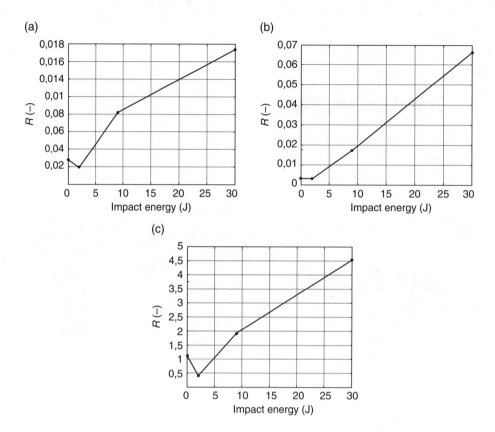

Figure 4.17 Analysis of modulation intensity R versus impact energy: (a) mode 1; (b) mode 2 and (c) mode 3

The modulation intensity was analysed using amplitudes of the carrier frequency and the first modulation sidebands, as described above. The resulting R parameter was calculated for all damage severities. The results are presented in Figure 4.17. Although, the intensity of modulation was smaller after the initial nondamaging 2 J impact, a significant increase in the R parameter can be observed for the 9 J and particularly 30 J impacts for all analysed mode shapes. This suggests that damage severity assessment is possible when nonlinear acoustic tests are performed.

A similar test was performed for a different composite structure. The composite plate was manufactured from carbon/epoxy (Seal HS160/REM) with unidirectional prepreg layers. The stacking sequence of the laminate was [03/903]. The average laminate thickness was equal to 2 mm. The plates were cured in an autoclave at a maximum temperature of $160^{o}C$ and a maximum pressure of 6 bar. The specimen was ultrasonically C-scanned prior to testing to assess the quality of the laminate and to exclude any presence of possible manufacturing defects. Impact tests were conducted using an instrumented drop-weight testing machine. The experimental arrangement was similar to the one used for the chiral panel. The intensity of the modulation parameter was calculated using amplitudes of the high frequency acoustic wave

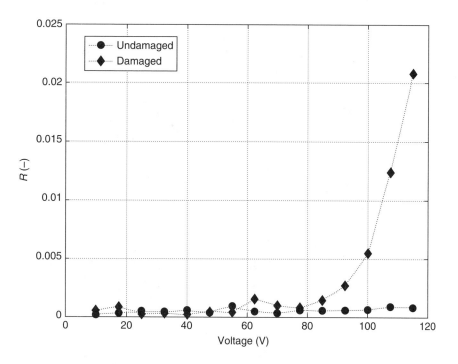

Figure 4.18   Analysis of modulation intensity R versus excitation level

and the first modulation sidebands, as described above. The analysis was performed for both damaged and undamaged plate. Figure 4.18 presents the R parameter versus amplitude of low-frequency excitation.

For a low level of excitation small values of modulation intensity parameters were observed for the damaged and undamaged specimen. A significant increased of R parameter could be observed for the damaged plate after the level of excitation exceeded 70 V. In the author's opinion, that may be due to a change in the dynamic behaviour of damage caused by the increase in low-frequency excitation. A possible explanation for this phenomenon is the change in the stress–strain characteristics from linear to bilinear. That makes the stiffness of the material higher for the compression than for the tensile phase and causes pulse-type stiffness modulation and wave deformation. This explanation is confirmed by the results of modelling. For the analysed mode shape, the contribution from the closing – opening action in the dynamics of damage is the largest and causes such a change in the stiffness characteristics. Figure 4.19 presents the amplitude of the second and third harmonics plotted against the amplitude of the fundamental 467 Hz harmonic.

A quadratic amplitude dependence (slope equal approximately to 2) for the second harmonic can be observed for the damaged plate (classical first-order nonlinearity), while for the undamaged specimen amplitude values are scattered. The third harmonic amplitude values are scattered until the amplitude of low-frequency excitation reaches -70 dB. This corresponds approximately to 70 V of low-frequency excitation, that is a level at which the intensity of modulation starts to increase significantly. This confirms the theory put forward during the

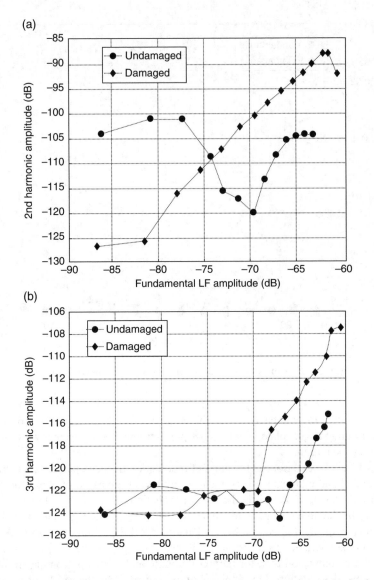

Figure 4.19 Analysis of the second (a) and third (b) harmonics as a function of the fundamental low-frequency (LF) amplitude

analysis of the modulation intensity factor that the nonlinear mechanisms caused by damage are revealed above a certain excitation level.

A very similar effect can be observed by analysing the amplitude of the first pair of sidebands (Figure 4.20). The amplitudes starts to increase significantly when the level of low-frequency excitation exceeds 70 V (70 dB). The results clearly show a classical first-order elastic perturbation nonlinearity caused by damage in a composite plate. This nonlinearity

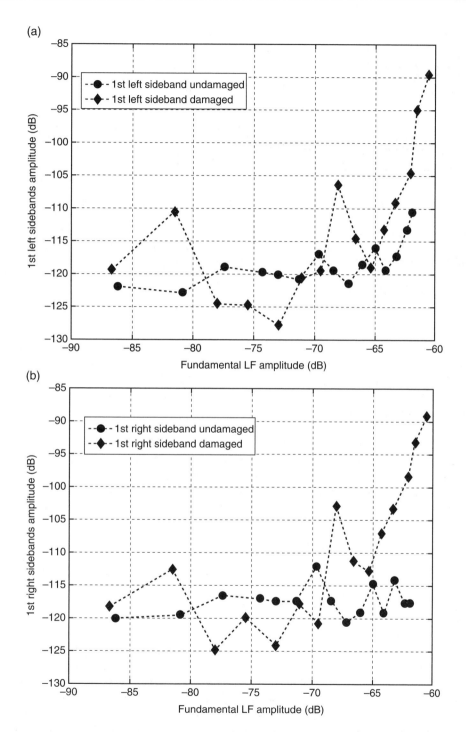

Figure 4.20 Analysis of the first left (a) and first right (b) sidebands as a function of the fundamental low-frequency (LF) amplitude

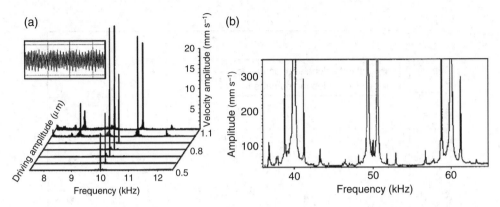

Figure 4.21 (a) Subharmonic and self-modulation spectra as functions of the driving amplitude for C/C-SiC-composite damaged structure. (b) UFP spectrum of an impacted GFR composite. Reproduced with permission from Solodov I, Wackerl J, Pfleiderer K and Busse G. 2004. Nonlinear self-modulation and subharmonic acoustic spectroscopy for damage detection and location. *Applied Physics Letters* **84**, 5386–5388, © 2004 American Institute of Physics

appears when a certain level of low-frequency excitation is reached. It follows that that in order to properly diagnose this type of structure it is necessary to obtain an adequate strain level that allows the nonlinear, dynamical behaviour of the damage to be released. Results of the analyses of higher harmonics and sidebands are also consistent with the results obtained in previous studies on fatigue damage.

Composite damages caused by mechanical impact can have a very complex nature. Depending on the fracture type different nonlinear mechanisms may appear, taking the form of delamination, crack, debonding or a combination of these. In the work by Solodov *et al.* (2004) examples of different nonlinear mechanisms are presented. Subharmonic self-modulation spectroscopy was used to explain nonlinear phenomena observed in a damaged structure. In Figure 4.21(a) the self-modulation of the signal response spectra for a delaminated C/C-SiC-composite structure are presented. As the authors noted, increasing driving frequency results in changes in structure dynamics. The self-modulation phenomenon observed for a low-amplitude input turns into a different nonlinear mechanism, which results in ultra-frequency pairs (UFP) components. The same components were observed for a 14-ply epoxy based glass-fibre reinforced (GFR) composite with a 9.5 J impact. The self-modulation spectrum [Figure 4.21(b)] shows UFP components resulting from nonlinear behaviour of the damaged structure.

The method based on local defect resonance (LDR) is presented in Solodov *et al.* (2012). A fundamental assumption of the method is that the damage causes a local change in stiffness for a certain mass of the structure. This is manifested by a particular resonant frequency. This frequency is a function of depth of delamination and its size. When the structure is excited with frequency equal to the LDR, the level of higher harmonics in the spectrum of signal response increases dramatically while for a different value of excitation frequency the spectrum of the signal response is close to linear. This effect is presented in Figure 4.22(a) and (b). Further nonlinear mechanisms arise if the driving frequency is fractional or an integer multiple of the

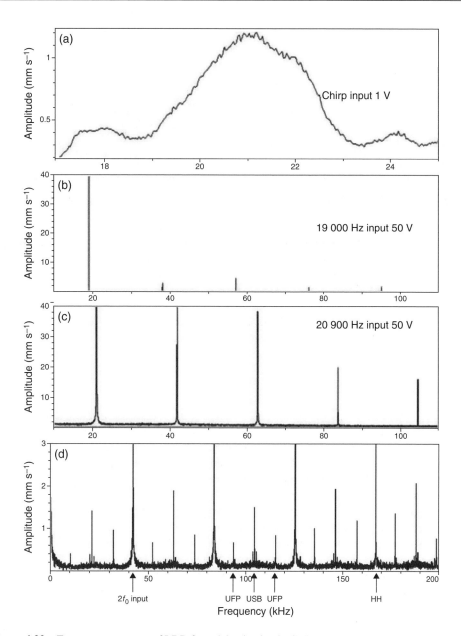

Figure 4.22 Frequency response of LDR for a delamination in GFRP (a), signal response spectrum for driving frequency outside LDR bandwidth (b), signal response spectrum for driving frequency equal to LDR (c) and higher harmonic USB-UFP resonance for a delamination in GFRP (d). Reproduced with permission from Solodov I, Bai J, Bekgulyan S and Busse G. 2012. A local defect resonance to enhance wave-defect interaction in nonlinear spectroscopy and ultrasonic thermography. *18th World Conference on Nondestructive Testing*

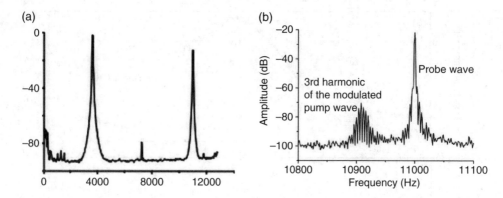

Figure 4.23   The spectra of the pump and probe waves used for excitation (a). The spectra demonstrated modulation transfer to the third harmonic of the pump wave (b). Reproduced with permission from Zaitsev V, Nazarov V, Gusev V and Castagnede B. 2006. Novel nonlinear-modulation acoustic technique for crack detection. *NDT & E International* **39**, 184–194. © 2006 Elsevier

LDR. Then the ultra-subharmonics (USB) and UFP can be observed on signal spectrum. An example of USB – UFP resonance is shown in Figure 4.22(c).

4.3.1.3   Glass, Plexiglass

Nonlinear damage detection methods are also used for materials with a structure different from solids. An example is glass, which is an inorganic material, cooled to a solid state without the crystallization process. Glass is not a plastic material: it can be deformed elastically or cracked (a very large Young's modulus, brittleness). Moreover, glass is an amorphous material with a nonperiodic atomic lattice. An example of a nonlinear technique applied for damage detection in glass is the cross-modulation technique. Pump and probe waves were used in (Zaitsev *et al.* 2006) to excite the structure. The frequency of these waves was tuned to a different resonance of the specimen. The result of the experiment is presented in Figure 4.23.

The study shows that the amplitude-dependent dissipative effect may play a dominant role in the modulation phenomena. The method based on nonlinear time reversal acoustics (NLR-TRA) is presented in Sutin *et al.* (2003). In this technique the classical time reversal acoustic method, which allows the focusing of an ultrasonic wave, was combined with a nonlinear elastic response. The characteristics of the second and third harmonics (Figure 4.24.) show nonclassical mechanisms in the damaged structure.

4.3.1.4   Concrete, Rocks, Geomaterials

Another group of materials to which nonlinear acoustics were applied are geomaterials. Their structures differ significantly from homogeneous materials such as liquids or solids. The dynamic behaviour of these materials is completely different. This difference is due to the mechanical properties. Monocrystalline structures have 'atomic elasticity' while geomaterials

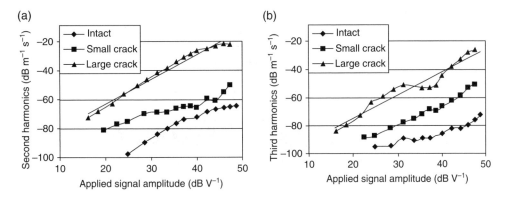

Figure 4.24 Characteristics of second (a) and third (b) harmonics focused on two different cracks. Reproduced with permission from Sutin A, Johnson P and TenCate J. 2003. Development of nonlinear time reversed acoustics (NLTRA) for applications to crack detection in solids. *Proceedings of the World Congress of Ultrasonics*, p. 121

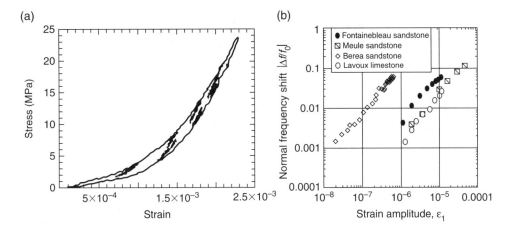

Figure 4.25 Stress–strain experimental characteristics for sandstone (a). Normalized frequency shift versus strain for various types of rocks (b). Reproduced with permission from Ostrovsky L and Johnson P. 2001. Dynamic nonlinear elasticity in geomaterials. *Rivista Del Nuovo Cimento*

have 'structural nonlinear elasticity' or 'mesoscpoic elasticity'. Nonlinear behaviour such as hysteresis and slow dynamic effects are characteristic for geomaterials. Many studies clearly demonstrate that the nonlinear behaviour of the micro-inhomogeneous material is released very quickly on damage occurrence. This allows the application of damage detection techniques in materials such as ceramics, microcrystalline materials, rocks and concrete. Nonlinear elasticity is evidently manifested during quasi-static tests. Figure 4.25(a) shows the stress–strain characteristics obtained for a sandstone sample (Ostrovsky and Johnson 2001).

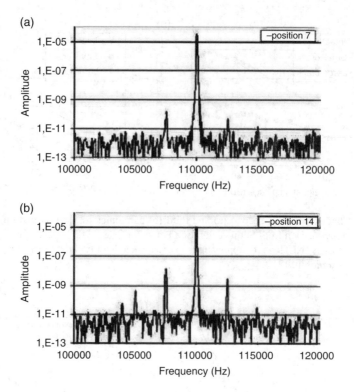

Figure 4.26 Modulated spectra of signal response for intact (a) and damaged (b) specimen. Reproduced with permission from Van Den Abeele K, Katkowski T, Wilkie-Chancellier N and Desadeleer W. 2006. *Universality of Nonclassical Nonlinearity. Applications to Non-Destructive Evaluations and Ultrasonics.* © 2006 Springer

Nonlinear behaviour resulting from the presented characteristics includes nonlinear stress–strain, hysteresis and the discrete memory effect. In the same work different nonlinear mechanisms observed in rock are also presented. In Figure 4.25(b) the frequency shift versus fundamental mode strain amplitude for different types of rock is presented. The relationship between frequency shift and the level of strain can identify the type of nonlinearity. In this case, the dependence has a form of power law, which means that it is nonclassical and hysteresis behaviour is present for both static and dynamic forcing.

In the work by Van Den Abeele *et al.* (2006) a sample of damaged and undamaged Balegem stone was investigated using nonlinear wave modulation spectroscopy (NWMS) in impact mode. The structure was excited simultaneously by impact hammer and piezoelectric transducer generating high frequency acoustic waves. The results are presented in Figure 4.26. The energy of the sidebands was calculated as a damage index.

Another study on the application of nonlinear acoustics for geomaterials is that by Van Den Abeele *et al.* (2000b). In this work the nonlinear resonance acoustic spectroscopy (NRUS) method was applied for damage detection in a slate beam with fatigue damage. Figure 4.27 shows the results of the experiment for intact and damaged sample.

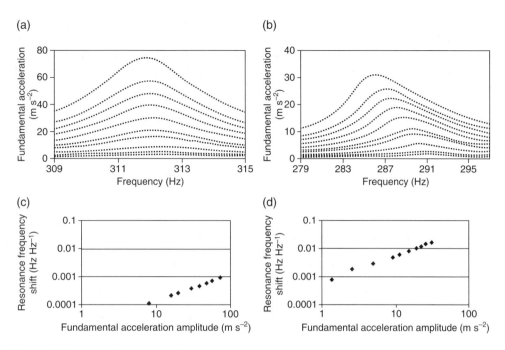

Figure 4.27 Acoustic spectra of signal response for intact (a) and damaged (b) sample. Frequency shift in resonant frequency for intact (c) and damaged (d) sample. Reproduced with permission from Van Den Abeele K, Carmeliet J, TenCate J and Johnson P. 2000b. Nonlinear elastic wave spectroscopy (news) techniques to discern material damage, Part II: Single-mode nonlinear resonance acoustic spectroscopy. *Research in Nondestructive Evaluation* **12**, 31–42. © 2000 Springer

The characteristics are typical for structures with hysteretic mesoscopic nonlinearity. For a damaged structure the softening effect is clearly visible together with nonclassical amplitude dependent behaviour. The analysis of higher harmonics and attenuation coefficient confirms that in this type of material the hysteretic nonlinearity completely dominates the classical atomic nonlinearity.

4.4   Conclusions

The presented nonlinear phenomena and effects and the application of the methods to damage detection do not exhaust the main issues related to nonlinear acoustics. Most of the presented mechanisms that cause nonlinear behaviour of the structures are theoretical models. Some of them were proved both analytically and experimentally whilst others are still under investigation. The biggest problem in nonlinear acoustics techniques is that most of the mechanisms can occur simultaneously and manifest themselves in a similar way. In many cases this complicates the interpretation of results. Studies clearly show that nonlinear methods are more sensitive than their linear equivalents. Nonlinear techniques also allow a precise analysis of damage dynamical behaviour on the basis of nonlinear phenomena generated by damage. This enables both SHM and material characterization. The experiments presented in this chapter show that these methods can be applied to various types of structures, both homogeneous and inhomogeneous. In recent years it has been shown that more and more methods of nonlinear acoustics are applicable and despite difficulties, studies on those methods are continuing and providing increasingly promising results.

References

Ambrozinski L, Stepinski T, Packo P and Uhl T 2012 Self-focusing Lamb waves based on the decomposition of the time-reversal operator using time-frequency representation. *Mechanical Systems and Signal Processing* **27**, 337–349.

Awrejcewicz J and Olejnik P 2005 Analysis of dynamic systems with various friction laws. *Applied Mechanics Reviews* **58**, 389–411.

Aymerich F and Staszewski W 2010a Experimental study of impact-damage detection in composite laminates using a cross-modulation vibro-acoustic technique. *Structural Health Monitoring* **9**, 541–553.

Aymerich F and Staszewski W 2010b Impact damage detection in composite laminates using nonlinear acoustics. *Composites Part A: Applied Science and Manufacturing* **41**, 1084–1092.

Ballada E, Vezirova S, Pfleiderer K, Solodov and Busse G 2004 Nonlinear modulation technique for NDE with air-coupled ultrasound. *Ultrasonics* **42**, 1031–1036.

Bateman T, Mason W and McSkimin H 1961 Third-order elastic moduli of germanium. *Journal of Applied Physics* **5**, 928–936.

Belyaeva I, Ostrovsky L and Zaitsev V 1997 Microstructure induced nonlinearity of unconsolidated rocks as related to seismic diagnostics problems. *Nonlinear Processes in Geophysics* **4**, 410.

Bentahar M, Marec A, El Guerjouma R and Thomas J 2009 Nonlinear acoustic fast and slow dynamics of damaged composite materials: correlation with acoustic emission. *Ultrasonics Wave Propagation in Non Homogeneous Media* **128**, 161–171.

Boller C, Chang FK and Fujino Y (eds) 2009 *Encyclopedia of Structural Health Monitoring*. John Wiley & Sons, Ltd.

Breazeale M and Ford F 1965 Ultrasonic studies of the nonlinear behaviour of solids. *Journal of Applied Physics* **36**, 3486–3490.

Bruneau M and Potel C 2009 *Materials and Acoustics Handbook*. Wiley-ISTE.

Buck O 1990 Nonlinear acoustic properties of structural materials – a review. *Review of Progress in Quantitative Nondestructive Evaluation* **9**, 1677.

Buck O and Morris W 1978 Acoustic harmonic generation at unbonded interfaces and fatigue cracks. *The Journal of the Acoustical Society of America* **64**, S33.

Cantrell JH and Yost WT 1994 Acoustic harmonic generation from fatigue-induced dislocation dipoles. *Philosophical Magazine A* **69**, 315–326.

Cantwell WJ and Morton J 1992 The significance of damage and defects and their detection in composite materials: A review. *The Journal of Strain Analysis for Engineering Design* **27**, 29–42.

Donskoy D, Sutin A and Ekimov A 2001 Nonlinear acoustic interaction on contact interfaces and its use for nondestructive testing. *NDT&E International* **34**, 231–238.

Duffour P, Morbidini M and Cawley P 2006 A study of the vibro-acoustic modulation technique for the detection of cracks in metals. *Journal of the Acoustical Society of America* **119**, 1463–1475.

Farrar CR, Worden K, Todd MD, Park G, Nichols J, Adams DE, Bement MT and Farinhol K 2007 *Nonlinear System Identification for Damage Detection*. National Nuclear Security Administration.

Friswell M and Penny J 2002 Crack modeling for structural health monitoring. *Structural Health Monitoring* **1**, 139.

Gherlone M, Mattone M, Surace C, Tassotti A and Tessler A 2005 Novel vibration-based methods for detecting delamination damage in composite plate and shell laminates. *Key Engineering Materials* **293–294**, 289–296.

Gros XE 1995 An eddy current approach to the detection of damage caused by low-energy impacts on carbon fibre reinforced materials. *Materials & Design* **16**, 167–173.

Guyer R, McCall K and Van Den Abeele K 1998 Slow elastic dynamics in a resonant bar of rock. *Geophysical Research Letters* **25**, 1585–1588.

Haller K and Hedberg C 2006 Frequency sweep rate and amplitude influence on nonlinear acoustic measurements. *The 9th Western Pacific Acoustics Conference (WESPAC)*. Seoul, Korea. http://www.bth.se/fou/forskinfo.nsf/all/e0a64b8c0e80a449c1257295002ed3ea/$file/HallerWespac2006.pdf (last accessed 8 January 2013).

Haron M and Adams DE 2008 Implementation of nonlinear acoustic techniques for crack detection in a slender beam specimen. *Proceedings of the SPIE, Health Monitoring of Structural and Biological Systems 2008* (ed. Kundu T), vol. 6935, pp. 69350P–69350P–12.

Hu HF, Staszewski WJ, Hu NQ, Jenal RB and Qin GJ 2010 Crack detection using nonlinear acoustics and piezoceramic transducers instantaneous amplitude and frequency analysis. *Smart Materials and Structures* **19**, 065017.

Iwaniec J 2011 *Selected Issues of Exploitational Identification of Nonlinear Systems*. AGH University of Science and Technology.

Jhang K and Kim K 1999 Evaluation of material degradation using nonlinear acoustic effect. *Ultrasonics* **37**, 39–44.

Kessler S, Spearing S and Soutis C 2002 Damage detection in composite materials using Lamb wave methods. *Smart Materials and Structures* **11**, 269–278.

Kim J, Yakovlev V and Rokhlin S 2003 Parametric modulation mechanism of surface acoustic wave on a partially closed crack. *Applied Physics Letters* **82**, 3203.

Kim JY and Rokhlin S 2002 Surface acoustic wave measurements of small fatigue cracks initiated from a surface cavity. *International Journal of Solids and Structures* **39**, 1487.

Klepka A, Jenal RB, Szwedo M, Staszewski WJ and Uhl T 2010 Experimental analysis of vibro-acoustic modulations in nonlinear acoustics used for fatigue crack detection. *Proceedings of the Fifth European Workshop Structural Health Monitoring*. DEStech Publications, pp. 541–546.

Klepka A., Staszewski WJ, Jenal R, Szwedo, M., Uhl T and Iwaniec J 2012a Nonlinear acoustics for fatigue crack detection experimental investigations of vibro-acoustic wave modulations. *Structural Health Monitoring* **25**, 197–211.

Klepka A, Staszewski WJ, Uhl T and Aymerich F 2012b Sensor location analysis for nonlinear-acoustics-based damage detection in composite structures. *Proceedings of Health Monitoring of Structural and Biological Systems 2012*, 8348.

Kolomenskii AA, Lioubimov VA, Jerebtsov SN and Schuessler HA 2003 Nonlinear surface acoustic wave pulses in solids: Laser excitation, propagation, interactions. *Review of Scientific Instruments* **74**, 448–452.

Kundu T 2003 *Ultrasonic Nondestructive Evaluation: Engineering and Biological Material Characterization*. CRC Press.

Landau L and Lifshitz E 1986 *Theory of Elasticity* (translated from the Russian by J. B. Sykes and W. H. Reid). Pergamon Press.

Meo M, Polimeno U and Zumpano M 2008 Detecting damage in composite material using nonlinear elastic wave spectroscopy methods. *Applied Composite Materials* **1**, 115.

Meo M and Zumpano M 2005 Nonlinear elastic wave spectroscopy, identification of impact damage on a sandwich plate. *Computers & Structures* **71**, 469–474.

Morris W, Buck O and Inman R 1979 Acoustic harmonic generation due to fatigue damage in high-strength aluminum. *Journal of Applied Physics* **50**, 6737.

Moussatov A, Gusev V and Castagnede B 2003 Self-induced hysteresis for nonlinear acoustic waves in cracked material. *Physical Review Letters* **90**, 124301.

Muller M, Sutin A, Guyer R, Maryline M, Talmant P, Laugier and Johnson P 2005 Nonlinear resonant ultrasound spectroscopy (NRUS) applied to damage assessment in bone. *The Journal of the Acoustical Society of America* **118**, 3946–3952.

Nagy PB 1994 Excess nonlinearity in materials containing microcracks. *Review of Progress in Quantitative Nondestructive Evaluation* **13**, 1987–1994.

Nagy PB 1998 Fatigue damage assessment by nonlinear ultrasonic material characterization. *Ultrasonics* **36**, 375–381.

Nazarov V, Radostin A and Soustova I 2002 Effect of an intense sound wave on the acoustic properties of a sandstone bar resonator. *Acoustical Physics* **48**, 76.

Nazarov VE 1988 Nonlinear acoustics of micro-inhomogeneous media. *Nonlinear Acoustics of Micro-inhomogeneous Media* **50**, 65.

Ostrovsky L and Johnson P 2001 Dynamic nonlinear elasticity in geomaterials. *Rivista Del Nuovo Cimento*. **24**, 1–46.

Pearson MR, Eaton MJ, Featherston CA, Holford KM and Pullin R 2011 Impact damage detection and assessment in composite panels using macro fibre composites transducers. *Journal of Physics: Conference Series, 9th International Conference on Damage Assessment of Structures* **305**, doi:10.1088/1742–6596/305/1/012049.

Pecorari C 2004 Adhesion and nonlinear scattering by rough surfaces in contact: Beyond the phenomenology of the Preisach–Mayergoyz framework. *The Journal of the Acoustical Society of America* **116**, 1938.

Pecorari C and Solodov I 2006 *Non-classsical Nonlinear Dynamics of Solid Interfaces in Partial Contact for NDE Applications*. Springer, pp. 307–324.

Pieczonka L, Staszewski W, Aymerich, F, Uhl T and Szwedo M 2011 Numerical simulations for impact damage detection in composites using vibrothermography. *IOP Conference Series: Materials Science and Engineering* **10**, Article 012062.

Raghvan A CCES 2007 Review of guided-wave structural health monitoring. *The Shock and Vibration Digest* **39**, 91–114.

Riviere, J, Renaud G, Haupert S, Talmant M, Laugier P and Johnson P 2010 Nonlinear acoustic resonances to probe a threaded interface. *Journal of Applied Physics* **107**, doi:10.1063/1.3443578.

Rizzo P, Cammarata M, Dutta D and Sohn H 2009 An unsupervised learning algorithm for fatigue crack detection in waveguides. *Smart Materials and Structures* **18**, 025016.

Scholey JJ, Wilcox, P. D, Lee CK, Friswell MI and Wisnom MR 2006 Acoustic emission in wide composite specimens. *Advanced Materials Research* **13-14**, 325–332.

Solodov I 1998 Ultrasonics of nonlinear contacts: Propagation, reflection and NDE-applications. *Ultrasonics* **36**, 383.

Solodov I 2010 Nonlinear acoustic NDT: Approaches, methods, and applications. http://www.ndt.net/article/Prague 2009/ndtip/proceedings/Solodov_10.pdf (last accessed 8 January 2013).

Solodov I, Bai J, Bekgulyan S and Busse G 2012 A local defect resonance to enhance wave-defect interaction in nonlinear spectroscopy and ultrasonic thermography. *18th World Conference on Nondestructive Testing*. Durban, South Africa. http://www.ndt.net/article/wcndt2012/papers/34_Solodov.pdf (last accessed 8 January 2013).

Solodov I, Wackerl J, Pfleiderer K and Busse G 2004 Nonlinear self-modulation and subharmonic acoustic spectroscopy for damage detection and location. *Applied Physics Letters* **84**, 5386–5388.

Staszewski W 2004 *Structural Health Monitoring Using Guided Ultrasonic Waves*. Springer.

Staszewski W and Parsons Z 2006 Nonlinear acoustics with low-profile piezoceramic excitation for crack detection in metallic structures. *Smart Materials and Structures* **15**, 1110–1118.

Staszewski WJ, Boller C and Tomlinson GR (eds) 2004 *Health Monitoring of Aerospace Structures*. John Wiley & Sons, Ltd.

Sutin A, Johnson P and TenCate J 2003 Development of nonlinear time reversed acoustics (NLTRA) for applications to crack detection in solids. *Proceedings of the Word Congress of Ultrasonics*. Paris, France. http://ees.lanl.gov/source/orgs/ees/ees11/geophysics/nonlinear/2003/NLTRA_wcu.pdf (last accessed 8 January 2013).

Van Den Abeele K, Campos-Pozuelo C, Gallego-Juarez J, Windels, F and Bollen B 2002 Analysis of the nonlinear reverberation of titanium alloys fatigued at high amplitude ultrasonic vibration. *Proceedings of Forum Acustica* http://digital.csic.es/bitstream/10261/7959/1/non02002.pdf (last accessed 8 January 2013).

Van Den Abeele K, Johnson P and Sutin A 2000a Nonlinear elastic wave spectroscopy (NEWS) technique to discern material damage. Part I: Nonlinear wave modulation spectroscopy. *Review of Progress in Quantitative Nondestructive Evaluation* **12**, 17–30.

Van Den Abeele K, Johnson P and Sutin A 2000b Nonlinear elastic wave spectroscopy (NEWS) technique to discern material damage. Part II: Single-mode nonlinear resonance acoustic spectroscopy. *Research in Nondestructive Evaluation* **12**, 31–42.

Van Den Abeele K, Katkowski T, Wilkie-Chancellier N and Desadeleer W 2006 Laboratory experiments using nonlinear elastic wave spectroscopy (NEWS): a precursor to health monitoring applications in aeronautics, cultural heritage, and civil engineering. In *Universality of Nonclassical Nonlinearity. Applications to Non-Destructive Evaluations and Ultrasonics* (ed. Delsanto PP). Springer, pp. 389–409.

Wilcox PD, Konstantinidis G, Croxford AJ and Drinkwater BW 2007 Strategies for guided wave structural health monitoring. In *Review of Progress in Quantitative Nondestructive Evaluation* (eds Thompson DO and Chimenti DE), vol. 894 of American Institute of Physics Conference Series. American Institute of Physics, pp. 1469–1476.

Woolfries S 1998 The effects of friction between crack faces on nonlinear wave propagation in microcracked media. *Journal of the Mechanics and Physics of Solids* **46**, 621.

Zaitsev V, Gusev V and Castagnede B 2002a Luxembourg-Gorky effect retooled for elastic waves: a mechanism and experimental evidence. *Physical Review Letters* **89**, 105502.

Zaitsev V, Gusev V and Castagnede B 2002b Observation of the "Luxemburg-Gorky effect" for elastic waves.. *Ultrasonics* **40**, 627–631.

Zaitsev V, Nazarov V, Gusev V and Castagnede B 2006 Novel nonlinear-modulation acoustic technique for crack detection. *NDT & E International* **39**, 184–194.

Zaitsev V and Sas P 2000 Dissipation in microinhomogeneous solids: Inherent amplitude-dependent losses of a non-hysteretical and non-frictional type. *Acta Acustica united with Acustica* **86**, 429.

5

Piezocomposite Transducers for Guided Waves

Michał Mańka, Mateusz Rosiek and Adam Martowicz
Department of Mechatronics and Robotics, Faculty of Mechanical Engineering and Robotics, AGH University of Science and Technology, Poland

5.1 Introduction

Ultrasonic guided waves (GWs) are the type of elastic waves that may propagate in structures confined and guided by their geometry. During propagation, the GWs interact with discontinuities in structures and produce reflected waves that may be detected by dedicated transducers. The nature of GWs is very complicated but on account of their advantages, such as ability of long distance propagation with little attenuation and selective sensitivity for different types of structural damage, GWs have gained importance in SHM applications. A review and the state of the art concerning their use in SHM systems was presented by Raghavan and Cesnik (2007).

The transducers used in SHM should be small and light in order to be integrated into the structure without significant influence on its behaviour. Additionally, to make the system economically justified, each transducer should cover a large area of the monitored structure and its costs should be low.

This chapter is concerned with piezoelectric transducers capable of sensing and generating surface and Lamb waves in SHM systems. Lamb waves can be excited in planar structures using different types of transducers, for instance, ultrasonic angle-beam transducers (Boonsang *et al.* 2006), electromagnetic acoustic transducers (EMATs) (Boonsang *et al.* 2006; Dutton *et al.* 2006; Murayama and Mizutani 2002) and laser ultrasound systems (Cosenza *et al.* 2007; Silva *et al.* 2003; Staszewski *et al.* 2004). Most of those transducers may work both as sensors and actuators, but many of them do not comply with the demands of SHM systems

Advanced Structural Damage Detection: From Theory to Engineering Applications, First Edition.
Edited by Tadeusz Stepinski, Tadeusz Uhl and Wieslaw Staszewski.
© 2013 John Wiley & Sons, Ltd. Published 2013 by John Wiley & Sons, Ltd.

since they are bulky (e.g. angle-beam transducers), brittle (piezo-wafers) or they generate low energy waves [e.g. capacitive micromachined ultrasonic transducers (CMUTs) and EMATs].

Piezoelectric transducers made of lead zirconate titanate (PZT) ceramics are a type of device that complies with many of the demands of SHM. They are affordable and their size is relatively small compared with the size of the monitored structures. The transducers most suitable for the SHM systems should be thin and flexible, capable of generating intensive Lamb waves with high degree of directionality. From the point of view of flexibility, two types of piezoelectric transducers may be distinguished: inflexible piezoceramic patches (PCs) with uniform electrodes; and flexible patches made of piezocomposite, i.e. active fibre composite (AFC) (Bent and Hagood 1997; Birchmeier *et al.* 2009; Brunner *et al.* 2005; Gentilman *et al.* 2003) and macro-fibre composite (MFC) (High and Wilkie 2003; Howard and di Scalea 2007; Salas and Cesnik 2008, 2009; Williams *et al.* 2002).

AFC and MFC transducers are normally optimized as actuators and for that reason provided with dense, comb-like electrodes. It appears, however, that sparse comb-like electrodes, similar to those used in interdigital transducers (IDTs) (Jin *et al.* 2005; Mamishev *et al.* 2004) create very interesting properties related to mode selectivity. Feasibility of the IDT for generation of Lamb waves in a structure monitoring application was investigated by Monkhouse *et al.* (1997, 2000). The authors used PVDF (polyvinylidene difluoride) as piezoelectric material to manufacture and investigate IDTs with different topologies for SHM applications. PVDF is a very flexible and cheap material, however, it exhibits considerably weaker piezoelectricity than PZT ceramics. Moreover, it cannot be used at high temperatures which may preclude its use in some structures.

In this chapter, after a brief introduction on piezoelectric transducers, three different transducers made of MFC substrate are investigated. Two of them are commercially available MFC actuators with dense electrodes. The third one is a specially designed IDT with double-sided sparse electrodes. Beam patterns of all three transducers are first calculated using numerical simulation and then compared with those measured using a laser vibrometer. Mode selectivity is also investigated using time and frequency analysis of the generated GWs. Finally, the transducer with the best directivity and mode selectivity is tested numerically and experimentally as the receiver, in order to identify the directional and frequency sensitivity for the approaching Lamb wave.

5.2   Piezoelectric Transducers for Guided Waves

5.2.1   Piezoelectric Patches

Despite their drawbacks, piezoelectric elements made of PZT ceramics are commonly used as transducers for Lamb wave generation and sensing in SHM applications (Badcock and Birt 2000; Diamanti *et al.* 2002; Dimitriadis *et al.* 1991; Klepka and Ambrozinski 2010; Schulz *et al.* 1999). Typical PZT transducers used for Lamb wave generation are built as piezoelectric patches attached to the inspected structure. Usually they take the form of a small disc that generates omni directional wave with relatively high amplitude. Frequency, mode selectivity and main directions of propagation of the wave are strongly related to the excitation frequency and patch shape (Collet *et al.* 2011). Multiple transducers forming ultrasonic phased arrays can be used to achieve directionality of the generated waves (Ochonski *et al.* 2010), however, rather

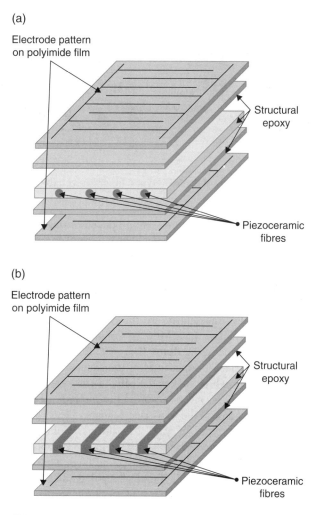

Figure 5.1   Structure of the piezocomposite transducers: (a) AFC; (b) MFC

complex electronics for signal generation and reception are required. A serious drawback of the PZT patch transducers in SHM applications is their fragility and limited life cycle which makes them very vulnerable to damage.

5.2.2   *Piezocomposite Based Transducers*

As mentioned in Section 5.1, two types of flexible piezoelectric composites suitable for patch transducers have been developed: the AFC and MFC. The AFCs have been developed by Bent and Hagood (1997) from MIT. The structure of an AFC transducer, presented in Figure 5.1(a), consists of three main layers: a single layer of piezoceramic fibres embedded in an epoxy matrix and sandwiched between two sets of dense interdigital electrodes (IDEs)

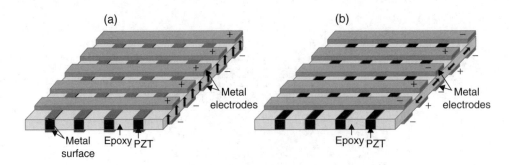

Figure 5.2   Examples of the MFC types: (a) d31 effect; (b) d33 effect

(Brunner *et al.* 2005). The piezoceramic fibres are produced through the injection moulding process of PZT powder mixed with wax binder that is heated to achieve the specific viscosity and then injected at high pressure into a cooled mould (Gentilman *et al.* 2003). The dense comb-like electrodes that are placed on the top and bottom surface of the fibres allow for the generation of directional wave and the epoxy layers make those transducers flexible (Birchmeier *et al.* 2009).

The other type of composite transducers are the MFC transducers invented by NASA (Williams *et al.* 2002). The structure of a MFC is similar to that of an AFC, as shown in Figure 5.1(b), except for the shape of the fibres. The AFC fibres are round whereas fibres in the MFC are rectangular. Piezoelectric fibres in MFC transducers are diced from piezoceramic wafers (High and Wilkie 2003), which makes the MFC a cost-effective version of the AFC. The electrode patterns in a MFC may vary according to the type of the piezoelectric effect used for the wave generation and/or to the transducer's shape.

Figure 5.2 presents two main types of MFC transducers (Collet *et al.* 2011). The first type, illustrated in Figure 5.2(a), uses the d31 effect in the piezoelectric fibres with the voltage applied through the fibres' thickness. The second type, exploits the d33 effect with the voltage applied along the fibre's length [Figure 5.2(b)]. Similarly to the AFC, the MFC transducers generate directional GWs and the use of piezocomposites makes the transducers flexible.

Piezocomposite transducers may be manufactured in different shapes and sizes but rectangular shapes are most common. Properties of the MFC transducer have been investigated in many studies, among them is an interesting one by Howard and di Scalea (2007). In this paper the authors investigate directional properties of the rectangular MFC transducers. One of the outcomes of that investigation is the design of the rosette MFC transducer [Figure 5.3(a)], which is able to detect damage in the structure not only by measuring time-of-flight (TOF) of the GWs, but also by detecting the direction of the approaching waves.

Another sensor topology that is sensitive to the GWs arrival azimuth is the transducer proposed by Salas and Cesnik (2008, 2009). The authors developed a new type of MFC structure [Figure 5.3(b)], in which a round shaped MFC transducer is divided into sectors that may be used for the detection of wave direction and/or directional wave generation.

It should be noted, however, that both transducer types, i.e. rosette and CLoVER, are provided with dense, comb-like electrodes with finger spacing much smaller than the wavelength of the generated GWs.

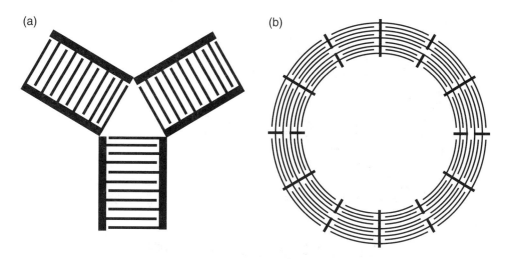

Figure 5.3   Examples of the MFC types: (a) rosette; and (b) CLoVER

5.2.3   Interdigital Transducers

5.2.3.1   **Structure of the IDT**

Contrary to the transducers presented in the previous section that were not optimized for any frequency, IDTs can be designed to excite modes with a specific wavelength. By adjusting the finger spacing in the comb electrodes accordingly to the specific wavelength in the inspected plate, the transducer will become a selective bandpass filter. The origin of the IDTs can be traced back to the surface acoustic wave (SAW) devices (Jin *et al.* 2005). However, in SHM applications the operating frequencies are normally below 1 MHz while SAW devices are designed to work with much higher frequencies in the range of tens of megahertz to several gigahertz.

A typical IDT design is presented in Figure 5.4. The IDT consists of three layers: two electrode layers separated by a piezoelectric layer. The piezoelectric layer may be made of piezoelectric polymer (i.e. PVDF; Bellan *et al.* 2005; Capineri *et al.* 2002; Monkhouse *et al.* 1997, 2000; Wilcox *et al.* 1998), piezoceramics (Luginbuhl *et al.* 1997) or piezoceramic composite (Hayward *et al.* 2001; Manka *et al.* 2010, 2011; Williams *et al.* 2002). The properties of the piezoelectric layer determine the transducer's features such as elasticity, maximal energy and frequency of the generated waves.

The most significant feature of the IDTs is the comb-finger shaped electrode pattern. The distance between the phase electrodes in the IDT (finger separation) determines the nominal wavelength of the Lamb waves generated by the transducer (Monkhouse *et al.* 1997, 2000). Similarly to the AFC and MFC transducers, IDTs generate bidirectional waves in the direction perpendicular to the finger electrodes (Na and Blackshire 2010; Na *et al.* 2008). The wave divergence angle (the main lobe width) depends on the number of fingers and their length.

Two main types of IDTs can be distinguished depending on the electrode pattern: a traditional single-sided (IDT-SS) and a double-sided (IDT-DS) (Jin *et al.* 2005). A typical

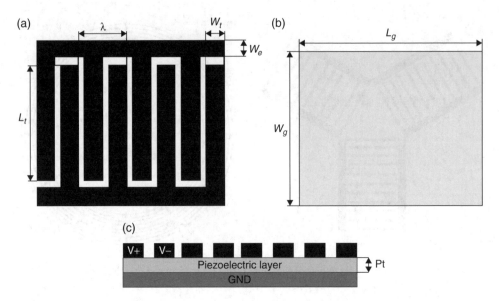

Figure 5.4   Structure of the single-sided IDT: (a) top electrode; (b) bottom electrode; and (c) cross section

IDT-SS, presented in Figure 5.4, has comb electrodes only on one side of the piezoelectric layer [Figure 5.4(a)]. The second side of the substrate is covered by ground plate electrode [Figure 5.4(b)]. Three wires are required to connect electrical signals to the electrodes (two opposite phases for the comb electrodes and ground) which means that antisymmetric sources of the phase signals are required.

   The IDT-DS, presented in Figure 5.5, has both sides covered by interdigital electrodes. To simplify the wiring, the top and the bottom electrodes are connected in pairs so that the opposite phase electrodes are placed over each other. Such design requires only two wires and does not need antisymmetric signal sources. An additional advantage of the IDT-DS compared with the IDT-SS is that the amplitude of GWs generated by the double-sided version is higher than that generated by the single-sided one (Jin *et al.* 2005).

5.2.3.2   Modifications of the Structures of the IDTs

Both IDT types presented above may be further modified to achieve some desired properties. Two types of modifications are presented here. The aim of the first group of modifications is to improve the transducer's frequency properties. The second one is aimed at achieving unidirectional GW generation.

Improving Frequency Properties of the Transducer
It is worth noting that basic function of the SAW IDT is its ability to delay analogue signals. If two sets of electrodes are placed on a piezoelectric substrate and one of those is fed with an input signal while the second is used to receive the propagating SAW, the output

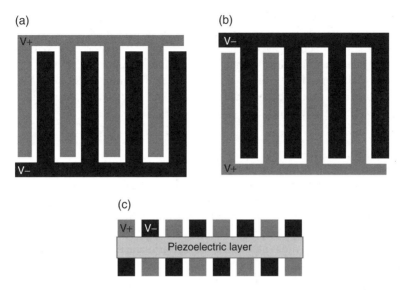

Figure 5.5 Structure of the double-sided IDT: (a) top electrode; (b) bottom electrode; and (c) cross section

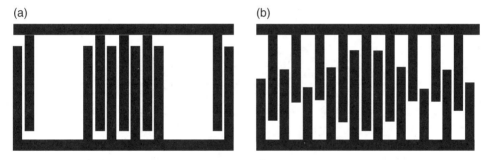

Figure 5.6 Example of structure modification of IDTs changing frequency characteristics: (a) withdraw weighting; and (b) apodization

will be delayed with respect to the input. The delay, created by the TOF, depends on the distance between the electrode sets and velocity of SAW in the substrate. It was soon discovered that the electrode sets can be modified to perform signal processing (filtering) of the signals (Morgan 1973).

One of the methods of improving frequency properties of the IDT (e.g. suppression of the sidelobe amplitude) was presented in Tancrell (1969). By controlling amplitudes and delays of each finger separately, he created a possibility to test advanced signal processing using SAW devices. Although this method provided flexibility useful in experimental work, its main disadvantage was a very complicated power supply. The second proposed method, called withdrawal weighting, consists of removing those finger electrodes where decreased coupling is required [Figure 5.6(a)]. The main drawback of this technique is that it is effective only for very long transducers.

Another method, presented in Tancrell (1969), is apodization. In this method a variation in the overlapping area of single finger electrodes is used to modify the amplitude of the generated GWs. In this method, the efficiency of each finger is proportional to its active length [Figure 5.6(b)] and therefore it does not require a multiphased power supply. Because of the simplicity and adaptivity of this technique, it is now commonly used in SAW devices. The appodization pattern may be symmetric or nonsymmetric and its shape may be changed according to particular requirements.

Unidirectional Wave Generation

Another feature of the IDT design is the ability to generate unidirectional GWs. A typical IDT (double-sided and single-sided version) generates Lamb waves in the direction perpendicular to the electrodes (Figure 5.7). In some applications, however, an unidirectional transducer is required, for instance, to reduce spurious edge reflections from the structure. Several methods have been proposed to achieve the unidirectionality in wave generation. The first of the presented methods involves multi phase structures (Hartmann *et al.* 1972). In this method, the voltage applied to the transducer's electrodes is provided by a multi phase signal generator. The transducer, also known as a unidirectional transducer (UDT), is powered in such a way that it generates waves only in the forward direction. An example of the three-phased transducer of this type is presented in Figure 5.8(a). In this design, three electrodes, separated by a

Figure 5.7   Example of the simulated GWs lobe generated by IDT

(a)   (b)

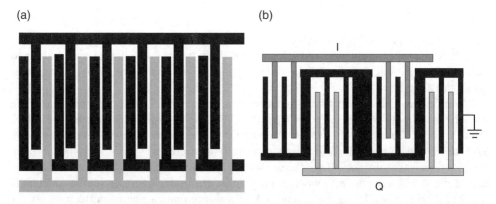

Figure 5.8   Example designs of the multi phase UDT: (a) three-phase UDT; and (b) GUDT

distance of λ/3 are used. The 120° electrical phase shift between the electrodes is required to achieve unidirectional generation of the emitted GWs. This type of transducer provides excellent performance but involves a complex construction, multilevel fabrication and the need for a three-phase power supply.

In order to overcome the disadvantages a group type of the UDT (GUDT) was introduced by Yamanouchi *et al.* (1975). Also in this design a three-phase transducer is used, [Figure 5.8(b)], but contrary to the previous device, the electrodes are split into two groups: I (in-phase) and Q (quadrature). They are powered by signals shifted in phase by 90°. Such design does not require multilevel fabrication and ensures unidirectional wave generation.

The second group of methods used to obtain unidirectional wave generation are passive methods. Contrary to the active methods, presented above, the passive methods do not require complicated and expensive multiphase power supply units.

The most straightforward among passive techniques is the method that involves resonator structures (Ash 1970). In this method reflectors are placed on the single, 'rear', side of the transducer and the rear propagating wave is trapped in an acoustic cavity. An example of this design is presented in Figure 5.9(a). Another variation of this method is placing the reflective structures in such a position that the reflected wave may be added to the generated one and practically cancel waves in the reverse direction. Because of the properties of the generated GWs a single line reflector does not ensure proper wave reflection and multiple lines are required. Moreover, such a structure has very narrow-band characteristics and only GWs at given frequencies may be reflected.

Other types of passive transducers, called single-phase UDTs (SPUDTs), are similar to the transducer proposed by Ash (1970). Also in this case the reflective elements are used but contrary to the transducers presented above, internal reflections are utilized. Over the years many different designs of the single-phase devices have been developed. Hartmann *et al.* (1982) proposed a transducer in which finger electrodes only partially overlap the reflective electrodes [Figure 5.9(b)] and the remaining part of the electrodes attenuates the rear part of the GWs.

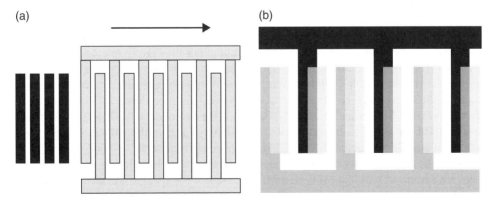

Figure 5.9 Example designs of the single-phase UDT: (a) IDT with reflectors; and (b) IDT with partially overlapping reflectors

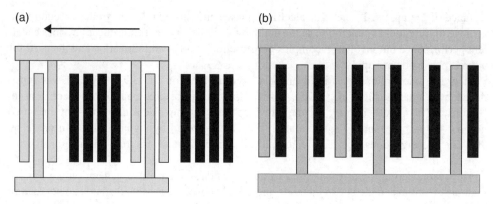

Figure 5.10 Example designs of the single-phase UDT: (a) IDT with internal reflectors; and (b) floating electrode UDT

Another design of the SPUDT was presented by Lewis (1983). In that design some regions of the IDT were removed and replaced by internal reflectors [Figure 5.10(a)]. The same principle was used in the SPUDT proposed by Yamanouchi and Furuyashiki (1984). In this design, a series of floating electrodes were used [Figure 5.10(b)]. Changes in segmentation of the reflecting electrodes, enabled variable reflectivity parameters to be achieved.

The example modifications, presented above, are only a small selection of the structure modifications of IDTs that have been used to adjust their properties to particular applications.

5.3   Novel Type of IDT-DS Based on MFC

Motivated by the specific needs of SHM applications, this chapter focuses on flat and flexible transducers capable of generating directional waves that can be easily integrated with monitored structures. AFC and MFC have similar properties, fulfil SHM's requirements and, in principle, both might be used for designing IDT for GWs. However, since the MFC is commercially available and cost effective, our attention in this chapter is focused on transducers based on this material.

Monolithic PZT offer a high piezoelectric effect (high values of electro-mechanical constant) and ability of omnidirectional wave generation, but they are fragile and have limited durability. Although these disadvantages are not so crucial in the case of small disc or plate transducers with dimensions below 10 mm, for larger transducers with size in the range of tens of millimetres, they become essential. Compared with monolithic PZT material, MFC offers increased flexibility and structural durability at the cost of decreased electro-mechanical constants. This, together with high degree of anisotropy (due to fibre structure) makes it an attractive material for structurally integrated GWs sensors/actuators.

In the following sections the commercially available MFC transducers with dense electrodes are compared with the MFC based IDTs designed for specific wavelength. The IDTs presented in the following sections are designed to excite the A_0 mode in a 4 mm thick aluminium plate. This mode exists at a broad band of frequencies, however, the most useful are

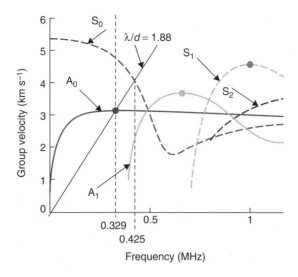

Figure 5.11 Group velocity of a 4 mm thick aluminium plate

the frequencies characterized by low dispersion, i.e. the frequency band for which the excited waves travel with an approximately constant speed.

Such a situation occurs at the points where group velocity is either stationary or almost stationary with respect to frequency, which takes place at the maxima indicated by dots in the dispersion plot shown in Figure 5.11. In the case presented here, the maximum of the group velocity of the A_0 mode in a 4 mm thick aluminium plate exists at approximately 330 kHz, and therefore this frequency has been chosen as the nominal frequency for the presented IDT. The wavelength of the Lamb waves in our case is

$$\lambda = \frac{c_p}{f} = \frac{246 \; (\text{km s}^{-1})}{330 \; (\text{kHz})} \simeq 7.5 (\text{mm}), \tag{5.1}$$

where c_p is phase velocity and f is frequency.

Another essential aspect of designing the IDTs is directionality of the generated/sensed wave. It is related to the finger length (L) and it can be approximately evaluated from

$$\gamma = \arcsin \frac{\lambda}{L}, \tag{5.2}$$

where γ is beam divergence angle.

In the transducer presented here, the finger length L is 15 mm, which yields the divergence angle 30°. The double-sided version of the IDT was chosen for our design due to the higher amplitude of the excited wave and simpler wiring. The designed transducer (subsequently referred to as IDT-DS4) had the following dimensions: finger separation equal to $\lambda = 7.5$ mm and thickness of the comb electrode $W_t = 1.9$ mm [Figure 5.4(a)].

Table 5.1   Geometric parameters of the transducers used during experiments

Transducer type	MFC-M2814-P2 (d31 effect)	MFC-M2814-P1 (d33 effect)	IDT-DS4 (d31 effect)
Active length (mm)	28	28	28.2
Active width (mm)	14	14	15

(a)   (b)

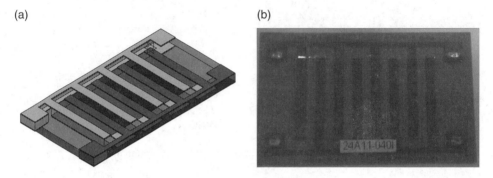

Figure 5.12   The IDT-DS4 prototype used in the laboratory tests: (a) 3D design; and (b) photograph

Two standard, commercially available MFC transducers have been chosen for the laboratory tests reported here (P1 with d33 effect and P2 with d31 effect) with dimensions close to those of the designed IDT. Both of them were provided by Smart Material GmbH, Dresden, Germany. This manufacturer also provided a number of MFC based IDTs with custom designed electrodes. Dimensions of the tested transducers can be compared in Table 5.1. The properties of the piezoelectric layers were similar for all the investigated transducers (fibre cross section 0.5×0.2 mm). The electrodes on both sides of the fibre substrate were made of a 0.18 mm thick copper layer. The 3D design and photograph of the manufactured IDT are Presented in Figure 5.12.

5.4   Generation of Lamb Waves using Piezocomposite Transducers

5.4.1   Numerical Simulations

To identify the properties of the designed IDT-DS4 and the MFC transducers acting as transmitters, a set of numerical simulations has been performed. The numerical models of the transducers and the aluminium plate have been created in ANSYS Multiphysics software. Each of the transducers was placed centrally on a 4 mm thick aluminium plate with dimensions 500×500 mm.

The model of the structure was built using 20-node brick finite elements. Fully coupled transient analysis was performed to simulate the piezoelectric effect. The transducers were excited with electrical signals (five-cycle tone burst modulated with Hanning window, amplitude of 100 V peak to peak and frequency of 100, 330 and 425 kHz) as shown in Figure 5.13. GW propagation was simulated in the time interval of 80 μs.

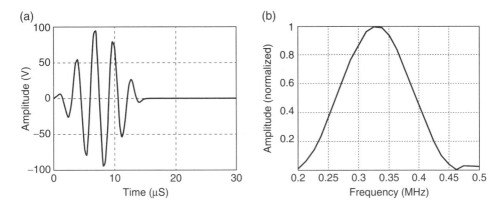

Figure 5.13   The excitation signal with frequency of 330 kHz: (a) time plot; and (b) frequency spectrum

Three excitation frequencies were chosen for the tests: 330 kHz which is the nominal frequency for the designed IDT (maximum on the dispersion plot for A_0 mode in 4 mm aluminium plate); 425 kHz, the frequency for which the S_0 mode's wavelength is equal to 7.5 mm (finger separation); and 100 kHz, low frequency excitation to verify the behaviour of the tested transducers at low frequencies.

The averaged values of the MFC parameters, provided by Smart Material GmbH, have been used as the properties of the piezoelectric layer (see Table 5A.1 and Table 5A.2 in Appendix).

5.4.1.1   Beam Pattern Calculation

The response surface method (RSM) (Box and Draper 1986; Myers and Montgomery 1995) was applied to calculate beam patterns in all the simulations performed, to analyse the dynamic properties of the modelled piezoelectric transducers. The RSM enabled the approximate maximal nodal velocity to be found for all the measurement points spread on the circle with radius of 150 mm where subsequently measurements were carried out using a scanning laser vibrometer. Since a direct comparison with the results of the measurements was impossible when rectangular FEs were used, a set of nodes in FE mesh located in close vicinity to the area with radius of 150 mm was selected. Transformation of the coordinates from the Euclidean to polar system was performed for the selected nodes to minimize geometric irregularities of the RSM input domain. The Hilbert transform was used to determine envelopes for time–history plots of the vertical nodal velocity for all the selected nodes. Finally, the maximal value for each envelope was found and assumed to represent the nodal velocity of the generated mode (either symmetric or antisymmetric). These data were used as an input for the RSM application. An example snapshot for a node is presented in Figure 5.14(a) for the IDT-DS4 excited with 100 kHz.

A polynomial metamodel was applied to approximate the relationship between input parameters (normalized radius and angle) and the maximal value of the envelope. The least square method was applied to determine the coefficients of 141 regressors by using 1405 known

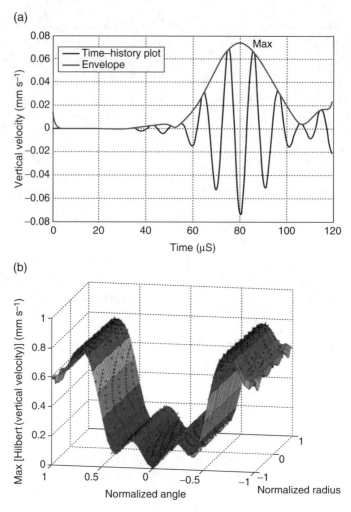

Figure 5.14 Example of the beam pattern calculation: (a) snapshot of the vertical velocity for a node; and (b) extremes of the envelopes

values of envelope extremes. An example of the response surface generated for the IDT-DS4 excited with 100 kHz is presented in Figure 5.14(b). The normalized values of angle and radius from the interval [−1, 1] represent the variation [−90, 90] and [148 mm, 152 mm], respectively.

5.4.2   Experimental Verification

To verify the numerically identified properties of the investigated transducers experimental tests were performed. The investigated transducers were mounted on the $1000 \times 1000 \times 4$ mm

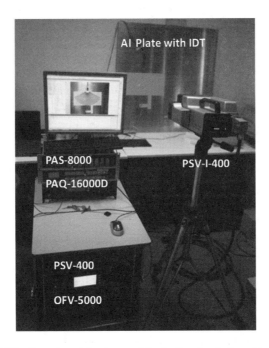

Figure 5.15   Experimental setup used for testing the designed transducers

aluminium plate using Loctite 3430 epoxy bond that was cured before the tests for at least 24 h. The excitation signals were generated by Piezo-Acquisition System PAS-8000 and PAQ-16000D, specially designed for SHM applications by EC-Electronics, Poland,

The measurements of the out-of-plane vibrations were performed using a Polytec PSV-400 scanning laser vibrometer located in front of the tested plate as shown in Figure 5.15. The vibrometer consists of two main parts: a PSC-I-400 sensor head and OFV-5000 vibrometer controller.

To improve the reflective properties of the surface, the measured area was covered with retroreflecting material dedicated for laser vibrometry tests. Sampling frequency of the vibrometer was set at 5.12 MHz, and the sensitivity at $20\,\mathrm{mm\,s^{-1}\,V^{-1}}$. To suppress the influence of the noise, the measurements were repeated 4 times in each point and subsequently averaged and filtered with a bandpass filter with bandwidth 100 kHz ($-/+$ 50 kHz round the excitation frequency).

The excitation signals used in the experiment consisted of five-cycle tone sine burst modulated with Hanning window. The burst frequencies used in the experiments were the same as in the FEM simulations: 100, 330 and 425 kHz. The transducers' beam patterns were calculated for each of the excitation frequencies on the basis of the vibrometer measurements (Figure 5.16). The Hilbert transform was used to determine the envelopes of the snapshots of the out-of-plane velocity for each of the measured points. Next, the maximal value of each envelope was found and used for the beam patterns calculation. Finally, the calculated beam paterns were normalized and smoothed by a moving average filter.

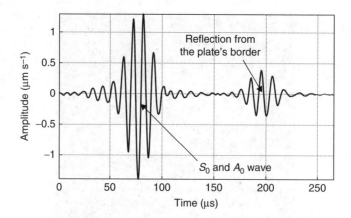

Figure 5.16   Snapshot of Lamb wave propagating in a 4 mm thick aluminium plate

5.4.3   Numerical and Experimental Results

5.4.3.1   Results for the MFC-P2 with d31 Effect

The numerical results of the simulations for the MFC-M2814-P2 transducer presented in Figure 5.17(a) show the transducer's beam patterns for different excitation frequencies. It can be seen that the GWs generated by that transducer propagate in four perpendicular directions simultaneously for all of the considered frequencies. All beam patterns, however, have relatively wide lobes. The main lobe width (divergence angle) calculation, defined as $-6\,dB$ amplitude drop, shows that for the MFC-M2814-P2 transducer two divergence angles may be distinguished: the main and the side. The calculated divergence angles for both directions are presented in Table 5.2.

Figure 5.17(a) shows that the highest amplitude of the propagating wave occurs for frequency of 425 kHz. For this frequency the main to side lobe amplitude ratio (MSLR) is 1.3. For excitation frequency of 100 kHz this ratio is 1.1. It is interesting that at excitation frequency of 330 kHz MSLR is 0.84 which means that the main radiation direction is parallel to the comb electrodes. The observed variations in the main radiation direction are similar to those observed by Collet *et al.* (2011).

Figure 5.17(b) presents the comparison of the experimentally identified beam patterns generated by the MFC-M2814-P2 transducer with different excitation frequencies. Similarly to the numerical results, both front and side lobes, can be observed for the experimental beam patterns. The front lobes have similar amplitudes for all three frequencies while the side lobes, amplitude varies with the frequency and achieves its largest value for 330 kHz. The main lobe width is the largest for the lowest frequency of 100 kHz and it decreases for higher frequencies (Table 5.2). In all of the tested frequencies side lobe width calculations give slightly different results for the 'left' and 'right' sides of the transducer and that is why two values are given in Table 5.2 for the side lobes. The variations observed in numerical data in the main radiation direction are also pronounced in the experimental data presented in Figure 5.17(b). At the excitation frequency of 100 kHz the MSLR is 1.06 while for higher frequencies this ratio is reduced to 0.4 for 330 kHz and 0.6 for 425 kHz, respectively. The experiments

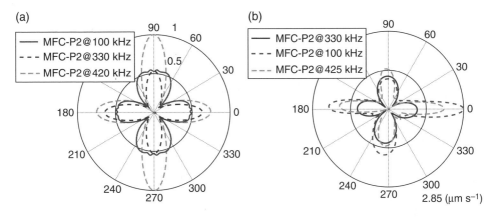

Figure 5.17   Normalized beam patterns of MFC-M2814-P2 at excitation frequencies of 100, 330 and 425 kHz: (a) numerical; and (b) experimental

Table 5.2   Identified main and side divergence angles for MFC-M2814-P2

| Excitation frequency | 100 kHz | 100 kHz | 330 kHz | 330 kHz | 425 kHz | 425 kHz |
Radiation direction	Main	Side	Main	Side	Main	Side
Numerical results (°)	79	51	35.28	24.7	37.62	29.76
Experimental results (°)	67.5	44/50	27.7	22/18.19	27.7	20.2/23.2

show that relatively high mode selectivity was achieved only for the lowest frequency of 100 kHz where only a single-mode wave was observed; for higher frequencies additional modes appear.

Figure 5.18(a–c) enables direct comparison of the simulated and experimental data for MFC-M2814-P2 for different frequencies. Note that the maximal experimentally obtained amplitudes, given in the lower-right corner of each diagram, correspond to the radius of the largest circle. Those amplitudes were used for the amplitude normalization of the respective simulated beam patterns. Figure 5.18(a–c) demonstrates a good agreement between the simulations and the experiments in terms of shape of the generated beam patterns.

5.4.3.2   Results for the MFC-P1 with d33 Effect

The second simulation series was performed for the MFC-M2814-P1 transducer, which exhibits d33 piezoelectric effect. The highest amplitude of the generated wave was observed again for excitation frequency of 425 kHz as can be seen in Figure 5.19(a). For this transducer only the main lobe exists. The divergence angles identified during simulation and experiments are presented in Table 5.3. The highest main lobe width is at 100 kHz and is above 71° while with increase in frequency it drops down to 31.6° at 425 kHz.

Also in this case the changes in amplitude with excitation frequency can be observed. The amplitudes of the waves generated for lower frequencies were approximately two times lower

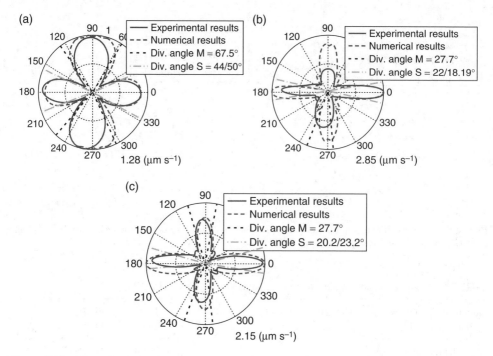

Figure 5.18 Normalized simulated (dashed) and experimental (solid) beam patterns of the MFC-M2814-P2 for different frequencies: (a) 100 kHz; (b) 330 kHz; and (c) 425 kHz. Note the scale factors (in μm s^{-1}) given in the lower-right corner of each diagram

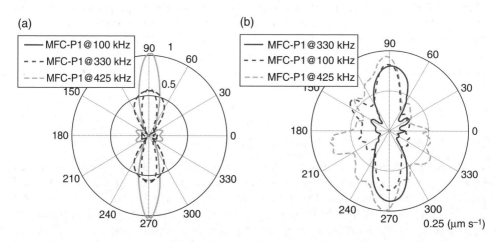

Figure 5.19 Normalized beam patterns of MFC-M2814-P1 at excitation frequencies of 100, 330 and 425 kHz: (a) numerical; and (b) experimental

Table 5.3 Identified divergence angles for MFC-M2814-P1

Excitation frequency	100 kHz	330 kHz	425 kHz
Numerical results ($^\circ$)	71.82	35.44	31.62
Experimental results ($^\circ$)	55	46	—

than those for 425 kHz and the beam patterns had much wider main lobes. It appears that the MSLR also depends on the excitation frequency: for 100 kHz this ratio was 10.8 and for the higher frequencies, 330 and 425 kHz, it was reduced to 6.5 and 7.3, respectively.

The normalized results of the experiments for this transducer, can be directly compared in Figure 5.19(b). The presented results confirm that the MFC-M2814-P1 transducer generates a bidirectional wave in directions perpendicular to the electrodes for all of the tested frequencies.

Similarly to the MFC-M2814-P2, also in this case better mode selectivity was observed mainly at the lowest frequency (100 kHz) where only a single mode could be distinguished with the frequency corresponding to the excitation frequency. With the change in the excitation frequency additional modes of GWs were generated. Contrary to the previous transducer, the shape of the generated beam pattern [Figure 5.19(b)] changed with the excitation frequency. For higher frequencies the bidirectional character of the wave was preserved, however, additional side lobes appeared with amplitudes and localizations changing with frequency. In this case the divergence angle could not be estimated for the 425 kHz frequency because of the high amplitudes of the side lobes. Results obtained during experiments are presented in Table 5.3. The best MSLR was observed at excitation frequency of 100 kHz with a value of 4.1.

Also in this case a good agreement between the experimental and the numerical data can be observed, especially at the lowest excitation frequency [Figure 5.20(a)]; for higher frequencies additional side lobes are visible in the experimental results that were not present in the numerical data. Note that the laser vibrometer velocities corresponding to the beam pattern amplitudes, shown in Figure 5.20, were approximately one order of magnitude lower than those measured for MFC-M2814-P2.

5.4.3.3 Results for IDT-DS with d31 Effect

The third investigated transducer was the custom-made IDT-DS with the d31 piezoelectric effect, made of the same substrate as the MFC-M2814-P2. The IDT-DS4 was designed to work at selected, nominal frequency (330 kHz) but for the comparison with the MFCs it was excited with the same frequencies as the transducers presented above.

The comparison between the simulation results for the IDT-DS4 presented in Figure 5.21(a) shows that the best performance was obtained for excitation frequency of 330 kHz. It can be seen that for this frequency the main lobes are characterized by excellent MSLR and narrow divergence angle equal to 30°. For higher excitation frequencies the beam pattern remains similar, but well pronounced side lobes appear at approximately 30° on both sides of the main lobe. For excitation frequency of 100 kHz, the simulated beam pattern takes the form of a butterfly with wide main lobes (Table 5.4) and considerable side lobes.

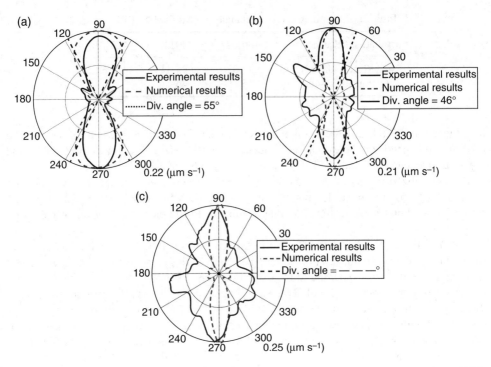

Figure 5.20 Normalized simulated (dashed) and experimental (solid) beam patterns of the MFC-M2814-P1 for different frequencies: (a) 100 kHz; (b) 330 kHz; and (c) 425 kHz. Note the scale factors (in $\mu m\,s^{-1}$) given in the lower-right corner of each diagram

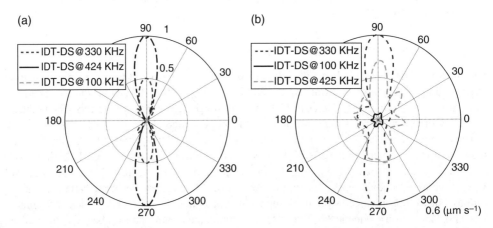

Figure 5.21 Normalized beam patterns of IDT-DS4 at excitation frequencies of 100, 330 and 425 kHz: (a) numerical; and (b) experimental

Table 5.4 Identified divergence angles for IDT-DS4

Excitation frequency	100 kHz	330 kHz	425 kHz
Numerical results (°)	86.6	30.34	31.36
Experimental results (°)	—	34	36/46

The best MSLR of the IDT-DS4, equal to 110, is observed at its nominal frequency of 330 kHz; for other frequencies it is much lower but still higher than that for the MFC-M2814-P1 transducer, i.e. 28 for 100 kHz and 39 for 425 kHz, respectively.

The experimental beam pattern characteristics of this transducer, shown in Figure 5.21(b), demonstrate its ability to generate a bi-directional wave in the direction perpendicular to its comb electrodes As in the simulation results, the best performance of the transducer can be observed at the 330 kHz frequency.

Time and frequency domain analysis of the results has shown that three modes were generated at this frequency: the A_0 and S_0 modes at 330 kHz, and a weak S_0 mode at 425 kHz, which was the symmetric mode corresponding to the electrodes' spacing as shown in Figure 5.11. This mode was excited by the small amount of energy present in the higher frequency band of the 330 kHz burst and was amplified by the finger separation that corresponds to the wave length of the S_0 mode at 425 kHz. The S_0 modes propagate with higher speeds (4.8 and 4.1 km s^{-1}, respectively) than the A_0 one, which has a velocity of 3.1 km s^{-1}. The calculated MSLR at the nominal frequency was 4.6 and the divergence angle was equal to 34° [Figure 5.21(b)].

The result obtained at the 100 kHz frequency is very difficult to interpret because the amplitude of the generated wave is very low. From Figure 5.22 it can be seen that the maximal amplitude for 100 kHz is an order lower than for 330 kHz and the beam pattern does not indicate any well pronounced directionality. At the highest frequency the main lobe is better pronounced but the amplitude is still lower than that observed at the nominal frequency and additional side lobes are also visible. Also at this frequency, besides the modes related to the excitation frequency, the A_0 mode related to the electrode spacing can be distinguished (A_0@330 kHz). The comparison between the simulated and the experimental results for the IDT-DS4, presented in Figure 5.22(a–c), shows a very good agreement between the simulations and the experiments for the nominal frequency. For the 425 kHz frequency the beam patterns are also similar, including the presence of additional side lobes generated at the edges of the transducer.

5.4.4 Discussion

Only single results from each type of transducer have been chosen for the comparison between the numerical and the experimental tests. In the numerical results of the MFC transducers the highest amplitudes, related to the lowest divergence ratio, were observed for the 425 kHz frequency and this frequency was chosen for comparison. The numerical tests of the IDT-DS4 transducers show that in this case the best results are obtained for the nominal frequency of the transducer, in this case 330 kHz. Figure 5.23(a) enables comparison between the beam

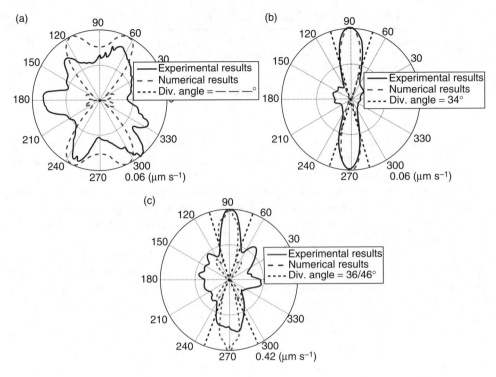

Figure 5.22 Normalized simulated (dashed) and experimental (solid) beam patterns of the IDT-DS4 for different frequencies: (a) 100 kHz; (b) 330 kHz; and (c) 425 kHz. Note the scale factors (in μm s^{-1}) given in the lower-right corner of each diagram

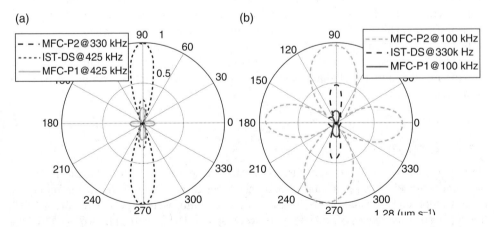

Figure 5.23 Comparison of the normalized beam patterns of the investigated transducers: (a) numerical; and (b) experimental

patterns normalized with reference to the maximal amplitude among those generated by the investigated transducers, in this case to the IDT-DS4@330 kHz.

It is visible at the first glance that the beam pattern generated by the IDT-DS4 transducer has not only the highest MSLR but also the highest main lobe amplitude and the smallest divergence angle. In the case of the experimental tests for the MFC-M2814-P2 the highest amplitude is observed at the 330 kHz frequency but this amplitude is generated in the direction parallel to the finger electrodes (side lobe) which leads to the lowest MSLR. The highest main lobe amplitude is observed at the 425 kHz frequency but at this frequency, the side lobe's amplitude is also very high, which results in a poor MSLR. In this case the best MSLR accompanied by a considerable amplitude is achieved at the 100 kHz frequency and this frequency has been chosen for the comparison. The experimental tests of the MFC-M2814-P1 have shown that also in this case the best MSLR accompanied by the best directionality of the excited wave is observed at the 100 kHz excitation frequency.

Contrary to the other two transducers, the IDT-DS4 shows the best directionality and MSLR for the nominal frequency of 330 kHz, which was chosen for the comparison. Figure 5.23(b) provides the beam patterns with the highest MSLR generated by the investigated transducers during the experiments. The scale factor for the amplitudes of the beam patterns is $1.28 \, \mu m \, s^{-1}$. From Figure 5.23(b) it can be seen that the best directionality among the inspected transducers in terms of side lobe level was achieved by the IDT-DS4, where the MSLR was approximately 5 while for the MFC-M2814-P1 it was approximately 4, and for the MFC-M2814-P2 only 1. Also the best directionality in terms of the divergence angle was observed for the IDT-DS4 that had the main lobe width 34°. This can be compared with 55° for the MFC-M2814-P1 and 67° (50° side wave) for the MFC-M2814-P2.

Based on the performed tests only a qualitative comparison can be done because only a single transducer of each type has been tested. The transducers were bonded manually in different localizations on the plate, which may have led to differences in bond quality between the transducers and the plate. To reduce the negative influence of these factors further multiple tests should be performed and then, based on a statistical analysis, a quantitative comparison could be performed. The experimental results confirm the numerical ones in terms of the shape of the beam patterns, however, contrary to the numerical results, the highest amplitude among the tested transducers was generated by the MFC-M2814-P2. In the performed experiments the amplitude was over 2 times higher than that generated by the IDT-DS4, and over 10 times higher than the amplitude for the MFC-M2814-P1. The source of this difference may be any of the factors mentioned above (the bonding quality, the material properties of the plate or of the piezocomposite) as well as the imperfections of the numerical model. The source of the discrepancy between the numerical and the experimental results is one of the issues investigated in the present work.

5.5 Lamb Wave Sensing Characteristics of the IDT-DS4

The results presented in the previous sections have proven that the IDT can be used as a high performance source of the Lamb wave. The mode and frequency selectivity allow for tuning the transducer to the desired frequency. In this section the directional sensing properties of the designed IDT-DS4 transducer are investigated. To identify the sensing capabilities of the designed transducer both numerical simulations and experimental tests have been performed.

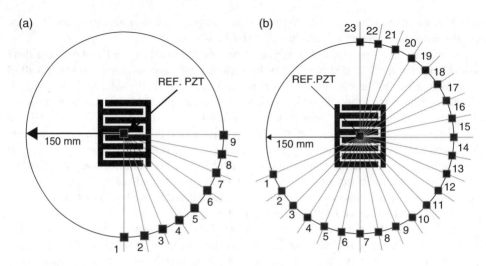

Figure 5.24 Test setups used for directionality evaluation during: (a) numerical tests; and (b) experiments

5.5.1 Numerical Simulations

The numerical simulations have been performed with ANSYS Multiphysics software to identify sensing capabilities of the designed IDT-DS4 transducer. Compared with the previously used models, additional PZT transducers were placed on the plate at a 150 mm radius from the centre and one reference PZT was placed at the centre of the plate.

To identify the influence of the angular position of the source of the Lamb waves on the electrical output signal generated by IDT-DS4, nine rectangular transducers were placed at a radius of 150 mm in the setup shown in Figure 5.24(a). The angular distance between the transducers was 11.25°. The transducers were modelled as $2 \times 2 \times 2$ mm PZT (similar to the Noliac CMAP12 transducer). Next, they were excited with electrical signal at a frequency of 330 kHz (the nominal frequency of the designed IDT-DS4).

The signals presented in Figure 5.25 were measured on the 100 Ω resistor (load) connected to the comb electrodes. Next, to determine the envelopes of the snapshots, the Hilbert transform was used for all measurements. Finally, the cubic spline interpolation from Matlab software was used to smooth the plot and increase the angular resolution (Figure 5.26).

The results obtained during the simulations, show that the tested IDT transducer is very sensitive to the direction of the approaching wave. The amplitude of the measured signal for the wave approaching from the direction parallel to the finger electrodes is two orders smaller than that obtained for the wave incident from the perpendicular direction (Figure 5.25). An angular sensitivity plot was determined on the basis of the measured signal (Figure 5.27). Next, the directional sensitivity was calculated determining the azimuth where signal drops below -6 dB. This angle, which was identified as approximately 30°, is equal to the divergence angle presented in the previous sections.

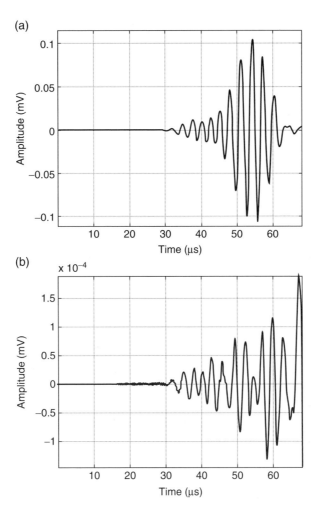

Figure 5.25 IDT-DS4 measurements of the approaching wave from two different directions: (a) perpendicular; and (b) parallel (note amplitude scale difference)

5.5.2 Experimental Verification

To verify the numerical results the experimental tests were performed in a similar setup as previously, presented in Figure 5.24(b). Additional PZT transducers (Noliac CMAP12) were mounted using wax at radius 150 mm with the same angular spacing as previously (11.25°). To reduce the influence of the bond quality between different sources, the reference signal was measured using a PZT transducer (Noliac CMAP12) mounted at the same place as the IDT, however, on the opposite side of the plate. PAS-8000 Piezo-Acquisition System was used to acquire the sensed signals from the tested IDT-DS4 and the referential PZT transducer.

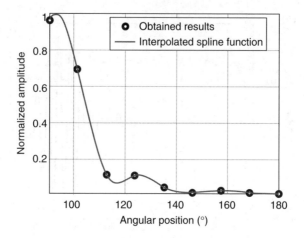

Figure 5.26   Numerical results and their interpolation

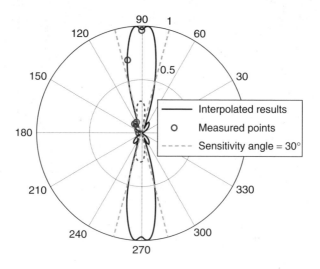

Figure 5.27   Numerical angular sensitivity of the IDT-DS4

In the experiments, Lamb waves were generated successively by each of the PZT transducers mounted on the 150 mm circle. The measurements were repeated several times for each of the transmitting transducers and an average value was calculated. The results significantly different from the average were excluded. Then, the new average values based on the remaining results were calculated and used for the envelope calculation. The final step of the data processing was normalization of the signals from the IDT-DS4 with the use of the signal measured by the middle PZT transducer. The aim of this procedure was reduction of the influence of the PZT bonding quality on the measured signals.

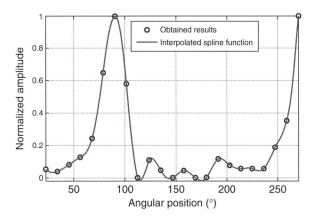

Figure 5.28   Experimental results and their interpolation

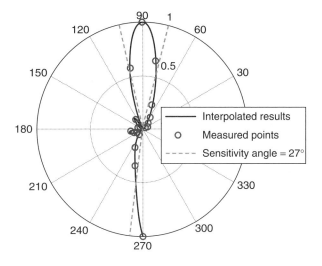

Figure 5.29   Experimental angular sensitivity of the IDT-DS4

Also in this case, to increase the angular resolution of the calculated results, the cubic spline data interpolation was used. The results of the interpolation and the measured results are presented in Figure 5.28.

A polar plot of the measured angular IDT-DS4 sensitivity is presented in Figure 5.29. Similarly to the numerical results, the amplitude of the signal arriving from the direction perpendicular to the comb electrodes is approximately two orders higher than of the signal incident from the direction parallel to the comb electrodes. The −6 dB peak sensitivity was 27°, which is slightly smaller than the values obtained from the numerical simulation as well as the divergence angle obtained during the experimental tests for IDT-DS4 working as the transmitter [Figure 5.22(b)].

5.6   Conclusions

In this chapter we compared, using numerical simulations and experiments, two transducer types built using the MFC substrate. The transducers had different electrodes' layouts: the standard dense electrodes and the IDT-DS electrodes. Both transducers were in the form of thin and flexible patches suitable for SHM applications. Two types of MFC substrate were used for the transducers with dense electrodes, the d33 polarized for MFC-P1 and the d31 for MFC-P2 while the IDT-DS transducer was made of the d31 MFC.

The results of the numerical simulations and the experiments presented in this chapter showed the superiority of the IDT transducer in comparison with the other two types, as it features, for its nominal frequency, both the narrowest main lobe with high amplitude and the lowest side lobe level. Moreover, the IDT-DS showed the mode selectivity related to the distance between its electrodes, matched to the wavelength of the selected wave mode (A_0@330 kHz).

This means that the MFC based IDT-DS is a better choice for SHM applications than the MFC-P1 and P2 with dense electrodes. Higher amplitude and the possibility of tuning up the operating frequency to the frequency in which dispersion of the excited mode is minimal, make the IDT very suitable for long range ultrasonic tests.

The results of the second simulation and experiment series have proven that the mode selectivity and high directivity at the nominal frequency are also preserved when the IDT-DS is operated as a sensor. Moreover, the directional properties of those transducers allow for the arrival azimuth detection. The IDT sensor, capable of Lamb wave angle of arrival detection may be built, for instance, in a way similar to the rosette MFC transducers proposed by Howard and di Scalea (2007).

Until now, the IDT transducers have been manufactured using PZT ceramics or PVDF as the piezoelectric layer, which resulted in brittle (PZT) or weak (PVDF) transducers. The results presented in this chapter have shown that the IDT-DS based on the MFC substrate unifies the advantages of IDT and MFC, since it enables the manufacture of flexible transducers capable of generating high amplitude waves characterized by a high directionality.

Appendix

Table 5A.1   Piezoelectric properties of 31 composite used in IDT-DS4 and MFC-M2814-P2

$c_{11}^E = 3.94 \times 10^{10}$ Pa	$c_{22}^E = 2.03 \times 10^{10}$ Pa	$c_{33}^E 3.25 \times 10^{10}$ Pa
$c_{12}^E = 1.29 \times 10^{10}$ Pa	$c_{12}^E = 0.83 \times 10^{10}$ Pa	$c_{23}^E = 0.53 \times 10^{10}$ Pa
$c_{44}^E = 0.55 \times 10^{10}$ Pa	$c_{55}^E = 0.55 \times 10^{10}$ Pa	$c_{66}^E = 1.31 \times 10^{10}$ Pa
$\epsilon_{31} = -7.12$ C m^{-2}	$\epsilon_{33} = 12.1$ C m^{-2}	$\epsilon_{32} = -4.53$ C m^{-2}
$\epsilon_{24} = -17.03$ C m^{-2}	$\epsilon_{15} = -17.03$ C m^{-2}	$\rho = 7000$ kg m^{-3}
$\varepsilon_{11}^\varepsilon = 237.2\varepsilon_0$ C V^1 m^{-1}	$\varepsilon_{22}^\varepsilon = 237.2\varepsilon_0$ C V^1 m^{-1}	$\varepsilon_{33}^\varepsilon = 143.4\varepsilon_0$ C V^{-1} m^{-1}

Table 5A.2 Piezoelectric properties of 33 composite used in MFC-M2814-P1

$c_{11}^E = 3.94 \times 10^{10}\,\mathrm{Pa}$	$c_{22}^E = 2.03 \times 10^{10}\,\mathrm{Pa}$	$c_{33}^E = 3.25 \times 10^{10}\,\mathrm{Pa}$
$c_{12}^E = 1.29 \times 10^{10}\,\mathrm{Pa}$	$c_{12}^E = 0.83 \times 10^{10}\,\mathrm{Pa}$	$c_{23}^E = 0.53 \times 10^{10}\,\mathrm{Pa}$
$c_{44}^E = 0.55 \times 10^{10}\,\mathrm{Pa}$	$c_{55}^E = 0.55 \times 10^{10}\,\mathrm{Pa}$	$c_{66}^E = 1.31 \times 10^{10}\,\mathrm{Pa}$
$\epsilon_{31} = 13.62\,\mathrm{C\,m^{-2}}$	$\epsilon_{33} = -4.1\,\mathrm{C\,m^{-2}}$	$\epsilon_{32} = 0.55\,\mathrm{C\,m^{-2}}$
$\epsilon_{24} = -17.03\,\mathrm{C\,m^{-2}}$	$\epsilon_{15} = -17.03\,\mathrm{C\,m^{-2}}$	$\rho = 7000\,\mathrm{kg\,m^{-3}}$
$\varepsilon_{11}^\varepsilon = 141.2\varepsilon_0\,\mathrm{C\,V^{-1}\,m^{-1}}$	$\varepsilon_{22}^\varepsilon = 141.2\varepsilon_0\,\mathrm{C\,V^{-1}\,m^{-1}}$	$\varepsilon_{33}^\varepsilon = 141.2\varepsilon_o\,\mathrm{C\,V^{-1}\,m^{-1}}$

References

Ash E 1970 Surface wave grating reflectors and resonators. *IEEE Proceedings of the 1970 G-MTT International Microwave Symposium*, pp. 385–386.

Badcock R and Birt E 2000 The use of 0–3 piezocomposite embedded Lamb wave sensors for detection of damage in advanced fibre composites. *Smart Materials and Structures* **9**, 291–297.

Bellan F, Bulletti A, Capineri L, Masotti L, Yaralioglu GG, Degertekin LF, Khuri-Yakub BT, Guasti F and Rosi E 2005 A new design and manufacturing process for embedded Lamb waves interdigital transducers based on piezopolymer film. *Sensors and Actuators A: Physical* **123–124**, 379–387.

Bent AA and Hagood NW 1997 Piezoelectric fiber composites with interdigitated electrodes. *Journal of Intelligent Material Systems and Structures* **8**(11), 903–919.

Birchmeier M, Gsell D, Juon M, Brunner AJ, Paradies R and Dual J 2009 Active fiber composites for the generation of Lamb waves. *Ultrasonics* **49**, 73–82.

Boonsang S, Dutton B and Dewhurst RJ 2006 Modelling of magnetic fields to enhance the performance of an in-plane EMAT for laser-generated ultrasound. *Ultrasonics* **44**, 657–665.

Box G and Draper N 1986 *Empirical Model Building and Response Surfaces*. John Wiley & Sons, Ltd.

Brunner AJ, Barbezat M, Huber C and Flueler PH 2005 The potential of active fiber composites made from piezo-electric fibers for actuating and sensing applications in structural health monitoring. *Materials and Structures* **38**, 561–567.

Capineri L, Gallai A, Masotti L and Materassi M 2002 Design criteria and manufacturing technology of piezo-polymer transducer arrays for acoustic guided waves detection. *IEEE International Ultrasonic Symposium*.

Collet M, Ruzzene M and Cunefare KA 2011 Generation of Lamb waves through surface mounted macro-fiber composite transducers. *Smart Materials and Structures* **20**, 1–14.

Cosenza C, Kenderian S, Djordjevic B, Green Jr RE and Pasta A 2007 Generation of narrowband antisymmetric Lamb waves usinga formed laser source in the ablative regime. *IEEE Transactions on Ultrasonics, Ferroelectrics, and Frequency Control* **1**(54), 147–156.

Diamanti K, Hodgkinson J and Soutis C 2002 Damage detection of composite laminates using PZT generated Lamb waves. *Proceedings of the First European Workshop on Structural Health Monitoring*. (ed. Balageas D). DEStech Publications, pp. 398–405.

Dimitriadis E, Fuller C and Rogers C 1991 Piezoelectric actuators for distributed vibration excitation of thin plates. *Journal of Vibration and Acoustics* **113**, 100–107.

Dutton B, Boonsang S and Dewhurst RJ 2006 A new magnetic confguration for a small in-plane electromagnetic acoustic transducer applied to laser ultrasound measurements: Modelling and validation. *Sensors and Actuators A: Physical* **125**, 249–259.

Gentilman R, McNeal K, Schmidt G, Pizzochero A and Rosetti GJ 2003 Enhanced performances active fibercomposites. *Smart Structures and Materials: Industrial and Commercial Applications of Smart Structures Technologies* **5054**, 350–359.

Hartmann C, Jones W and Vollers H 1972 Wideband unidirectional inderdigital surface wave transducers. *IEEE Transactions on Sonics and Ultrasonics* **SU-19**(2), 372–381.

Hartmann C, Wright P, Kansy R and Garber E 1982 An analysis of SAW interdigital transducers with internal reflections and the application to the design of single-phase unidirectional transducers *IEEE International Ultrasonics Symposium* **1**, 40–45.

Hayward G, Hailu B, Farlow R, Gachagan A and McNab A 2001 The design of embedded transducers for structural health monitoring applications. *Proceedings of SPIE* **4327**, 312–323.

High JW and Wilkie KW 2003 Method of fabricating NASA-standard macro-fiber composite piezoelectric actuator. Technical report, NASA.

Howard MM and di Scalea FL 2007 Macro-fiber composite piezoelectric rosettes for acoustic source location in complex structures. *Smart Materials and Structures* **16**, 1489–1499.

Jin J, Quek ST and Wang Q 2005 Design of interdigital transducers for crack detection in plates. *Ultrasonics* **43**, 481–493.

Klepka A and Ambrozinski L 2010 Selection of piezoceramic sensor parameters for damage detection and localization system. *Diagnostyka / Polskie Towarzystwo Diagnostyki Technicznej* **4**, 17–22.

Lewis M 1983 Low loss saw devices employing single stage fabrication. *IEEE International Ultrasonics Symposium* **1**, 104–108.

Luginbuhl P, Collins SD, Racine G, Gretillat MA, de Rooij NF, Brooks KG and Setter N 1997 Microfabricated Lamb wave device based on PZT sol-gel thin film for mechanical transport of solid particles and liquids. *Journal of microelectromechanical systems* **6**(4), 337–346.

Mamishev AV, Sundara-Rajan K, Yang F, Du Y and Zahn M 2004 Interdigital sensors and transducers. *Proceedings of the IEEE* **92**(5), 808–845.

Manka M, Rosiek M, Martowicz A, Uhl T and Stepinski T 2010 Design and simulations of interdigital transducers for Lamb-wave based SHM systems. *Proceedings of 11th IMEKO TC 10 Workshop on Smart Diagnostics of Structures*. Krakow, Poland.

Manka MM, Rosiek M, Martowicz A, Uhl T and Stepinski T 2011 Properties of interdigital transducers for Lamb-wave based SHM systems. The 8th International Workshop on Structural Health Monitoring. Stanford University, p. 1488.

Monkhouse R, Wilcox P and Cawley P 1997 Flexible interdigital PVDF transducers for the generation of Lamb waves in structures. *Ultrasonics* **35**, 489–498.

Monkhouse R, Wilcox P, Lowe M, Dalton R and Cawley P 2000 The rapid monitoring of structures using interdigital lamb. *Smart Materials and Structures* **9**, 304–309.

Morgan DR 1973 Surface acoustic wave devices and applications: 1. Introductory review. *Ultrasonics* **11**(3), 121–131.

Murayama R and Mizutani K 2002 Conventional electromagnetic acoustic transducer development for optimum Lamb wave modes. *Ultrasonics* **40**, 491–495.

Myers R and Montgomery D 1995 *Response Surface Methodology Process and Product Optimization using Designed Experiments*. John Wiley & Sons, Ltd.

Na JK and Blackshire JL 2010 Interaction of Rayleigh surface waves with a tightly closed fatigue crack. *NDT&E International* **43**, 432–439.

Na JK, Blackshire JL and Kuhra S 2008 Design, fabrication and characterization of single-element interdigital transducers for NDT applications. *Sensors and Actuators A* **148**, 359–365.

Ochonski J, Ambrozinski L, Klepka A, Uhl T and Stepinski T 2010 Choosing an appopriate sensor for the designed SHM system based on Lamb waves propagation. *Proceedings of 11th IMEKO TC 10 Workshop on Smart Diagnostics of Structures*. Kraków, Poland.

Raghavan A and Cesnik CE 2007 Review of guided-wave structural health monitoring. *The Shock and Vibration Digest* **2**(39), 91–114.

Salas KI and Cesnik CE 2008 Design and characterization of the CLoVER transducer for structural health monitoring. *Proceedings of SPIE* **6935**.

Salas KI and Cesnik CE 2009 Guided wave excitation by a CLoVER transducer for structural health monitoring: theory and experiments. *Smart Materials and Structures* **18**, 1–27.

Schulz M, Pai P and Inman D 1999 Health monitoring and active control of composite structures using piezoceramic patches. *Composites: Part B* **30**, 713–725.

Silva M, Gouyon R and Lepoutre F 2003 Hidden corrosion detection in aircraft aluminium structures using laser ultrasonics and wavelet transform signal analysis. *Ultrasonics* **41**, 301–305.

Staszewski WJ, Lee BC, Mallet L and Scarpa F 2004 Structural health monitoring using scanning laser vibrometry: I. Lamb wave sensing. *Smart Materials and Structures* **13**, 251–260.

Tancrell R 1969 Acoustic surface wave filters. *IEEE International Ultrasonics Symposium* **1**, 48–64.

Wilcox PD, Cawley P and Lowe MJ 1998 Acoustic fields from PVDF interdigital transducers. *IEE Proceedings Science, Measurement and Technology* **145** (5), 250–259.

Williams BR, Park G, Inman DJ and Wilkie KW 2002 An overview of composite actuators with piezoceramic fibers. *20th International Modal Analysis Conference*, Los Angeles pp. 421–427

Yamanouchi K and Furuyashiki H 1984 Low-loss SAW filter using internal reflection type of new single phase unidirectional transducer. *IEEE International Ultrasonics Symposium* **1**, 68–71.

Yamanouchi, K. and Nyffeler F and Shibayama K 1975 Low insertion loss acoustic surface wave filter using group-type unidirectional interdigital filter transducer. *IEEE International Ultrasonics Symposium* **1**, 317–321.

6

Electromechanical Impedance Method

Adam Martowicz and Mateusz Rosiek
Department of Mechatronics and Robotics, Faculty of Mechanical Engineering and Robotics, AGH University of Science and Technology, Poland

6.1 Introduction

The electromechanical impedance based method emerges as a promising local SHM technique for mechanical structures. It is based on the variations of indirectly measured mechanical impedance, which contains the information concerning structural properties of the mechanical system. Considering the difficulties with registering mechanical impedance, the electrical impedance of the piezoelectric transducers mounted on the monitored structure is measured. Due to the presence of electromechanical coupling, the electrical response of the PZT is directly related to the mechanical impedance. Thus, it can be easily used for assessing the state of the structure and is called the electromechanical impedance (Liang and Rogers 1994; Park *et al*. 2003; Sun *et al*. 1995).

The described method has been successfully applied to fault detection in aircraft components. One of the first attempts was performed by Chaudhry *et al*. (1995). They studied the possibility of damage detection in the tail section of a light twin-engine airplane. Two damages of joints connecting the structural elements were introduced in two different locations of the structure: the first one in close vicinity to the piezoelectric transducer and the second one in a farther location. The experiments proved considerable sensitivity of the method to the close-field failure and insensitivity to the distant one.

The impedance based method was also used to detect disbonds in the rear rotor blades of a military helicopter (Giurgiutiu 2007). During operation rotor blades are exposed to in-flight vibrations which may result in disbonds between their structural elements. To investigate this kind of damage several piezoelectric sensors were placed on the rotorblade section. Failure

Advanced Structural Damage Detection: From Theory to Engineering Applications, First Edition.
Edited by Tadeusz Stepinski, Tadeusz Uhl and Wieslaw Staszewski.
© 2013 John Wiley & Sons, Ltd. Published 2013 by John Wiley & Sons, Ltd.

was introduced by making local disbonds with a sharp knife. Occurrence of the damage was manifested by frequency shifts of the resonance peaks in the impedance plot (due to the decrease in the local stiffness), an increase in peak amplitudes (caused by the decrease of local damping) and emergence of new peaks corresponding to the new local mode shapes.

In order to repair ageing aircraft structures, repair patches are commonly used. Such elements, made mainly of composite materials, are placed in the degraded areas of the construction, i.e. cracked or corroded regions. However, after a patch has been applied, it is difficult to monitor the propagation of damage located underneath the composite, as well as to assess the quality of bonding. An experimental study assessing the feasibility of damage detection in a structure repaired by a composite patch is presented in Koh et al. (1999). Boron/epoxy patches were bonded to aluminium specimens to reproduce a typical composite-repaired structure. Teflon inserts, simulating the disbonds, were placed in the bonding layer during the fabrication process. Each of the specimens was subsequently equipped with three piezoelectric transducers, one attached on the composite repair and two on the host structure. Due to the presence of disbonds, significant vertical shifts of the entire impedance plots were reported. Koh and his team were able to determine the size of the failure on the basis of the registered data. They have also proven that the selection of the frequency range for impedance measurements is a key issue.

Roach and Rackow (2006) compared the efficiency of the described technique with other damage detection methods. They tested the possibility of detecting cracks and corrosion flaws in aluminium riveted panels representing the aircraft components. The monitored elements were covered with square piezoelectric transducers and interrogated using a laboratory impedance analyzer. The obtained results confirmed the suitability of the electromechanical impedance measurements to observe structural alterations of such elements. Emerging application areas of SHM determines the necessity of continuous development of monitoring systems, including both hardware and software contributions. Effort is put into increasing the quality of the monitoring process by improving the sensitivity to incipient damages as well as to prevent it from false alarms. On the other hand, reduction of energy consumption as well as installation and maintenance costs may be a key issue when designing a new SHM system.

In this chapter, after the theoretical background for the electromechanical impedance method, the measurement setups and signal processing algorithms used for damage detection will be introduced. The method will be illustrated with the results of numerical FE simulations performed for simple mechanical structures, such as beams and plates. The results of experiments performed in the laboratory conditions for simple structures (e.g. aluminium plate and pipeline section) will be presented. Finally, the method verification on real-life structures subjected to damage detection tests, performed using a specially designed multiple-channel instrument will be presented. Results of the electromechanical impedance measurements performed on two aircraft structures (bolted joint in the main undercarriage bay and riveted fuselage panel) will be discussed in terms of the method feasibility for SHM.

6.2 Theoretical Background

For the electromechanical impedance based method piezoelectric transducers, mounted on a monitored construction, employ both direct and converse piezoelectric effect. The structure is excited to vibrate by the use of the converse piezoelectric effect in which the strain is produced in the presence of an electric field. Conversely, electric impedance or admittance can be measured with the direct piezoelectric effect since strain can produce an electric response. The dynamic response obtained from the area of interest is transferred back to the sensor and then represented in the form of an electric signal. Eventually, the occurrence of damage causes changes in the amplitude and phase of electric response.

6.2.1 Definition of the Electromechanical Impedance

For a linear piezoelectric material the relationship between mechanical and electric properties can be defined as follows (Park and Inman 2007):

$$
\begin{bmatrix} S \\ D \end{bmatrix} = \begin{bmatrix} s^E & d_t \\ d & \epsilon^T \end{bmatrix} \begin{bmatrix} T \\ E \end{bmatrix},
\tag{6.1}
$$

where S is mechanical strain, T is mechanical stress, E is electric field, D is electric displacement field (i.e. charge density in the case of a capacitor which represents a piezoelectric transducer in an electric circuit), s is mechanical compliance, d is the piezoelectric strain constant and ϵ is the electric permittivity. The superscripts E and T mean that the quantities are, respectively, measured with electrodes connected together and at zero stress. The subscript t indicates transposition. The direct piezoelectric effect is described by the second part of the equation whereas the first part describes the converse effect.

An example of the derivation of the formula used to determine electromechanical admittance (the inversion of electromechanical impedance) can be found in Bhalla and Soh (2003). In the cited reference the model of a PZT patch was bonded to the surface of the monitored structure using a high-strength epoxy adhesive. The PZT patch vibrates along its longitudinal axis in direction '1'. An alternating electric field E_3 is applied along transversal direction '3'. The behaviour of the piezoelectric transducer is governed by the following constitutive equation derived from Equation (6.1) Bhalla and Soh 2003):

$$
\begin{bmatrix} S_1 \\ D_3 \end{bmatrix} = \begin{bmatrix} s_{11}^E & d_{31} \\ d_{31} & \epsilon_{33}^T \end{bmatrix} \begin{bmatrix} T_1 \\ E_3 \end{bmatrix},
\tag{6.2}
$$

where S_1 is the strain in direction '1', D_3 is the electric displacement field (i.e. charge density) over the PZT transducer, d_{31} is the piezoelectric strain coefficient and T_1 is the axial stress in the PZT patch in direction '1'. The complex mechanical compliance s_{11}^E is the inverse of complex stiffness, directly defined as the complex Young's modulus Y_{11}^E:

$$
s_{11}^E = \left(Y_{11}^E \right)^{-1}.
\tag{6.3}
$$

Both Young's modulus Y_{11}^E and electric permittivity ϵ_{33}^T are considered as complex quantities and allow for the introduction of mechanical and electric loss factors.

For the elaborated model the following assumptions were made:

- The vibrating patch is infinitesimally small compared with the host structure and therefore has negligible mass and stiffness.
- The structure can be assumed to possess uniform dynamic stiffness over the entire bonded area.
- The two end points of the mounted PZT patch encounter equal mechanical impedance Z_M by the structure.

The electromechanical admittance Y depends on mechanical impedances of the monitored structure Z_M and the piezoelectric transducer Z_a according to the following equation (Park et al. 2003; Sun et al. 1995):

$$Y = j\omega a \left[\epsilon_{33}^T - \frac{Z_M}{Z_M + Z_a} d_{31}^2 Y_{11}^E \right],$$
(6.4)

where a is the geometric constant of the piezoelectric transducer.

Any damage in the structure will change the mechanical impedance Z_M. When the assumption is made that mechanical properties of the piezoelectric transducer do not change, the resultant electromechanical admittance Y and impedance Z should represent any changes in mechanical properties of the monitored structure.

Equation (6.4) involves two components. The first one is the capacitive admittance of the free PZT. It describes the linear growth of the admittance Y with frequency. This component represents the majority in the imaginary part of both Y and Z. The second term represents the variation of the frequency characteristics of Y resulting from the mechanical properties of both the monitored structure and the transducer. It defines the majority of the real part of electromechanical admittance and impedance. Therefore real parts of both impedance and admittance, i.e. $Re(Z)$ and $Re(Y)$ are most often used for monitoring as more sensitive to the variation of structural parameters. The measurements of real components provide information about both the electrical behaviour of the piezoelectric element and the mechanical behaviour of the whole structure (Sun et al. 1995).

6.2.2 Measurement Techniques

There are two possible configurations that are applicable for measuring electromechanical impedance:

- Point Frequency Response configuration, as shown in Figure 6.1(a);
- Transfer Frequency Response configuration, as shown in Figure 6.1(b).

In the case of the Point Frequency Response configuration it is assumed that only one PZT is used, which simultaneously acts as both actuator and sensor (Park et al. 2003). However, the drawback of the application is the reduced detection area and sensitivity of the measurement, because of the compromise on the electromechanical characteristics of PZT which is expected to be both an effective actuator of vibration and a sensitive sensor. The circuit in Figure 6.1(a)

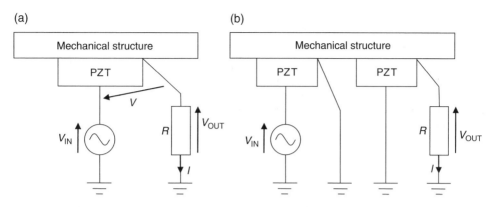

Figure 6.1 Point Frequency Response configuration (a) and Transfer Frequency Response configuration (b)

is powered by applying alternating voltage V_{IN}. The output voltage V_{OUT} depends on the voltage V, which is generated in PZT. It corresponds to the mechanical impedance. Thus, any variation of the mechanical impedance resulting from structural damage should be reflected in the measured complex electrical impedance Z of the piezoelectric transducer (Park and Inman 2007). The electromechanical impedance is determined with the voltage and current measured directly for PZT, and can be found by the following equation:

$$Z = \frac{(V_{IN} - V_{OUT})\, R}{V_{OUT}} \tag{6.5}$$

The latter configuration determines the use of two PZT for separate excitation and measurement (Koh *et al.* 1999; Peairs *et al.* 2002; Sun *et al.* 1996). The advantages of the described approach are: the increase of measurement sensitivity and vibration energy transmitted into a structure and resulting from the fact that each of the PZT used can be separately chosen according to the task it accomplishes. The main drawback, however, is the increase of the number of bonded PZT and a more complex control electronic circuit, which must contain additional systems including charge amplifiers. The electromechanical impedance is calculated as follows:

$$Z = \frac{V_{IN} R}{V_{OUT}} \tag{6.6}$$

During the impedance measurements high-frequency vibrations are induced in the structure in order to excite local mode shapes. A typical frequency range for such experiments is from 10 up to 500 kHz, depending on the examined construction and the measurement device (Ayres *et al.* 1998; Naidu and Soh 2004; Park and Inman 2007; Yan *et al.* 2007). High frequency measurement characterizes significant sensitivity to local changes in the mechanical properties of the monitored structure. It results from the fact that these changes mostly interfere with high frequency normal modes characterizing small wavelengths, i.e. of damage size. It must be noted, however, that the impedance based SHM may by effectively performed only within the vicinity of the mounted PZT, i.e. with distances of millimetres and centimetres rather than metres (Chaudhry et al. 1995). Moreover the localization effect is due to present damping, e.g. friction damping in surrounding joints for truss structures (Sun et al. 1995), that significantly disturbs the propagation of wave induced by the piezoelectric transducer in joints and in the material (Ayres et al. 1998). Additionally, in the range of high frequencies a small voltage is required to excite the structure properly, usually less than 1V (Park and Inman 2007). The measurement method is suitable for structures which are exposed to low-frequency vibrations, static loads and displacements during the normal operation. Global changes of mass, stiffness and boundary conditions that occur beyond the intimate neighbourhood of the mounted sensor do not influence the registered impedance plots (Ayres et al. 1998).

6.2.3 Damage Detection Algorithms

The process of damage assessment requires definition of the baseline impedance plots when the structure is considered healthy. As the time passes the currently measured frequency characteristics are gradually collected and their changes with respect to the stored baseline plot are monitored. To qualitatively assess the changes of the registered impedance plots, i.e. symptoms of damage, the following damage metrics can be applied (Park and Inman 2007; Rosiek et al. 2010b):

$$DI_1 = \sum_{i=1}^{n} \left| \frac{Re(Z_{0,i}) - Re(Z_i)}{Re(Z_{0,i})} \right|, \tag{6.7}$$

$$DI_2 = \sqrt{\sum_{i=1}^{n} \left(\frac{Re(Z_{0,i}) - Re(Z_i)}{Re(Z_{0,i})} \right)^2}, \tag{6.8}$$

$$DI_3 = \sum_{i=1}^{n} \sqrt{\left| \frac{Re(Z_{0,i}) - Re(Z_i)}{Re(Z_{0,i})} \right|}, \tag{6.9}$$

$$DI_{4,q} = \left(1 - \frac{1}{1-n} \frac{\sum_{i=1}^{n} (Re(Z_{0,i}) - Re(Z_0)) (Re(Z_i) - Re(Z))}{s_0 s} \right)^q, \tag{6.10}$$

where $Z_{0,i}$ and Z_i are, respectively, the referential and the current value of the electromechanical impedance for the ith frequency. Z_0, s_0 and Z, s are the mean values and the standard deviations of the referential and current impedances, n is the number of considered frequencies and q is the order of the damage metric. Damage index DI_2 is the root mean square deviation (RMSD) widely used by various researchers (Naidu and Soh 2004; Rosiek et al. 2010b), while DI_4 is a statistical metric known as cross correlation (Park and Inman 2007; Raju et al. 1998). Metrics DI_1 and DI_3 are proposed by the authors of the chapter (Rosiek et al. 2011b). The relative deviation is used in Sun et al. (1995) for the truss structure with joints loosening. In Park *et al.* (1999) the sum of the real admittance change squared is proposed as a damage index. In Bhalla and Soh (2003) it is proposed that both real and imaginary components of measured admittance can be used in damage quantification. Other damage metrics were investigated in Zagrai and Giurgiutiu (2001), e.g. mean absolute percentage deviation, covariance change, correlation coefficient deviation. The impedance based SHM is characterized by application versatility dealing with both the type of monitored mechanical structure and the construction material. Most known applications of the described type of SHM are (Ayres *et al.* 1998; Koh *et al.* 1999; Park and Inman 2007; Peairs *et al.* 2002; Roach and Rackow 2006; Sun *et al.* 1996): bolted and screw joints, welded and spot-welded joints, glued joints, pipelines and railroad tracks, performed for metallic and composite materials, concrete and reinforced concrete. The most known kinds of damages that can be detected are e.g. cracks, fatigue fracture, corrosion, loosening of joints, disbonds and delaminations.

6.3 Numerical Simulations

Although the impedance based damage detection method is not a model based technique, a model of the electromechanical response of the monitored structure may be useful for predicting the damage growth or planning the locations of piezoelectric transducers in the construction. Analytical modelling of the electromechanical impedance can be performed only for non complicated structures (beams, plates, rods) (Giurgiutiu and Rogers 1999; Giurgiutiu and Zagrai 2001). Attempts were also made to use the spectral element method with electric circuit analysis for modelling the piezoelectric systems (Peairs *et al.* 2005). This technique is very efficient in terms of computational cost, however, it is difficult to implement due to the lack of software appropriate for modelling structures with complex geometry. Besides the methods mentioned above, the coupled FEM seems to be the most accurate and easiest to apply for modelling more complex engineering structures, although it requires high computational power (Liu and Giurgiutiu 2007; Yang *et al.* 2008). The following subsection presents a study utilizing the coupled FEM for understanding the phenomenon of electromechanical impedance in conjunction with damage detection in mechanical structures.

6.3.1 Modelling Electromechanical Impedance with the use of FEM

The first examined object was a freely suspended aluminium beam with bonded PZT transducer, as shown in Figure 6.2(a). The dimensions of the aluminium beam were 100 mm × 16 mm × 1 mm. PZT of size 10 mm × 10 mm × 0.3 mm was made of PIC151 material (PI Ceramic) and was located on the left side of the structure (5 mm from the beam end). The second tested object was a cantilever steel beam of size 120 mm × 15 mm × 1 mm with two bonded piezoelectric transducers made of S1 and S2 materials (Noliac), as shown in Figure 6.2(b). The smaller PZT of size 5 mm × 5 mm × 2 mm was located 24.5 mm from the clamped edge of the beam and was acting as an actuator. The larger one (7 mm × 7 mm × 2 mm) was placed 64 mm from the fastening and treated as a sensor. A thin layer of epoxy adhesive was also taken into account in both models.

The FE models were built in the ANSYS software using 3D 20-node parabolic SOLID95 and SOLID226 elements for structure modelling and CIRCU94 elements for voltage source and resistor modelling. The SOLID 226 is a coupled-field element which has thermoelectric,

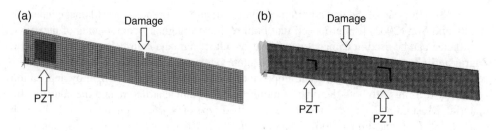

Figure 6.2   FE models of the simply suspended aluminium beam (a) and the cantilever steel beam (b)

piezoelectric and piezoresistive capabilities, thus it is well suited for modelling several complex multiphysics problems. In particular, it can incorporate full electromechanical coupling which is present in piezoelectric materials (Yang *et al.* 2008). Piezoelectric transducers were excited by applying alternating voltage V_{IN} with magnitude of 1V to one of their electrodes. For both structures a 100Ω referential resistor was used to evaluate the electromechanical impedance. In order to reach the convergence of the solution and obtain sufficient accuracy of the results, it was essential to prepare an appropriate mesh. According to the criterion recommended by Yang *et al.* (2008), an element size of 1 mm was chosen with respect to the selected frequency range. The damage was modelled as a vertical notch with the depth varying from 1 to 4 mm. Subsequent damage cases are denoted in Figure 6.3 as Damage 1 to Damage 4. As a result of harmonic analysis, the Point Frequency Response Functions (Point FRFs) of the electromechanical impedance were obtained for the aluminium beam and the Transfer Frequency Response Functions (Transfer FRFs) were acquired for the steel beam. Figure 6.3(a) and (b) present the absolute values of these impedance characteristics. It can be seen, that in conjunction with growing failure, most of the resonance peaks are shifted towards lower frequencies due to a decrease of resultant global stiffness of the beam. Some minor changes in the amplitude of peaks can also be observed. For larger notches (i.e. 3 mm and 4 mm) disturbance in the structure is so significant, that new normal modes appear.

Figure 6.4 shows a close-up view of the real part of the Point FRFs obtained for the aluminium beam. Most of the natural frequencies can be identified during the impedance measurement, excluding those which basically cannot be excited using the PZT patch considered in the model. If no significant piezoelectric effect d_{31} is present for a particular mode shape, or if one half of the transducer exhibits negative stress, while the second half is exposed to positive stress, the overall, resultant, charge induced in the PZT is close to zero. In that case, the current flowing through the referential resistor is negligible and there is no peak present in the plot. However, as it was mentioned before, some of those natural frequencies may appear in the impedance plot evaluated for a damaged structure, due to the unsymmetrical character of the mode shape.

For the calculation of damage metrics only the real part of impedance was taken into account, because of its greater sensitivity to incipient damage (Park *et al.* 2003). All four previously defined damage indices DI_1, DI_2, DI_3 and DI_4 were calculated according to Equation (6.7), Equation (6.8), Equation (6.9) and Equation (6.10) and were based on the impedance plots presented in Figure 6.3(a) and (b). In all cases a monotonic relationships between failure size and metric value have proven that it is possible to track the growth of damage in the monitored structure (Figure 6.5).

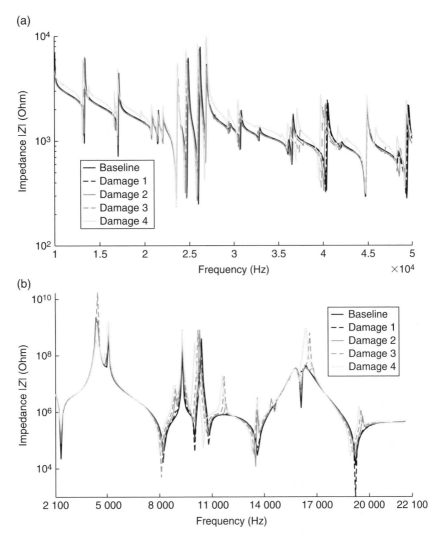

Figure 6.3 Point Frequency Response Functions simulated for the aluminium beam (a) and Transfer Frequency Response Functions simulated for the steel beam (b)

The accuracy of the modelling technique chosen to simulate the dynamical behaviour of the discussed structures was verified experimentally. The Agilent 4395A and Analog Devices AD5933 impedance analyzers were used to perform measurements for the previously described beams. The impedance plots obtained from experiments were compared with the simulations and there was a good agreement between numerical and experimental data as shown in Figure 6.6(a) and (b). Some minor changes in resonance peaks and amplitudes can be observed due to the fact that the FE model was not updated after performing the measurements.

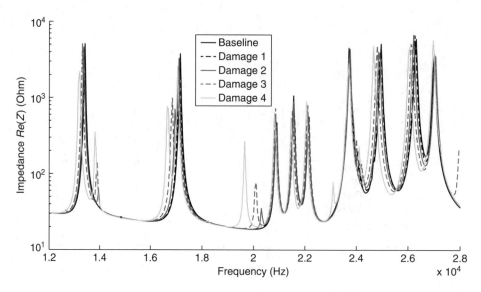

Figure 6.4   Close-up view of the Point Frequency Response Functions

6.3.2   Uncertainty and Sensitivity Analyses

In order to examine the effect of the variation of structural parameters on the impedance characteristics, uncertainty and sensitivity analysis was performed using the Monte Carlo method (Rosiek *et al.* 2010a). Parameters which determine the electrical and mechanical properties of the piezoelectric transducers and the mechanical properties of the bonding adhesive were chosen as random input variables. It was assumed that all input parameters follow the normal distribution, with corresponding mean values and standard deviations shown in Table 6.1.

FE models of an aluminium beam with no damage and with a 2 mm notch were considered for probabilistic treatment, as presented in Figure 6.2. The analysis consisted of 250 simulation loops arranged according to the applied Monte Carlo method. The Latin hypercube sampling technique was used for better coverage of the input domain with generated samples and to decrease the computational time (Helton and Davis 2003).

As a result of the numerical simulations a scatter of the Point FRF was obtained. The real part of the electromechanical impedance was taken into consideration [Figure 6.7(a)]. Figure 6.7(b) shows the envelopes of scatter plots for damaged and undamaged beams. The general trend of shifting resonance peaks of the damaged structure to a lower frequency range was maintained though the whole frequency domain.

The histogram plots presented in Figure 6.8 show the uncertainty propagation of four output parameters, i.e. damage metrics. As it can be seen, the statistical index DI_4 was the least sensitive for the parameters variation. Almost all values of that index are within the bound of $\pm 20\%$ of its mean value whereas e.g. for DI_1 and DI_2 their extent of variation is in the interval of $\pm 50\%$ of mean values. The results of the sensitivity analysis carried out for all damage indices are presented in Figure 6.9.

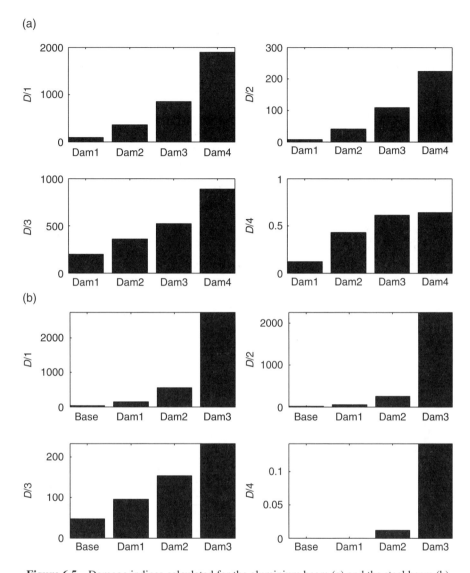

Figure 6.5 Damage indices calculated for the aluminium beam (a) and the steel beam (b)

All results of the sensitivity analysis were first ordered according to the absolute values and then collected considering their signs. The shown orders help to determine which input uncertain parameters are of the greatest influence on damage metrics and those which, in turn, can be neglected as their impact is relatively small. The analysed significance level was defined as a ratio between percentage change of damage index with respect to its mean value and, similarly, percentage change of input parameter with respect to its nominal value. All sensitivities were defined as coefficients of linear regressors in the task of linear approximation available in the response surface method (Myers and Montgomery 1995).

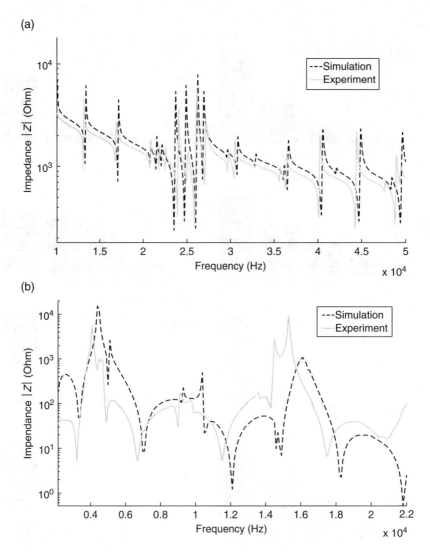

Figure 6.6 Comparison between simulated and experimental Point Frequency Response Functions (a) and Transfer Frequency Response Functions (b)

The parameter numbers in Figure 6.9 correspond to those in Table 6.1. Parameter w_p, i.e. the transducer width, is the most influential for all damage indices. This observation can be explained by the small difference between width of beam and transducer. PZT covers almost the whole beam at the transversal cross-section in the position where it is placed. Considering this significant contribution in the cross-section area of the beam, even a small percentage change of parameter w_p may cause large changes in normal modes, natural frequencies and finally the value of electromechanical impedance. A similar conclusion can be formulated in the case of y_p, i.e. the parameter defining transducer localization. A group of parameters

Table 6.1 Random input variables

No.	Input variable	Nominal value	Standard deviation with respect to nominal value (%)
1	Piezoelectric material permittivity $\epsilon^S_{31}/\epsilon_0 (-)$	1110	2.5
2	Piezoelectric material permittivity $\epsilon^S_{33}/\epsilon_0 (-)$	852	2.5
3	Piezoelectric material density $\rho_p (\mathrm{kg\,m^{-3}})$	7.76×10^3	2
4	Piezoelectric material stiffness $c^E_{11} (\mathrm{N\,m^{-2}})$	10.76×10^{10}	5
5	Piezoelectric material stiffness $c^E_{12} (\mathrm{N\,m^{-2}})$	6.31×10^{10}	5
6	Piezoelectric material stiffness $c^E_{13} (\mathrm{N\,m^{-2}})$	6.38×10^{10}	5
7	Piezoelectric material stiffness $c^E_{33} (\mathrm{N\,m^{-2}})$	10.04×10^{10}	5
8	Piezoelectric material stiffness $c^E_{44} (\mathrm{N\,m^{-2}})$	1.96×10^{10}	5
9	Piezoelectric constant $e_{31} (\mathrm{C\,m^{-2}})$	-9.6	5
10	Piezoelectric constant $e_{33} (\mathrm{C\,m^{-2}})$	15.1	5
11	Piezoelectric constant $e_{15} (\mathrm{C\,m^{-2}})$	12	5
12	Transducer length $l_p (m)$	0.01	0.5
13	Transducer width $w_p (m)$	0.01	0.5
14	Transducer thickness $h_p (m)$	3×10^{-4}	5.7
15	Transducer centre location $x_p (m)$	0.01	1.7
16	Transducer centre location $y_p (m)$	0.008	2.1
17	Adhesive thickness $h_a (m)$	6×10^{-5}	5.5
18	Adhesive Young's modulus $E_a (Pa)$	2500	2
19	Adhesive Poisson's ratio $v_a (-)$	0.4	2
20	Adhesive density $\rho_a (\mathrm{kg\,m^{-3}})$	1150	2

characterizing smaller sensitivities stands for material properties. As reported for uncertainty analysis, the smallest absolute values of sensitivities were found for the statistic damage index DI_4. In the case of damage index DI_4 the absolute values of the found ratios are less than 3 whereas in the case of other indices their values extends to 10.

6.3.3 Discussion

In this chapter the usability of FEM for modelling the dynamical behaviour of piezoelectric systems is discussed. The FE models of simple beams with bonded piezoelectric transducers were elaborated and used in coupled simulations. The results of numerical simulations proved that the considered modelling technique is suitable for simulating electromechanical interactions in the mechanical structures with bonded piezoelectric transducers. With the use of FEM it was possible to reproduce changes of the impedance characteristics corresponding to the growing damage induced in the structure. FEM was also used to discover which types of mode shapes have the biggest influence on the resulting impedance characteristics. Next, a problem of robust design of the electromechanical impedance based SHM system was discussed. An

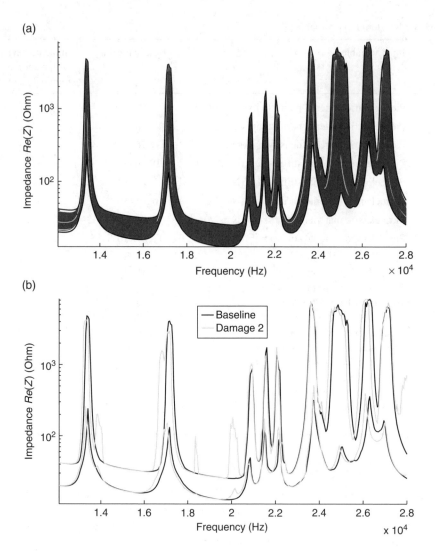

Figure 6.7 Scatter of impedance plots obtained for the model without damage (a) and envelope plots for the undamaged and damaged structure (b)

uncertainty and sensitivity analysis was performed using the Monte Carlo method to evaluate dispersion of the impedance plots and damage metrics in the presence of variation of electrical and mechanical parameters of transducers used in the system. In the case of deterministic analyses monotonic relationships between size of damage and value of damage index were found. Amongst all parameters only the geometric ones play a significant role in the case of variation of damage indices. The obtained results can be used to improve the calibration procedure applied in the impedance based monitoring system.

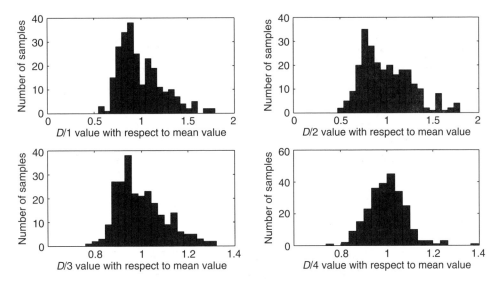

Figure 6.8 Histogram plots of damage metrics

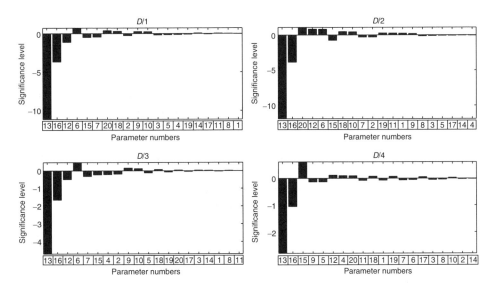

Figure 6.9 Sensitivity plots for damage metrics

6.4 The Developed SHM System

Based on the theoretical considerations, an SHM system incorporating the electromechanical impedance principle was developed at the AGH University of Science and Technology, Kraków, Poland. The designed system has a modular architecture and can be easily adopted to different applications, i.e. monitoring of civil, industrial, marine or aerospace structures.

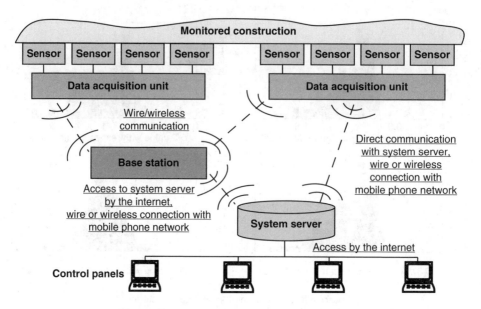

Figure 6.10   Structure of developed SHM system

The overall structure of the system is presented in Figure 6.10. The system consists of data acquisition units (DAUs), base stations and a system server which enables the connection with control panels. The DAUs and base stations are localized in the area of the monitored structure. The main task of the DAUs is to check the condition of the construction by the impedance measurements performed with mounted piezoelectric sensors. The data which are gathered in the DAU is sent to the base station or directly to the system server either by wire or wireless connection. All the information is stored in the system server and can be acquired at any time with control panels. Control panels are mobile or desktop computers which have access to the Internet and SHM system software already installed. The configuration settings of the whole system can be reprogrammed remotely. The measurements can be triggered automatically according to the programmed time intervals or manually with control panels. The main component of the system is the DAU, shown in Figure 6.11(a). The presented hardware works as a sensor node and can be used to interrogate the structure and register the impedance characteristics. The DAU is capable of performing the pre-processing of the measured impedance characteristics, calculating damage index values and evaluating the state of the monitored construction. The information gathered by the sensor node can be afterwards sent to the master unit, i.e. avionic system of the aircraft or external base station, shown in Figure 6.11(b). The designed DAU has the capability of registering impedance signatures using 16 piezoelectric transducers in a frequency range up to 100 kHz. The system is designed to allow for measurements of both Point FRF using one PZT (self-sensing actuation) and Transfer FRF between two locations on a structure in an arbitrary configuration of the transducers, amongst all connected PZT (Rosiek *et al.* 2011a,b). The sensor nodes can operate independently or may be combined in a network. The main hardware board of a DAU can be equipped with an additional processing module based on the ARM7 processor, which increases the overall computational power of the system. The environmental conditions can

(a)   (b)

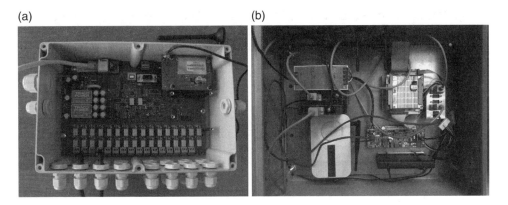

Figure 6.11   Data acquisition unit (a) and base station (b)

Table 6.2   Power consumption and data transmission time

Communication medium	Power consumption (W)	Transmission time (s)
WiFi-Ethernet	5.5	< 1
ZigBee	1	60
GSM-GPRS	3	5
Ethernet (wire)	4.5	< 1

be assessed by a DAU with additional temperature and humidity sensors. The variation of the two parameters mentioned above should be assessed as it significantly influences the frequency characteristics determined for electromechanical impedance (Park *et al.* 1999). In the system there are several possible wire or wireless transmission techniques to enable for the data transfer between all system components. Various communication modules can be easily replaced depending on the selected communication technique. Between the DAU and base station the following communication technologies have been implemented: wireless (ZigBee and WiFi) and wire connection with the Ethernet protocol. The data connection between the base station and system server can also be possible either by wire, i.e. by the Internet, or by using mobile phone network infrastructure (GSM technology). For applications where there is a direct communication between the DAU and system server, i.e. applications which do not require base stations, applicable communication techniques are the same as previously mentioned for the connection between the base station and system server. The DAU may work either independently or create a network. The approximated power consumption of the DAU during communication and transmission time of one data package (2213 bytes) for each communication medium are presented in Table 6.2. The collected data can be stored for further evaluation in a SQL database installed on a personal computer. Dedicated software was developed to provide access to the collected data and remote configuration of elements of the monitoring system. A graphical user interface of the software is presented in Figure 6.12.

The developed system may operate properly in a wide range of environmental conditions: in the temperature range −40 to +85°C (from −20 up to +50°C for the system equipped with additional high capacity accumulators) and humidity up to 100%. The housings used allow for

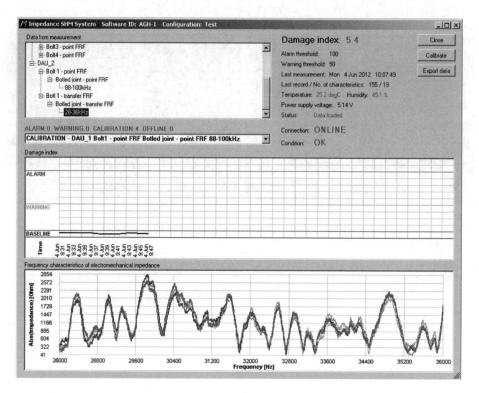

Figure 6.12   Dedicated software for acquiring and processing the measured data

industrial applications and ensure the protection is compliant with the IP66 protection level. All wire outlets from housing boxes are insulated. The DAU is supplied with accumulators optionally equipped with photovoltaic panels used to extend the operating period without any maintenance activity. The other system components are powered by external voltage 230 V/ 50 Hz. The base station is equipped with an uninterruptible power supply. The DAU used for the measurements requires calibration to evaluate true values of the impedance. However, the monitoring technique used is based on the relative changes of the characteristics registered for healthy and degraded structure. If DAU settings remain unchanged, then the calibration procedure can be omitted and raw data from the device can be compared in order to calculate the damage metrics. The impedance characteristics considered as the dimensionless quantities can still be used to calculate the damage indices and perform assessment of the structure's condition.

6.5 Laboratory Tests

To verify the properties of the developed monitoring system a series of experiments was performed under laboratory conditions. The tests incorporated measurements of the electromechanical impedance for various types of structures with different types of damage.

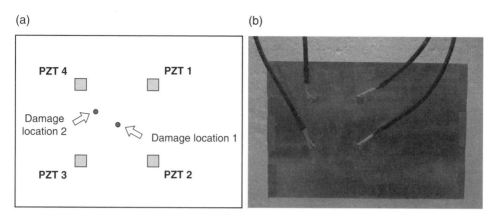

Figure 6.13 Dimensions of monitored structure with indicated damage locations (a) and examined specimen (b)

6.5.1 Experiments Performed for Plate Structures

The first structure tested was a freely suspended aluminium plate panel with four piezoelectric patches made of PIC151 material (PI Ceramic) permanently bonded using epoxy glue (Rosiek *et al.* 2011a). Dimensions of the tested object and damage localizations are presented in Figure 6.13(a) and the examined specimen is shown in Figure 6.13(b). Two locations of damage introduced to the structure were considered. In the first case, the damage was placed at equal distances from the transducers. In the second case, in turn, it was moved towards PZT no. 4. Damage was simulated as an additional mass and stiffness. Thick steel washers of two different sizes were attached to the panel using wax to induce local changes in the dynamical properties of the structure. Two sets of baseline measurements were performed for the undamaged structure to check the repeatability of experiments and to determine the initial values of damage metrics. Next, damage was introduced and measurements were repeated for the same frequency ranges. As an outcome of experiments a set of electromechanical impedance plots was obtained. Both Point and Transfer FRFs were evaluated for all piezoelectric transducers working as actuators and sensors. The measurements were performed for the frequency range from 24 up to 28kHz with a frequency step of 10Hz. The relatively small frequency step was chosen to ensure sufficient resolution of the measurements due to high modal density of the structure for the chosen high-frequency range.

The exemplary results obtained for Point FRF configuration for PZT no. 1 and no. 3 are shown in Figure 6.14(a) and (b), respectively. It can be seen that the appearance of damage causes shifts of the resonance peaks and changes of their amplitude. The values of damage metrics calculated on the basis of the recorded impedance data for the failure placed in location 1 are presented in Figure 6.15. Monotonic relationships between damage size and the value of *DI* for all PZT were obtained. In the case of the second damage location the proportions between metrics calculated for different damage sizes remained similar. Selected impedance plots evaluated for Transfer FRF configuration are shown in Figure 6.16. The corresponding *DI* for exemplary damage location 2 are presented in Figure 6.17.

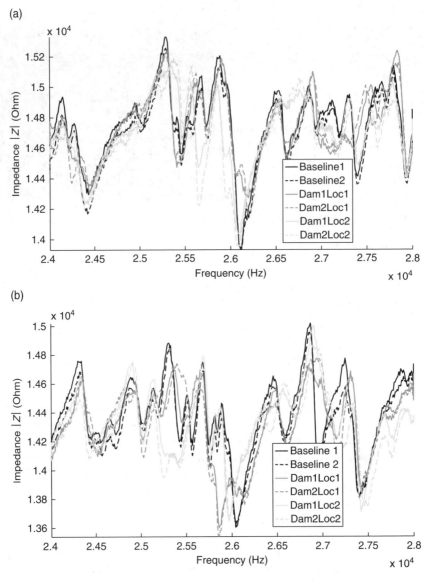

Figure 6.14   Point Frequency Response Functions obtained for the aluminium plate for transducer no. 1 (a) and transducer no. 3 (b)

In all considered cases the greatest differences between particular failure sizes were observed for the statistical metric DI_4. Moreover better repeatability for impedance baseline signals was found for the Transfer Frequency Response case. In accordance with the obtained experimental results a conclusion can be made that the tested application of impedance based SHM allows the presence of damage to be detected in a mechanical structure and track its growth.

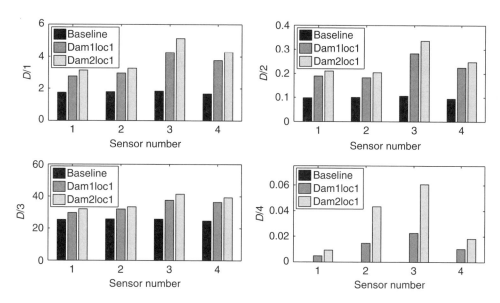

Figure 6.15 Damage indices calculated for the aluminium plate on the basis of Point Frequency Response Functions: damage location 1

6.5.2 *Condition Monitoring of a Pipeline Section*

The subject of the damage detection experiments was a section of a pipeline connected with four screws and nuts (Figure 6.18) (Rosiek *et al.* 2011b). It was chosen to verify the capability of the system to detect a loosening of a screw in the bolted joint. The tested object was equipped with a special type of steel washers with bonded PZT transducers placed between screw heads, pipe flanges and nuts. Such an approach was first proposed in Mascarenas *et al.* (2007).

The damage was introduced in the structure by loosening one of the screws after evaluating the baseline impedance plots. First, all four screws connecting the pipeline elements were tightened to a torque of 20N m. This case was treated as a baseline configuration for which the initial values of the damage indices were evaluated. Subsequently, one of the bolts was loosened and retightened to 15N m, which corresponded to the first damage case. The procedure was repeated for a torque of 10 N m (second damage case). For all tested configurations Point and Transfer FRFs were measured.

The results of measurements of Point FRFs are shown in Figure 6.19(a). The measurements were performed in a limited frequency range 80 to 100kHz due to the fact that only above 80kHz significant peaks were present in the impedance signature. Figure 6.19(b) presents electromechanical impedance plots obtained for the Transfer Frequency Response configuration, for which the measurements were performed over frequency range from 30 to 80kHz. Figure 6.20(a) and (b) presents, respectively, damage metrics calculated on the basis of Point and Point and Transfer FRFs.

It can be seen that the large changes of the amplitude and frequency of the resonance peaks can be observed in the impedance plots for the structure with a loosened bolt (Figure 6.19).

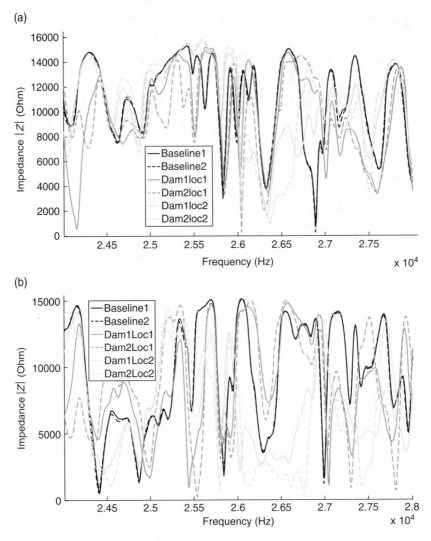

Figure 6.16 Transfer Frequency Response Functions obtained for the aluminium plate for transducer no. 2 (a) and transducer no. 4 (b)

The occurrence of the damage causes substantial shifts of the resonance peaks towards the lower frequencies and alteration of their amplitude. In all cases it was possible to find a mono-tonic relationship between the defect size and the values of the damage indices (Figure 6.20).

Among all analysed metrics the index DI_4 was found to be the most effective one, which was based on the statistical parameters and was characterized by the smallest initial values for the undamaged structure.

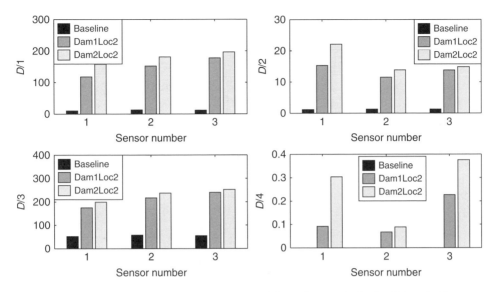

Figure 6.17 Damage indices calculated for the aluminium plate on the basis of Transfer Frequency Response Functions: damage location 2

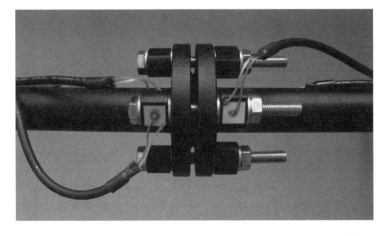

Figure 6.18 Section of a pipeline with mounted steel washers with bonded PZT patches

6.5.3 Discussion

The results of experiments performed for the aluminium plate panel showed that the described measurement method is sensitive to incipient and growing damage. Due to its properties, impedance based SHM appears to be a promising technique for local monitoring of

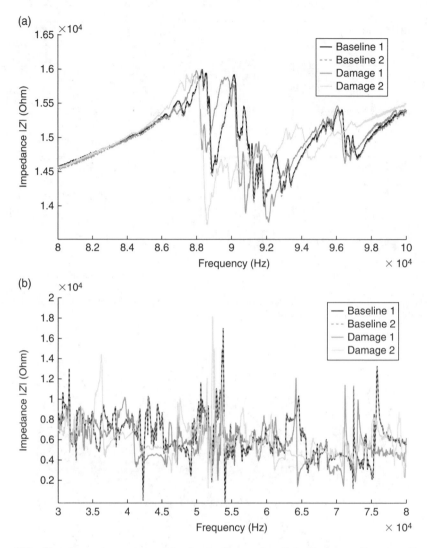

Figure 6.19 Impedance plots measured for the pipeline section. Point Frequency Response Functions (a) and Transfer Frequency Response Functions (b)

mechanical constructions. The results obtained for the pipeline section confirmed the ability to detect a loosened screw in the joint. It should be noticed that the method using electromechanical impedance measurements provides mainly a qualitative assessment of damage. The event of damage can be detected and tracked, however, the obtained results do not allow for the location of failure. It should be noticed that even the best damage indices may be useless for relatively large defects (i.e. missing screw) due to the large structural changes that can occur in the construction.

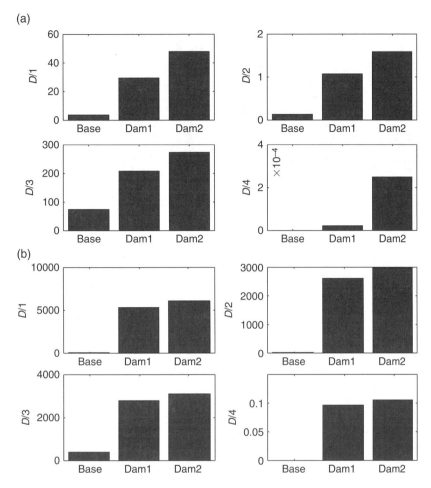

Figure 6.20 Damage indices calculated for the pipeline section on the basis of Point Frequency Response Functions (a) and Transfer Frequency Response Functions (b)

6.6 Verification of the Method on Aircraft Structures

In order to evaluate the usability of the developed SHM system to detect damage in aircraft structures, a series of tests was carried out on jet planes taken out of service. The experiments described below comprised measurements performed for two different types of fuselage element joints of two different aircraft (Rosiek *et al.* 2012).

6.6.1 Monitoring of a Bolted Joint in the Main Undercarriage Bay

The monitored section of the structure was a bolted connection of a fuselage in the main landing gear bay of a fighter jet aircraft [Figure 6.21(a)]. Four circular PZT patches, each with

(a)　　　　　　　　　　　　　　　　(b)

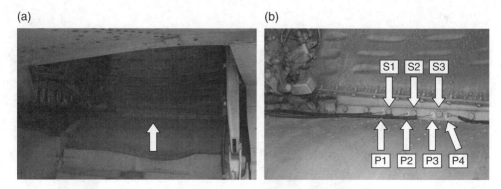

Figure 6.21 Monitored area of the fuselage (a) and piezoelectric patches bonded to the undercarriage bay (b)

Table 6.3 Values of the tightening torques for various damage cases

Damage case	Bolt S1 (N m)	Bolt S2 (N m)	Bolt S3 (N m)
No damage	25	25	25
Damage 1	25	15	25
Damage 2	25	0	25
Damage 3	25	0	0

diameter of 15 mm, were bonded to the construction using an epoxy glue [Figure 6.21(b)]. Monitored bolts are marked in Figure 6.21(b) as S1, S2 and S3, while piezoelectric transducers are indicated by P1, P2, P3 and P4, respectively.

Before performing the measurements, all four screws were tightened to torque of 25 N m using a torque spanner. For the considered structure the settings of the sensor node electrical circuit were selected and a set of baseline impedance plots was registered. The obtained results, which corresponded to the healthy state of the bolted joint, were used to calculate the initial values of the damage metrics. Afterwards, damage was introduced to the structure according to the following scheme:

- Damage 1: screw S2 was loosened and retightened with torque 15N m,
- Damage 2: screw S2 was fully loosened,
- Damage 3: screw S3 was fully loosened.

All values of the tightening torques corresponding to different damage cases are given in Table 6.3.

For the examined construction the Point FRFs were measured in the frequency range 20–60 kHz [shown in Figure 6.22(a)], and the Transfer FRFs in the range 20–52 kHz and 52–100 kHz [as an example the impedance plots obtained for the first range are shown in Figure 6.22(b)]. The occurrence of the incipient damage resulted in vertical shifts of the impedance plots without significant changes in its shape. Such a phenomenon was especially apparent for the Point

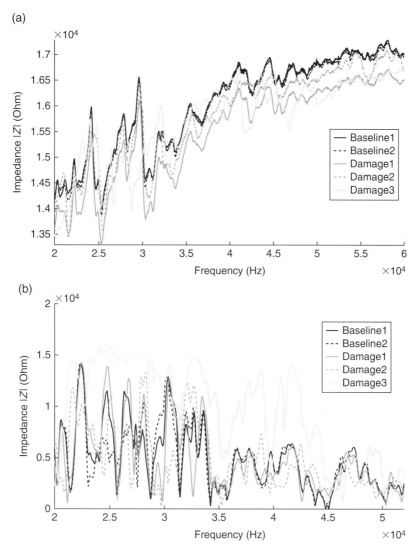

Figure 6.22 Impedance plots obtained for the fuselage joint: Point Frequency Response Function measured for the configuration P4-actuator/sensor (a) and Transfer Frequency Response Function measured for the configuration P3-actuator P4-sensor (b)

FRFs [Figure 6.22(a)]. The growth of the damage caused horizontal shifts of the resonances and emergence of new peaks. Significant changes of the transfer FRFs can be observed as a result of large structural changes present for the extensive damage 3 (complete loosening of bolts S2 and S3). Moreover, the Transfer FRFs were characterized by higher dynamics of the signal, better repeatability and higher signal to noise ratio than the Point FRFs as seen in Figure 6.22(b).

Bar charts shown in Figure 6.23, Figure 6.24 and Figure 6.25 present the values of applied damage metrics (see Equations 6.7, 6.8, 6.9 and 6.10), calculated on the basis of real parts of measured Point and Transfer FRFs. For the cases when a monotonic relation between the damage index value and failure size was identified, the bar graphs are indicated with dark shading. The obtained The obtained results show that it is difficult to infer about the location of damage because loosening of one bolt increases the damage metrics for due to their dense arrangement on the structure. This leads to the conclusion that monitoring of the area marked in Figure 6.21 might be possible with less piezoelectric elements.

The statistical metrics DI_4 turned out to be the least sensitive to the distant damage, as well as to the damage in the initial phase. Another important factor, which could hinder the damage assessment process, was that the measurements were conducted in ambient conditions with varying temperature, which may have influenced the obtained impedance characteristics.

6.6.2   Monitoring of a Riveted Fuselage Panel

The second tested object was a part of a riveted engine housing of a training jet aircraft [Figure 6.26(a)]. A construction node shown in Figure 6.26(b) was the subject of investigation. Two MFC d31-type piezoelectric transducers, marked in Figure 6.27(a) as P1 and P2, were bonded in the vicinity of the monitored region. Three rivets, between which a notch simulating a fatigue crack was made, are marked in Figure 6.27(b) as R1, R2 and R3.

The following damage scenario was assumed:

- Damage 1: notch between rivets R2 and R3 was made using a sharp tool,
- Damage 2: notch between rivets R2 and R3 was deepened,
- Damage 3: notch between rivets R1 and R2 was made using a sharp tool.

Due to the lack of resonance peaks in the point impedance signature only the transfer characteristics were registered during the experiment. In this configuration transducer P1 was used to excite the structure, while P2 was acting as a sensor.

The measurements were carried out in the frequency range 30–100 kHz. Significant changes in the amplitude of the resonance peaks can be observed in Figure 6.28(a), which presents Transfer FRFs registered for both undamaged and damaged structure. For the calculated damage metrics the growth of their values corresponding to damage propagation was observed [Figure 6.28(b)]. It was found that for the deterministic metrics DI_1, DI_2 and DI_3 the most significant increments corresponded to the deepening of the notch between rivets R2 and R3. The statistical damage metric DI_4, which is distinguished by the smallest initial values for the healthy structure, did not give satisfactory results for the first damage case. There was no significant change in DI_4 after making the first notch (damage 1), so the index did not indicate the appearance of incipient damage. However, the presence of the notch between rivets R1 and R2, denoted as damage 3, caused a very large variation in the DI_4 value, which was not observed for the remaining damage metrics. This leads to the conclusion that for robust monitoring of failure propagation in the structure, it is necessary to use multiple damage indices.

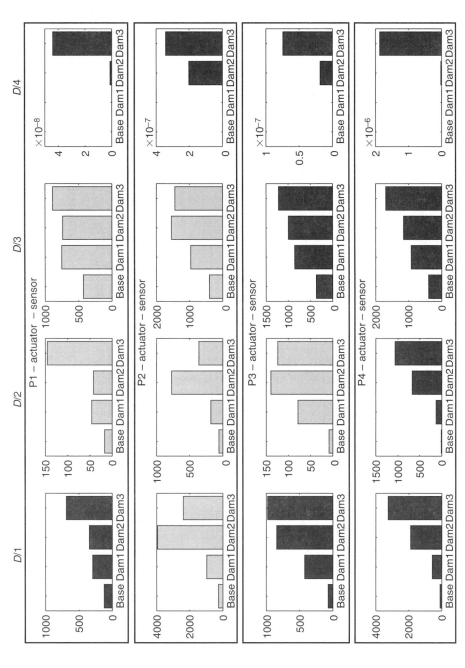

Figure 6.23 Damage indices calculated for the fuselage joint on the basis of Point Frequency Response Functions measured in the frequency range 20–60kHz

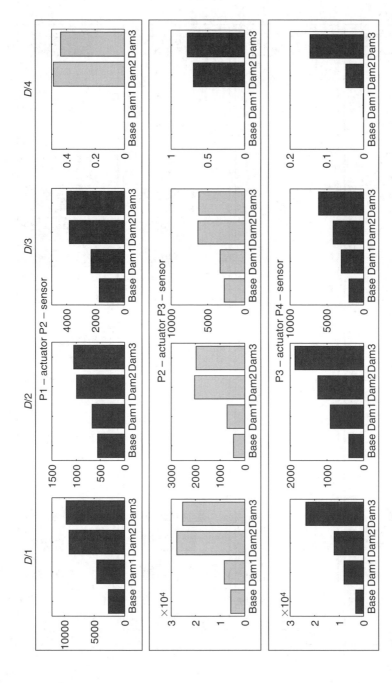

Figure 6.24 Damage indices calculated for the fuselage joint on the basis of Transfer Frequency Response Functions measured in the frequency range 20–52kHz

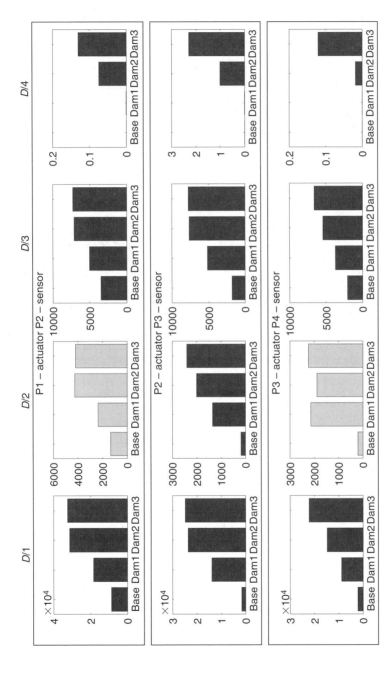

Figure 6.25 Damage indices calculated for the fuselage joint on the basis of Transfer Frequency Response Functions measured in the frequency range 52–100kHz

(a)   (b)

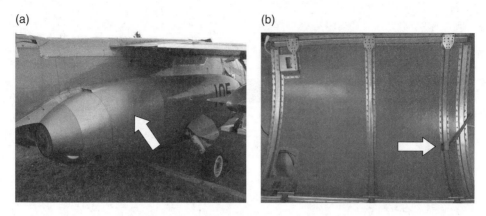

Figure 6.26   Examined part of the plane (a) and inner side of the monitored engine housing (b)

(a)   (b)

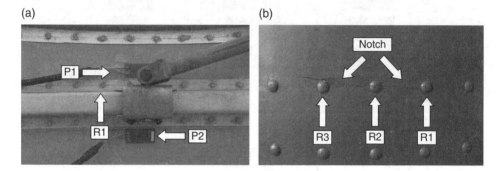

Figure 6.27   Close-up view of the engine housing: inner side with visible MFC transducers (a) and outer side with visible notch between rivets (b)

6.6.3   Discussion

The obtained results proved the qualitative character of the method. Using the damage index approach for the state assessment of a monitored structure, it is difficult to determine the size and location of the fault, especially when a multiple damage scenario occurs. However, the proposed technique allows for detection of damage in various stages with the use of different damage metrics. The deterministic indices make it possible to detect damage at the early stage of propagation. Although the statistical index is less sensitive to the incipient damage, it is also less susceptible to the shifts of impedance plots without alteration of its shape, which may result from temperature variations. A substantial matter influencing the efficiency of the method is the layout of the PZT transducers on the monitored structure. Due to the local character of the impedance technique, piezoelectric transducers should be placed in the locations where the occurrence of damage is most probable, i.e. places with a significant stress concentration. Nevertheless, too dense spacing of the sensors does not improve the efficiency of the fault detection and can increase the overall costs of implementing the SHM system. Thus, further research aimed at determining the sensing region for different types of structures should

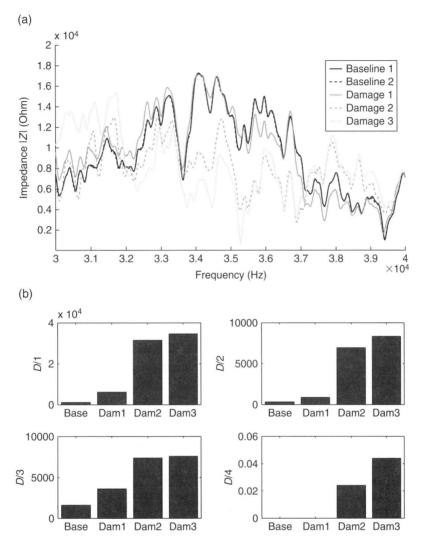

Figure 6.28 Transfer Frequency Response Functions measured for the engine housing (a) and damage indices calculated on their basis (b)

be carried out. Future work should also focus on investigating the influence of environmental effects such as temperature changes and in-flight static and dynamic loadings.

6.7 Conclusions

The electromechanical impedance based SHM method allows the effective detection of a damage and its growth by the utilization of the electromechanical coupling feature of piezo-electric transducers mounted in the monitored mechanical structure. Any significant change of mechanical properties caused by the presence of damage can be assessed with numerical metrics – damage indices. The measured electromechanical impedance contains information on both the monitored structure and piezoelectric transducers. Therefore, the procedure of self-monitoring applied for mounted transducers is also feasible. Contrary to the mechanical impedance based methods which appeared as first, all sensors used in the electromechanical impedance based methods are thin and do not interfere with the structure while it is in operation. This fact allows for the permanent assemblage of transducers to a monitored structure. Piezoelectric patches are characterized by small dimensions and weights, so their impact on the properties of the host structures is negligible. They can be placed almost everywhere. Piezoelements can be used both as actuators and sensors simultaneously which is crucial in the context of impedance based methods because it reduces the number of piezoelectric elements as well as electric wiring and therefore required hardware. A simple transducer enables measurement of the electromechanical impedance. Sensitivity of impedance based methods is comparable with sensitivity of ultrasonic methods but the former methods do not require an experienced technician to localize damages. The required hardware is also less expensive in the case of impedance based techniques. These methods act as a compromise between global structural methods and ultrasonic methods. Moreover, the mounted piezoelectric elements can also be used by other NDE approaches. A critical aspect of the piezoelectric active sensing technologies is that large numbers of distributed sensors and actuators are usually needed to perform the required monitoring process. The issue is also the detection of the scope of each piezotransducer. As presented in Sun et al. (1995), in the case of a truss structure it is experimentally proven that the detection range of a bonded PZT sensor is highly constrained to its immediate vicinity. Therefore damage location can be determined accurately, however with enough dense net of sensors. Moreover, in the case of composite laminates there is a possibility to include piezoelectric transducers in between subsequent plies of the monitored structure without degrading its integrity (Bois et al. 2007). Since the described method is sensitive to the variation of temperature, humidity, and additionally to boundary conditions and operational loads, but only when they interfere with the frequencies of PZT actuation, these parameters should be monitored in order to evaluate impedance correctly. However, the most important parameter that disturbs the measurement process is temperature. Variations of the temperature cause frequency shifts in the measured electromechanical impedance characteristics. Modified damage metrics which introduce the influence of the temperature are used to deal with the problem. In Park et al. (1999) the empirical approach is proposed for the compensation of varying temperature effect. The damage metric defined as the sum of the real admittance change squared is used with a coefficient calculated while minimizing the value of the applied damage metric. The impedance based methods are qualitative measures because various types of damage such as cracks, corrosion, delamination and loosened connections change the mechanical impedance similarly, which makes the distinction between each type of damage very difficult, if not impossible.

References

Ayres J, Lalande F, Chaudhry Z and Rogers C 1998 Qualitative impedance-based health monitoring of civil infrastructures. *Smart Materials and Structures* **7**, 599–605.

Bhalla S and Soh C 2003 Structural impedance based damage diagnosis by piezo-transducers. *Earthquake Engineering and Structural Dynamics* **32**, 1897–1916.

Bois C, Herzog P and Hochard C 2007 Monitoring a delamination in a laminated composite beam using in-situ measurements and parametric identification. *Journal of Sound and Vibration* **299**, 786–805.

Chaudhry Z, Joseph T, Sun F and Rogers C 1995 Local-area health monitoring of aircraft via piezoelectric actuator/sensor patches. *Smart Structures and Integrated Systems SPIE 2443, SPIE Conference*, San Diego, USA.

Giurgiutiu V 2007 *Structural Health Monitoring with Piezoelectric Wafer Active Sensors*. Elsevier Academic Press.

Giurgiutiu V and Rogers CA 1999 Modeling of the electro-mechanical (E/M) impedance response of a damaged composite beam. *ASME Winter Annual Meeting, ASME Aerospace and Materials Divisions, Adaptive Structures And Material Systems Symposium*. Nashville, USA, pp. 39–46.

Giurgiutiu V and Zagrai AN 2001 Electro-mechanical impedance method for crack detection in metallic plates. *Proceedings of SPIE, Advanced Nondestructive Evaluation for Structural and Biological Health Monitoring*, vol. 4335 of *Society of Photo-Optical Instrumentation Engineers (SPIE) Conference Series*. Newport Beach, USA, pp. 131–142.

Helton JC and Davis FJ 2003 Latin hypercube sampling and the propagation of uncertainty in analyses of complex systems. *Reliability Engineering and System Safety* **81**, 23–69.

Koh Y, Rajic N, Chiu W and Galea S 1999 Smart structure for composite repair. *Composite Structures* **47**, 745–752.

Liang C and Rogers C 1994 Coupled electro-mechanical analysis of adaptive material systems-determination of the actuator power consumption and system energy transfer. *Journal of Intelligent Material Systems and Structures* **5**, 12–20.

Liu W and Giurgiutiu V 2007 Finite element simulation of piezoelectric wafer active sensors for structural health monitoring with coupled-filed elements. *Proceedings of SPIE, Sensors and Smart Structures Technologies for Civil, Mechanical, and Aerospace Systems 2007*, vol. 6529 of *Society of Photo-Optical Instrumentation Engineers (SPIE) Conference Series*. San Diego, USA.

Mascarenas D, Todd M, Park G and Farrar C 2007 Development of an impedance-based wireless sensor node for structural health monitoring. *Smart Materials and Structures* **16**(6), 2137–2145.

Myers R and Montgomery D 1995 *Response Surface Methodology Process and Product Optimization Using Designed Experiments*. John Wiley & Sons, Ltd.

Naidu A and Soh C 2004 Damage severity and propagation characterization with admittance signatures of piezo transducers. *Smart Materials and Structures* **13**, 393–403.

Park G and Inman D 2007 Structural health monitoring using piezoelectric impedance measurements. *Philosophical Transactions of the Royal Society A* **365**, 373–392.

Park G, Kabeya K, Cudney H and Inman D 1999 Impedance-based structural health monitoring for temperature varying applications. *JSME International Journal* **42**(2), 249–258.

Park G, Sohn H and Farrar C 2003 Overview of piezoelectric impedance-based health monitoring and path forward. *The Shock and Vibration Digest* **35**, 451–463.

Peairs D, Gyuhae P and Inman D 2005 Simplified modeling for impedance-based health monitoring. *Key Engineering Materials* **293–294**, 643–652.

Peairs D, Park G and Inman D 2002 Self-healing bolted joint analysis. *Proceedings of 20th International Modal Analysis Conference*. Los Angeles, USA, pp. 1272–1278.

Raju V, Park G and Cudney H 1998 Impedance-based health monitoring technique of composite reinforced structures. *Proceedings of 9th International Conference on Adaptive Structures and Technologies*. Cambridge, USA, pp. 448–457.

Roach D and Rackow K 2006 Health monitoring of aircraft structures using distributed sensor systems. Technical report, Sandia National Laboratories, FAA Airworthiness Assurance Center Albuquerque.

Rosiek M, Martowicz A and Uhl T 2010a Uncertainty and sensitivity analysis of electro-mechanical impedance based SHM system. *IOP Conference Series: Materials Science and Engineering* **10**, 1–9.

Rosiek M, Martowicz A and Uhl T 2011a SHM system based on impedance measurements. *Diagnostyka-Diagnostics and Structural Health Monitoring* **59**(3), 3–8.

Rosiek M, Martowicz A and Uhl T 2011b Structural health monitoring system based on electromechanical impedance measurements. *Proceedings of 8th International Workshop on Structural Health Monitoring*. Stanford, USA, pp. 314–321.

Rosiek M, Martowicz A and Uhl T 2012 Electromechanical impedance based SHM system for aviation applications. *Key Engineering Materials* **518**, 127–136.

Rosiek M, Martowicz A, Uhl T, Stepinski T and Lukomski T 2010b Electromechanical impedance method for damage detection in mechanical structures. *Proceedings of 11th IMEKO TC 10 Workshop on Smart Diagnostics of Structures.* Kraków, Poland.

Sun F, Chaudhry Z, Liang C and Rogers C 1995 Truss structure integrity identification using PZT sensor-actuator. *Journal of Intelligent Material Systems and Structures* **6**, 134–139.

Sun F, Rogers C and Liang C 1996 Structural frequency response function acquisition via electric impedance measurement of surface-bonded piezoelectric sensor/actuator. *Proceedings of 36th AIAA/ASME/ASCE/AHS/ASC Structures, Structural Dynamics, and Materials Conference.* New Orleans, USA, pp. 3450–3461.

Yan W, Lim C, Chen W and Cai J 2007 A coupled approach for damage detection of framed structures using piezoelectric signature. *Journal of Sound and Vibration* **307**, 802–817.

Yang Y, Lim Y and Soh C 2008 Practical issues related to the application of the electromechanical impedance technique in the structural health monitoring of civil structures: II. Numerical verification. *Smart Materials and Structures* **17**(3), 1–12.

Zagrai A and Giurgiutiu V 2001 Electro-mechanical impedance method for crack detection in thin wall structures. *Proceedings of 3rd International Workshop on Structural Health Monitoring.* Stanford, USA.

7

Beamforming of Guided Waves

Łukasz Ambroziński

Department of Mechatronics and Robotics, Faculty of Mechanical Engineering and Robotics, AGH University of Science and Technology, Poland

7.1 Introduction

SHM of plate-like structures is a hot topic of research since many of such constructions must meet high safety standards. Lamb waves are a promising tool for these applications on account of their ability to propagate over long distances and sensitivity to various types of damages (Raghavan and Cesnik 2007). A critical factor for SHM of plate-like structures is the design of transducers and their distribution over the investigated plate. A well known approach in this field is the use of transducers distributed over the structure (Michaels 2008). The sensitivity of a SHM system can be enhanced by means of active ultrasonic phased arrays (PAs) due to their superior signal to noise ratio and beam-steering capability (Giurgiutiu 2008).

The concept of beam-steering of ultrasonic waves with the use of a PA is well-established both in medical (Karaman *et al.* 2009) and NDT (Drinkwater and Wilcox 2006) imaging. It appeared, however, that implementing the method for SHM of plate-like structures requires arrays with 2D topologies to avoid equivocal damage localization (Giurgiutiu 2008). Therefore, a number of reports have been published concerning various shapes of arrays, such as, star (Ambrozinski *et al.* 2011), square (Engholm and Stepinski 2010b), circular (Wilcox 2003a) and spiral (Yoo *et al.* 2010) shaped arrays.

The quality of an image created with the use of an array of a defined shape depends also, among other factors, on the array's aperture and imaging technique. A moving transducer can be used in mechanized NDT applications to increase the aperture of an array with the use of synthetic aperture focusing technique (SAFT) (Doctor *et al.* 1986). Since this solution would be infeasible in most SHM applications, this chapter is concerned with the static arrays that can be permanently attached to the monitored structure and used in SHM applications. These arrays can take advantage of multiple emitting transducers capable of illuminating

Advanced Structural Damage Detection: From Theory to Engineering Applications, First Edition.
Edited by Tadeusz Stepinski, Tadeusz Uhl and Wieslaw Staszewski.
© 2013 John Wiley & Sons, Ltd. Published 2013 by John Wiley & Sons, Ltd.

a defect from a set of diverse localizations, which results in an extension of the array's effective aperture (Moreau *et al.* 2009).

Another critical factor that affects the performance of a PA based SHM system is the signal processing technique implemented in the beamforming scheme. The most common scheme used for processing snapshots captured by an array, mostly due to its simplicity and robustness, employs delay and sum (DAS) operations in the time domain. More advanced methods, capable of angular (Engholm and Stepinski 2010a) and radial resolution (Liu and Yuan 2010; Wilcox 2003b) improvement, can also be implemented to the Lamb waves sensed using PAs. However, advanced processing techniques require precise information on material properties, which can create limitations in practical applications, especially, to anisotropic materials.

Self-focusing methods were introduced to solve such problems, for instance, time reversal mirrors were introduced for ultrasonic waves (Fink 1992; Fink *et al.* 2000). Time-reversal mirrors are a powerful tool that enabled signal to noise ratio improvement (Ing and Fink 1995) and reduction of the effect of dispersion (Ing and Fink 1996, 1998). The iterative time reversal mirror enables selective focusing on the strongest reflector in the region of interest to be obtained (Prada *et al.* 1991). In the SHM applications, however, focusing on targets with lower reflexivity is also often desired. The decomposition of time reversal operator (DORT) method has been introduced (Prada *et al.* 1996) with the aim of detecting and focusing waves on the multiple scatterers. DORT is a signal processing technique that is able to estimate, on the basis of the received data, time delays required for selective focusing waves on a target. The method has been successfully applied in NDT applications (Kerbrat *et al.* 2002) as well as for Lamb wave characterization (Prada and Fink 1998).

It appeared, however, that in some cases resolving multiple scatterers with this method may fail. To deal with this problem an extension of the method employing time – frequency representation (TFR) of the signals was proposed (Ambrozinski *et al.* 2012a). Lamb waves propagating in a thin plate normally generate nonstationary responses due to the dispersive phenomenon and therefore, the continuous wavelet transform (CWT) was used to calculate the TFR of the signals. The scatterers that could not be resolved correctly in the frequency domain could be resolved in the time domain (Ambrozinski *et al.* 2012a).

The extended DORT method was applied to damage imaging using a 2D star-shaped active array (Ambrozinski *et al.* 2011). The array consisted of a transmitting subarray, which could be self-focused on a target, and a receiving subarray used to capture echoes reflected from scatterers. In this chapter, the modified DORT is analysed and compared with the multi-static imaging (MSI) based on the standard DAS beamforming. Both methods were applied for damage detection in an aluminium plate using the 2D star-shaped active array.

This chapter is organized as follows: Section 7.2 gives a theoretical background for the presented approaches. First, principles of synthetic and effective aperture concepts are given in Sections 7.2.1 and 7.2.2, respectively. Imaging schemes based on the presented approaches are described in detail in Section 7.2.3. Finally, theoretical grounds of the self-focusing technique and its application for imaging are given in Section 7.2.4. The theoretical overview is followed by numerical investigations of the presented methods. The effective aperture concept is examined in Section 7.3.1, next, an application of various imaging schemes is presented using a star-shaped array in Section 7.3.2. Finally, the self-focusing DORT-CWT method is verified in Section 7.3.3. The experimental results obtained with the presented methods are given in Section 7.4. Evaluation of sensing arrays and effective aperture can be found in Sections 7.4.1 and 7.4.2, respectively. Next, an example of damage imaging using an array and synthetic aperture is given in Section 7.4.3. Finally, the self-focusing, DORT-CWT, method is validated in Section 7.4.4 and damage imaging results obtained using the technique are presented in Section 7.4.5. Discussion and conclusions are given in Sections 7.5 and 7.6, respectively.

7.2 Theory

In order to provide solid theoretical grounds for the presented approaches, a brief overview of the concepts of synthetic aperture and effective aperture is outlined below. The theoretical background is then used to present imaging schemes in the context of the synthetic aperture concept and to present a design method for 2D ultrasonic arrays and to evaluate their directivity characteristics.

In the following section damage imaging with the use of self-focusing technique will be presented. First, the theoretical background of the DORT technique and its extension to DORT-CWT will be outlined. Next, numerical and experimental evaluation of the extended technique will be given. Finally, imaging results obtained with the use of a self-focused transmitting array, operating in PA mode, will be presented.

7.2.1 *Imaging Using Synthetic Aperture*

Application of array technique to the SHM of plate-like structures involves a set of transducers that are permanently attached to or embedded into the structure to enable performing inspection of a large region of interest (ROI) from a fixed location. In an active pulse-echo setup the transmitting elements of the array generate elastic waves in the plate that are scattered from the discontinuities present in the structure and the scattered waves are subsequently sensed by the array's receiving elements.

If an array is used in the PA mode, the transmitting and receiving subarrays normally consist of the same elements. In transmission the elements are excited by time-shifted pulses to obtain a steered wavefront, which is received by the same elements in the reception mode and amplified by the introduction of respective time delays which bring the received pulses in phase. Beamformers implementing this technique, referred to as DAS, are commonly used in ultrasound applications. Using steered and possibly focused beam improves angular (azimuth) resolution in ultrasonic image but it requires scanning of the whole ROI. This means that the send – receive cycle has to be repeated for a number of azimuths and ranges within the ROI, which results in a low acquisition speed.

An alternative to PA is the synthetic aperture (SA) approach that has been developed for radar (SAR, synthetic aperture radar) and more recently applied in medical ultrasound. Generally, SA beamforming uses different transmitting and receiving subarrays for imaging. In the classical SA technique, the image is formed by successive excitation of all transmiting elements and reception of the scattered waves by all other array elements. The final image is formed as a superposition of beamformed energy from all the transmitted cycles. A SA image is generated as a result of post-processing of all received data which requires rather intensive computations. A SA image is essentially equivalent to an image produced by conventional PA using a number of transmit – receive focused beam cycles equal to the number of pixels in the ROI.

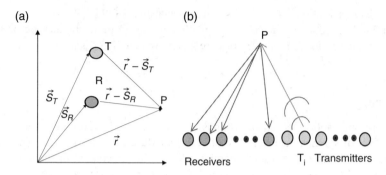

Figure 7.1 Beamforming using single receiver – transducer pair (a). Principle of synthetic aperture imaging (b)

For a linear array with N elements a full matrix of N^2 transmit – receive data can be acquired if all the elements are fired in N transmission cycles. There are, however, many other ways of gathering data for SA processing that depend on the array geometry and topology of the transmitting aperture. Most of these ways are designed to reduce the number of transmissions with minor loss of resolution and an acceptable signal to noise ratio. Generally, this can be achieved by using sparse transmitting apertures. If elements of a sparse transmitting aperture are distributed across the full aperture of the array, there will be no loss in the field of view and only a minor loss in lateral resolution. Below, we will present principles that can be applied to evaluate different combinations of 2D transmitting – receiving apertures used in SA imaging.

Consider a transmitter – receiver pair located at a homogeneous 2D medium (e.g. a plate-like structure) in the coordinate system shown in Figure 7.1(a). The point transmitter T located at $\vec{s}_T$ emits elastic waves that are scattered by the point reflector P at $\vec{r}$, and received by the point receiver R at $\vec{r}_R$. The wave front of the wave with frequency ω at the point P located at the distance $|\vec{r} - \vec{s}_T|$ from the transmitter T can be expressed as (Giurgiutiu 2008):

$$f(\vec{r}, \vec{s}_T, t) = \frac{A}{\sqrt{|\vec{r} - \vec{s}_T|}} e^{j[\omega t - \vec{k} \cdot (\vec{r} - \vec{s}_t)]}, \tag{7.1}$$

where A is a constant, $\vec{\eta} = \frac{\vec{r}}{|\vec{r}|}$, $\vec{k} = \vec{\eta}\omega/c$ is the wavenumber vector of nondispersive wave-mode propagating in the direction $\vec{r}$, t is time, and $\vec{k} \cdot (\vec{r} - \vec{s}_t)$ denotes dot multiplication. Assume that a transmitting aperture consists of M point sources emitting waves with angular frequency ω. In such case the individual wave front of the mth transmitter located at $\vec{s}_{Tm}$ becomes

$$f(\vec{s}_{Tm}, t) = \frac{A}{\sqrt{|\vec{r}_m|}} e^{j(\omega t - \vec{k}_m \vec{r}_m)}, \tag{7.2}$$

where $\vec{r}_m = \vec{r} - \vec{s}_{Tm}$ and $\vec{k}_m = \vec{\eta}_m \omega/c$.

The combined wave front transmitted by the aperture will be a superposition of the effects of M elements

$$
z_T(\vec{r}, t) = \sum_{m=0}^{M-1} w_m \frac{A}{\sqrt{|\vec{r}_m|}} e^{j\left(\omega t - \vec{k}_m \vec{r}_m\right)}
$$

$$
= f\left(\vec{r}, t - \frac{|\vec{r}|}{c}\right) \sum_{m=0}^{M-1} w_m \sqrt{\frac{|\vec{r}|}{|\vec{r}_m|}}\, e^{j\omega \frac{\vec{r}-\vec{r}_m}{c}},
$$

(7.3)

where w_m are weighting coefficients applied to the aperture elements.

Note that the overall characteristics of the imaging setup is defined by the point spread function (PSF), which is its response to a point-like target given by the convolution of the functions, Equation (7.3), in transmission and reception

$$
Z_{TR}(\vec{r}, \vec{s}_T, \vec{s}_R, t) = z_T(\vec{r}, t) * z_R(\vec{r}, t).
$$

(7.4)

It can be shown that for far-field conditions, i.e. for $|\vec{r}| \gg |\vec{r}_m|$, the respective vectors become almost parallel, i.e. $\vec{k}_m \approx \vec{\eta}\omega/c_p = \vec{k}$ and $\sqrt{r_m} \approx \sqrt{r}$ and Equation (7.2) can be approximated by

$$
f(\vec{r}_m, t) \approx f\left(\vec{r}, t - \frac{|\vec{r}|}{c}\right) e^{j\frac{\omega}{c}\vec{\eta}\cdot\vec{s}_{Tm}}.
$$

(7.5)

To generate a steered beam the desired azimuth ϕ_s we have to apply delays $\delta_m(\phi_s)$ and apodization coefficients w_m. The beam pattern (BP) of the transmitting aperture will be (Giurgiutiu 2008):

$$
\mathrm{BP}(w_m, \eta_m, \phi_s) = \frac{1}{M} \sum_{m=0}^{M-1} w_m e^{j\frac{\omega}{c}[\vec{\eta}\cdot\vec{r}_m - \delta(\phi_s)]}.
$$

(7.6)

A similar relation will also apply for the receiving aperture. Note that the beam pattern defines the array's characteristics in the frequency domain for a given angular frequency while DAS beamformers operate in the time domain.

The principle of DAS algorithms is to compensate the phase shifts that occur for pulse signals emitted by the transmitter, first in the image point $\vec{r}$ and then at the receiver point $\vec{r}_R$. This operation is applied in the time domain by compensating time delays equivalent to the phase shifts in the broadband signals acquired by the receiving element. After performing this operation the intensity of the image at the point P is given by (Moreau *et al.* 2009):

$$
I(\vec{r}) = y_{TR}\left(\frac{|\vec{r} - \vec{r}_T| + |\vec{r} - \vec{r}_R|}{c}\right) = y_{TR}(\tau_T + \tau_R),
$$

(7.7)

where c is the velocity of the propagating wave mode, τ_T and τ_R are propagation times from the transmitter and to the receiver from the point P, respectively, and $y_{TR}()$ is the temporal signal received by the receiver when the transmitter is firing. Consider now the 2D configuration of a linear ultrasonic array consisting of N transmitters and receivers shown in Figure 7.1b. The intensity of the image produced by the classical SA setup using N transmissions will be a sum of the intensities obtained from the combination of each element

$$I(\vec{r}) = \sum_{k}^{N} \sum_{l}^{N} y_{k,l} \left(\frac{|\vec{r} - \vec{r}_T^k| + |\vec{r} - \vec{r}_R^l|}{c} \right) = \sum_{k}^{N} \sum_{l}^{N} y_{k,l} \left(\tau_T^k + \tau_R^l \right). \tag{7.8}$$

Equation (7.8) can be used in the situation when the transmitting and receiving apertures do not have weighting functions, i.e. no apodization function is used to correct the aperture characteristics.

Apodization is implemented when sparse transmitting and receiving apertures have to result in a desired final PSF. Assume that the weighting coefficients are used for both apertures, that is, functions w_T and w_R represent the weight functions applied to the transmit and receive aperture in order to control the relative contribution of their elements. Then the intensity of the resulting image will be (Moreau *et al.* 2009):

$$I(\vec{r}) = \sum_{k \in T} \sum_{l \in R} w_T^k w_R^l y_{k,l} \frac{|\vec{r} - \vec{r}_T^k| + |\vec{r} - \vec{r}_R^l|}{c}. \tag{7.9}$$

Processing the temporal signals $y_{k,l}(\cdot)$ according to Equation (7.9) yields an image focused in all points (pixels) of the ROI. Note, however, that an interpolation is normally required due to the time-discrete character of the temporal signals $y_{k,l}(\cdot)$.

7.2.1.1 Examples of a 2D PA Design

As presented in the previous section, a PA of sensors can be considered in the transmission mode as a directional source that allows the waves to be enhanced in a desired azimuth and suppresses those arriving from other directions, while in the reception mode, it acts as a spatial filter enabling isolation of the waves arriving from a desired azimuth. The radiation and filtering, however, are almost never perfect, therefore an investigation of the array angular performance is required in the array's designing process. The beam pattern plot, which illustrates the power output from the array as a function of radiation angle, is a very useful tool for this purpose.

To illustrate this let us consider a linear array consisting of M elements spaced at d with coordinate origin in the middle, as shown in Figure 7.2(a). The location vector of the mth point transducer is $\vec{s}_m = [(m - \frac{M-1}{2})d, 0]$.

Then, the BP of this array in the far field will be (Giurgiutiu 2008)

$$BP\left(w_m, \frac{d}{\lambda}, M\right) = \frac{1}{M} \sum_{m=0}^{M-1} w_m e^{j2\pi \frac{d}{\lambda} \left[\left(m - \frac{M-1}{2}\right) \cos\phi \right]}, \tag{7.10}$$

where λ is the wavelength of the propagating mode, ϕ is the azimuth of the transmitted/received waves, and w_m is the weight coefficient belonging to the apodization function.

A normalized beam pattern of a linear unapodized array with $M = 9$ and $d = \lambda/2$, plotted in polar coordinates as a function of azimuth ϕ is presented in Figure 7.2(b). It can be seen that

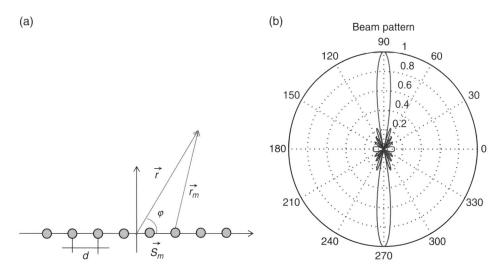

Figure 7.2 Linear uniform array (a). Normalized beam pattern of a linear uniform array with $M = 9$ elements (b)

the beam pattern is symmetric with respect to the array axis, which is a source of ambiguity in imaging. Furthermore, it can be also seen that the unapodized array has a considerable side lobe level. Another undesired feature of linear arrays is their azimuth dependent angular resolution – the main lobe width of a steered array increases with azimuth and achieves its maximum at the array's fire ends. To avoid the shortcomings of linear arrays we will consider a number of 2D array topologies characterized by uniform element distribution with high degree of symmetry.

Three example 2D topologies are presented in Figure 7.3(a), (b) and (c), respectively: cross-shaped, square, and star-shaped arrays of a different number of elements but the same pitch. The beam patterns, evaluated for the investigated arrays for two exemplary steering angles of 70° and 180° are shown in Figure 7.3(d–f). It can be seen that in terms of the main-beam width and the side-lobes level, the best performance features the square matrix [Figure 7.3(b) and (e)], which is not surprising since it is composed of the largest number of elements. Moreover, it can be seen that the star-shaped array offers considerably lower side lobes than the cross-shaped array and a slightly wider main beam than the square matrix. Therefore, the star-shaped array was selected for further evaluation as a compromise between simplicity and performance.

7.2.2 Effective Aperture Concept

In general, any combination of sparse transmitting/receiving apertures can be used in every measurement cycle to provide data for imaging algorithms that can take advantage of numerous snapshots acquired for a range of different emitters'/sensors' configurations. If the beamforming instrumentation is capable of switching between the transmission and reception of the elements in the same excitation cycle, a given array consisting of N elements can

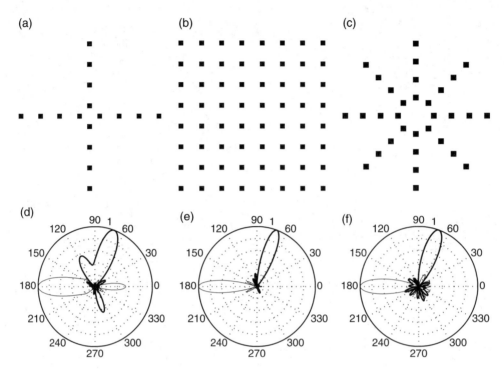

Figure 7.3 Comparison of three candidate 2D array configurations with their beam patterns obtained for steering angles of 70° and 180°: cross-shaped, 16 elements (a and d); square matrix, 64 elements (b and e); and star-shaped, 32 elements (c and f)

collect up to N^2 time traces, i.e. the full matrix of transmit – receive data. Although the full matrix offers the largest amount of information which can be gathered from a fixed position with the use of an array, its collection can be time-consuming and hence impractical in many applications (Lockwood *et al.* 1998; Moreau *et al.* 2009).

Sparsely populated apertures can be used as a compromise between image quality and acquisition time expressed as the number of images per time unit in the SA approach. The main idea is to effectively use transducers grouped into transmitting and receiving subarrays. The elements of a transmitting aperture can be fired in the same transmission cycle to obtain a steered wave front. Alternatively, individual transmitters can be excited separately while multiple responses are acquired by the elements of the receiving aperture. The snapshots of the acquired backscattered signals can be then used for synthetic focusing during post-processing.

If the array is divided into two parts, their individual beam patterns can be evaluated in the way presented in the previous section. However, when two subarrays contribute to imaging, the resulting radiation pattern of the whole active array can be evaluated using the effective aperture concept.

The effective aperture of an array is defined as an equivalent receive aperture that would produce an identical two-way radiation pattern if a point source was used for the transmission (Lockwood *et al.* 1996). Thus, in the case of a single element excitation the effective aperture is simply a receiving aperture, as the effective aperture is a convolution of the transmission and reception apertures (Lockwood *et al.* 1998; Moreau *et al.* 2009). Assuming that the weight functions $w_T(s_T)$ and $w_R(s_R)$ represent the transmission and reception apertures, respectively, then the effective aperture $a_{TR}(s_R, s_T)$ is their spatial convolution

$$a_{TR}(s_T, s_R) = w_T(s_T) * w_R(s_R) \tag{7.11}$$

where all three apertures are expressed as functions of vectors defining element location s_T and s_R in the transmitting and receiving aperture, respectively.

7.2.3   Imaging Schemes

Damage imaging in SHM applications normally relies on representing the ROI as a bitmap and processing the acquired signals to evaluate the intensity of the image. The processing usually involves simply time-shifts of the responses. However, depending on the measurement technique, the processing of the captured snapshots can differ in details. In the following section a practical implementation of damage imaging schemes with the use of Lamb waves will be presented. First, imaging in a monostatic setup using a single transmitter and multiple receiver array will be introduced. The method will be then generalized to MSI. Finally, a description of the SA concept using an active transmitting array used for sweeping the ROI will be given.

7.2.3.1   Single Transmitter, Multiple Receivers Array

Consider a setup consisting of uniform linear sensing array of M transducers, presented in Figure 7.4(a), and a point-like emitter localized in point $(0,0)$. The emitter is an omnidirectional point-like source of a single dispersive Lamb mode. Due to the dispersive character, the individual waves travel at a different velocity than the wave packet. Velocity of the individual waves is determined by phase velocity, which directly affects the phase of the wave. Group velocity, related to the speed of the group of individual packets, determines the velocity of wave energy transportation (Giurgiutiu 2008).

Assume now a point-like scatterer P in the array's far-field at the distance $|\vec{r}|$ and azimuth ϕ. The excited wave propagates through the medium and when it hits the target it is scattered and propagates towards the sensing array. The time-shift between the excited wave and the reflected wave packet, which reached the point $(0,0)$ is simply

$$\tau_g(|\vec{r}|) = \frac{2|\vec{r}|}{c_g}, \tag{7.12}$$

where c_g denotes group velocity of the Lamb wave.

If the scatterer is in the array's far field, the reflected wave can be assumed planar and its wavefront in $(0,0)$ forms a line perpendicular to the direction of arrival. Since the distances between the subsequent receivers and the wavefront differ, the captured snapshots will be shifted in phase. Under the assumptions specified above, the differences in wave

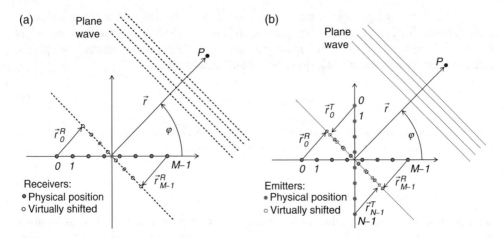

Figure 7.4 Imaging under far-field assumption using uniform linear sensing array: (a) synthetic aperture; and (b) reflections form a target localized at angle ϕ acquired by sensors of the uniform linear array

propagation and also time-shifts occurring due to these differences can be easily predicted using the following trigonometric expression

$$\tau_m(\phi) = \frac{d(m - \frac{M-1}{2})\cos\phi}{c_p},\tag{7.13}$$

where c_p is the phase velocity of the wave, and d is the sensor spacing.

As presented in the previous sections, the beamforming algorithm compensates for these time shifts when the array is steered in the direction of wave arrival. The compensation can be considered a virtual shift of the emitting/receiving elements (Wilcox 2003a).

In the approach presented here, imaging is performed in the form of a bitmap in a polar coordinate system where each pixel can be defined with its angular position α and the distance from the centre $|\vec{r}|$. Then, the intensity of the pixel can be evaluated using

$$I(|\vec{r}|, \alpha) = \sum_{m=0}^{M-1} w_m y_m(\tau_m(\alpha) + \tau_g(|\vec{r}|)),\tag{7.14}$$

where y_m is the temporal signal captured by the mth sensor contributing to imaging with an apodization weight w_m.

To localize the scatterer, the image has to be created for a set of azimuths. If α equal the angle of incidence ϕ the time delays are compensated, which results in the alignment of the snapshots shown schematically in Figure 7.4(a). Thus, the aligned signals will be added coherently which results in the enhanced contrast of the resulting image.

7.2.3.2 Synthetic Aperture in Multistatic Configuration

Assume now that an array of M sensing N emitting elements is used for imaging, as presented in Figure 7.4(b). In multistatic SA imaging transmitters are subsequently fired and the responses are captured by the sensing elements. Damage imaging can be performed similarly to the previous case, however, the time shifts for the emitting elements should be also compensated. A separate image is produced for each emission – reception cycle and the final result is formed as a sum of those images. Then, the intensity of the pixel $I(|\vec{r}|, \alpha)$ can be evaluated as

$$I(|\vec{r}|, \alpha) = \sum_{n=0}^{N-1} \sum_{m=0}^{M-1} w_T^n w_R^m y_{m,n} \left(\tau_m^R(\alpha) + \tau_n^T(\alpha) + \tau_g(|\vec{r}|) \right), \qquad (7.15)$$

where the terms $\tau_n^T(\alpha)$ and $\tau_m^R(\alpha)$ denote the time delays required to steer the transmitting and receiving subarray, respectively, with the angle of α, and $y_{m,n}$ is the temporal signal captured by the mth sensor due to the excitation by the nth transmitter.

If the hardware is capable of simultaneous excitation of the time-shifted signals from the emitting elements, the transmitting array can be used to steer the wave at a predefined set of azimuths in a sequence of transmissions. A response to each transmitted wavefront is captured by the receiving elements and since the wave steering is already done in the transmission the expression for imaging simplifies to

$$I(|\vec{r}|, \alpha) = \sum_{m=0}^{M-1} y_{m,\alpha_T} \left(\tau_m(\alpha) + \tau_g(|\vec{r}|) \right), \qquad (7.16)$$

where y_{m,α_T} is the response of the mth sensor due to steered transmission of the wavefront at the azimuth α_T.

If imaging is performed with an angular resolution of the transmission azimuths $\Delta\alpha$, the angle α in Equation (7.16) can be replaced by $\alpha_T = m\Delta\alpha + \alpha_0$. In the general case, however, imaging can be performed with different resolution steps in transmission and reception. In such a case the data obtained for the transmission angle α_T is used to generate the image at a larger set of angles α, e.g., defined by the relation $|\alpha_T - \alpha| \le \Delta\alpha/2$.

7.2.4 Self-Focusing Arrays

The theoretical background, outlined in this section, starts with principles of the self-focusing DORT method. Next, extension of the technique with the use of CWT will be presented. Finally, an imaging scheme using self-focusing of the transmitting array will be proposed.

7.2.4.1 DORT

In the DORT method an array of N transmit – receive transducers is considered a linear time invariant system with N inputs and N outputs. Signal $r_l(t)$, received by the lth element, $1 \le l \le N$, can be expressed as

$$r_l(t) = \sum_{m=1}^{N} k_{lm}(t) \otimes e_m(t), \qquad (7.17)$$

where $e_m(t)$ is the transmitted signal by element m, and k_{lm} is the impulse response function from the element m to l. In the Fourier domain Equation (7.17) can be presented using matrix notation

$$R(\omega) = K(\omega)E(\omega) \qquad (7.18)$$

where $R(\omega)$ and $E(\omega)$ are vectors of the Fourier transformed received and transmitted signals, respectively, and $K(\omega)$ is the $N \times N$ transfer matrix of the system. Based on Equation (7.18) an expression for the transmitted signal after two sequential time-reversal operations can be derived. If the initial input vector signal is E^0 then the output signal is $R^0 = KE^0$. The new input signal is a time-reversed output signal, which in the frequency domain is equivalent to the phase conjugate

$$E^1 = K^*E^0. \qquad (7.19)$$

The received signal due to the emitted signal given by Equation (7.19) is then

$$R^1 = K^*E^1 = \left(K * KE^0\right)^*. \qquad (7.20)$$

The matrix K^*K is called the time reversal operator (TRO). In accordance with the reciprocity theorem $k_{l,m}(t)$ and $k_{m,l}(t)$ should be equal. This implies that the matrix K is symmetrical and that K^*K can be diagonalized (Prada and Fink 1998). As the name of the DORT method suggests, it is based on the decomposition of the TRO. The detailed procedure of the DORT is presented schematically in Figure 7.5. First, the interelement impulse responses are collected

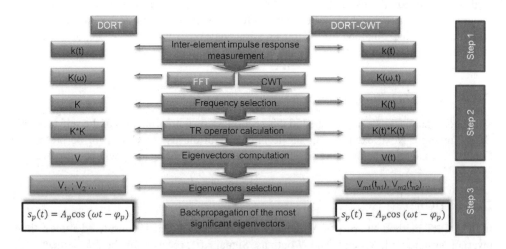

Figure 7.5   The DORT (left) and the DORT-CWT (right) algorithm flow diagram

to create the K matrix. In the next step, Fourier transform is applied to the temporal signals and the transfer matrix $K(\omega)$ is obtained. In the second step the diagonalization of the TRO is performed at a chosen frequency. It has been proven that the number of significant eigenvalues of the TRO is equal to the number of the point-like scatterers existing in the medium. Lamb waves, however, have a multimodal nature and for each mode reflected from a scatterer the corresponding eigenvector exists (Prada and Fink 1998). The final step of the DORT method is the backpropagation of the most significant eigenvectors in the structure. The eigenvector V_m calculated with the DORT method contains information about a phase and amplitude of a signal that should be applied to an array in order to focus a beam on the mth scatterer. In a monochromatic case this signal can be calculated as follows

$$V_m = \left[A_1 e^{i\phi_1}, A_2 e^{i\phi_2}, \cdots, A_n e^{i\phi_n} \right]^T \tag{7.21}$$

then the signal

$$s_p(t) = A_p \cos(\omega t - \phi_p) \tag{7.22}$$

is applied to the transducer p to obtain the beam focused on the damage (Mordant *et al.* 1999).

7.2.4.2   DORT-CWT Algorithm

The DORT method enables selective focusing of ultrasonic waves on scatterers in different media. The number of scatterers that can be resolved with the algorithm depends on the size of the TRO matrix K^*K, which is determined by the number of transducers. In the works mentioned above (Fink 1992; Fink *et al.* 2000; Ing and Fink 1995, 1996, 1998; Kerbrat *et al.* 2002; Prada and Fink 1998; Prada *et al.* 1996) arrays with a large number of transducers were used. Therefore, an attempt to implement the DORT method to the setup with a small array consisting of only 8 transducers was made (Ambrozinski *et al.* 2012a). This array, used for damage detecting in an aluminium plate, was theoretically capable of resolving up to 8 damages in the plate.

Since the snapshots produced by the propagating Lamb waves are nonstationary due to the dispersion phenomenon, in the proposed algorithm CWT is used to obtain the TFR of the signals. A flow diagram of the proposed DORT-CWT method can be compared with the classical DORT scheme in Figure 7.5. In the algorithm implementation presented here the complex Morlet wavelet was used as the mother wavelet. It enabled signals decomposed in the complex domain with a very good time and frequency localization to be obtained. Next, the TRO, $K(t)^*K(t)$, was calculated at a chosen frequency for the successive time samples. In the final step of the DORT-CWT, likewise in the classical DORT, the most significant eigenvalues were selected and backpropagation was performed. In the proposed method the distribution of eigenvalues versus time samples, $V_m(t_n)$, is obtained. The backpropagation step can be performed, as in the classical approach, with the use of Equation (7.22). Although each of the eigenvectors calculated for the successive time instants could be backpropagated, only the vectors corresponding to the significant eigenvalues should be taken into account. Even if a threshold value to separate the relevant eigenvalues is applied, still a large number of eigenvectors that should be backpropagated remains. Thorough analysis of the eigenvalues' distributions, performed on our simulated and experimental data revealed that the number of local maxima of the significant eigenvalues is equal to the number of damages existing in the plate. It will be shown below that backpropagation of the vectors corresponding to those maxima enables wave focusing on the damages.

7.2.4.3 Imaging with Self-focused Transmitting Array

In the approaches presented in Section 7.2 the imaging relies on steering the array at a set of angles and on analysing the time-shifted responses. Therefore, in the preliminary sweeping step the array has to be steered both in the azimuths where damages are expected as well as in the azimuths where they are not present. However, using self-focusing techniques allows to illuminate a large ROI and based on the obtained response, detect azimuths where strong scatterers are present.

Here, we propose the imaging algorithm that consists of two steps: self-focusing of the transmitting array using the DORT-CWT method and DAS of the acquired data to produce damage images.

The self-focusing algorithm yields a set of significant eigenvalues that can be backpropagated to obtain selective focusing of a beam at a scatterer's location. Note that regardless of the near- or far-field location of the target no parallel-ray approximation is used in the self-focusing step. In the second step, the physical backpropagation of the eigenvalues is performed. The backpropagation consists of simultaneous emission of time-delayed signals and the responses, captured by the receiving subarray, are processed to obtain a damage image. The imaging is performed with the use of Equation (7.14), however, for each backpropagation a separate image is obtained.

7.3 Numerical Results

In this section numerical evaluation of the presented methods is given. First, examples of effective apertures of cross- and star-shaped arrays are presented. Next, imaging using a star-shaped array and mono and multistatic setups is presented and compared. Finally, numerical evaluation of the self-focusing technique is given.

7.3.1 Examples of Effective Aperture

To illustrate the concept of effective aperture, in Figure 7.6(a) we present an effective aperture of a cross-shaped array, consisting of a linear transmitting subarray intersected at an angle of $90°$ with the receiving subarray. The transmit and receive apertures took the form of linear arrays that consisted of 8 point elements spaced at a distance of $d = 5\,\text{mm}$, whereas the wavelength of the assumed propagating mode was $\lambda = 12\,\text{mm}$.

The examples of the three transmit – receive configurations of star-shaped matrix topology considered in this chapter are presented in Figure 7.7(a), (b) and (c). In all cases the same number of transmitters and receivers was used and different results were obtained depending on the transmitters' location. Note that neither transmitter nor receiver apertures were apodized, i.e., w_T and w_R were unity vectors.

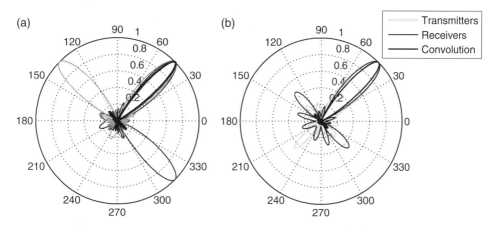

Figure 7.6 Example of beam pattern function calculated for: (a) cross matrix, in the horizontal line there were transmitters, in the vertical line receivers and the wavefront was directed at 45°; and (b) star matrix, transmitters were placed in horizontal and vertical lines, the receivers were set in both diagonals

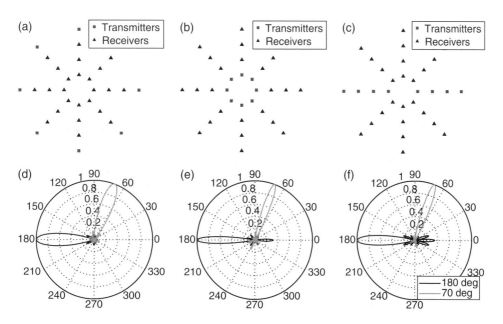

Figure 7.7 Example of transmitters and receivers layout in a star matrix (a–c); and beam pattern function calculated for wave front angles 70° and 180° (d–f)

The effective beam patterns calculated using Equation (7.6) and Equation (7.11) (the latter in the frequency domain) for the configurations shown in Figure 7.7(a–c), are presented in Figure 7.7(d–f). The BPs were calculated for the wavelength $\lambda = 2d$. Main lobe width, side lobe level and back lobe level were chosen as parameters for rating these configurations.

Table 7.1 Beampattern parameters calculated for three layouts

	Small circle	Large circle	Linear
Angle of wavefront steering (°)	Main lobe width (°)		
0	12.97	9.37	12.97
70	12.61	9.01	9.37
90	12.61	9.01	9.37
The worst parameter in 0–360° scan	12.97	9.37	13.33
	Side lobe level (dB)		
0	−16.87	−17.05	−13.47
70	−16.79	−16.99	−19.24
90	−16.87	−17.05	−18.9
The worst parameter in 0–360° scan	−16.56	−10.66	−10.09
	Back lobe level (dB)		
0	−13.8	−9.92	−9.23
70	−21.03	−15.37	−17.71
90	−13.87	−9.92	−9.24
The worst parameter in 0–360° scan	−13.8	−9.92	−9.23

The results are summarized in Table 7.1. Both circle configurations [Figure 7.7(a) and (b)] were symmetrical in all four quadrants and the respective parameters listed in Table 7.1 were relatively constant for all azimuth angles. Due to technical reasons, however, the configuration with the linear transmitter layout [Figure 7.7(c)] was chosen for further experiments. The simulation results in Table 7.1 show that this configuration generates the lowest level of side lobes in front of the transmitters line, while the main lobe width is only slightly broader than that in the large circle case [Figure 7.7(a)]. Note, that the widest main lobe is observed when the wave front was steered at one of the transmitters' line endfire angles. However, the effect of the broad transmitters' endfire lobes is compensated by the respective narrow lobes of the inclined receiver lines, which improves the overall azimuth characteristics of this configuration.

7.3.2 Imaging Using Star-like Array

In Section 7.2.1 simulated array directivity characteristics for different array topologies were presented. The simulations were performed for monochromatic continuous wave excitation. In real SHM applications, however, tone-burst or broad-band pulses are used to enable damage localization by the time of flight extraction of the damage-related reflections. Since the shape of the excitation signal plays a vital role in the mechanism of Lamb wave propagation and their beamforming, in this section simulations of Lamb wave propagation for a broadband excitation signal will be presented. The responses processed with beamforming algorithms will be used to present and discuss the performance of the selected 2D arrays.

Lamb waves are dispersive and multimodal. The relation between the phase velocity and the product of the plate thickness and excitation frequency can be evaluated by numerical solution of the Rayleigh–Lamb equation. When the dispersion curves are available, it is possible to evaluate the response of the structure $g_r(t)$ due to the excitation $g_e(t)$ using the following relation

$$g_r(t) = F^{-1}(g_e(\omega) \cdot G(k, x, \omega)), \tag{7.23}$$

where $G(k, x, \omega)$ is the frequency-dependent transfer function, k is the wavenumber of the propagating mode, x is the propagation distance, and F^{-1} stands for the inverse Fourier transform (Xu and Giurgiutiu 2007). In the case when only two basic A_0 and S_0 modes are present, the function $G(k, x, \omega)$ can be expressed as follows

$$G(k, x, \omega) = S(\omega)e^{-ik_S x} + A(\omega)e^{-ik_A x}, \tag{7.24}$$

where the terms $S(\omega)$, $A(\omega)$ correspond to the amplitude of the received S_0 and A_0 modes, respectively. It has been shown that these frequency-dependent amplitudes are related to the transducer type, its size and the plate thickness (Xu and Giurgiutiu 2007). Single-mode propagation can be assumed in the simulations if one of the terms $S(\omega)$ or $A(\omega)$ is set to 0.

The method for evaluation of dispersive signals was used to simulate wave propagation in a 2 mm thick aluminium plate. In the presented approach a set of point-like transducers arranged in a star-shaped array, acting as transmitters or receivers, was assumed. Moreover, a point-like scatterer localized in the far field of the array was assumed. The simulations of wave propagation were conducted with the use of Equation (7.23), in which the length of the propagation path was equal to the transmitter – scatterer – receiver distance. In order to pay more attention to the dispersion phenomenon than to the multimode nature of the Lamb waves, the coefficient $S(\omega)$ in the transfer function was set to 0.

7.3.2.1 Imaging Using a Single Transmitter and a Multiple Receiver Array

The first simulation series concerned the monostatic setup consisting of a single transmitter and multiple receivers (STMR) array. The receiving elements were distributed according to the topology of the star-shaped array, presented in Figure 7.3(b), whereas the transmitting element was placed in the centre of the array. A point-like scatterer was localized at a distance of 250 mm from the array centre at an azimuth of 110°.

Two simulations were conducted for this setup: first, a tone-burst signal consisting of 3 cycles of sine at a frequency of 100 kHz, modulated with Hanning window was used as an excitation. In the second simulation the number of cycles was increased to 30. The spacing of the array's elements was equal to half of wavelength of the A_0 mode at the excitation frequency. The responses captured by the consecutive sensors were evaluated using Equation (7.23). Both sets of the simulated snapshots were processed with the DAS beamformer assuming far-field approximation; the resulting images are shown in Figure 7.8(a) and (b). It can be seen that the longer burst signal results in poor range resolution. The beam patterns obtained from the maximum amplitudes of the sum of the beamformed signals as a

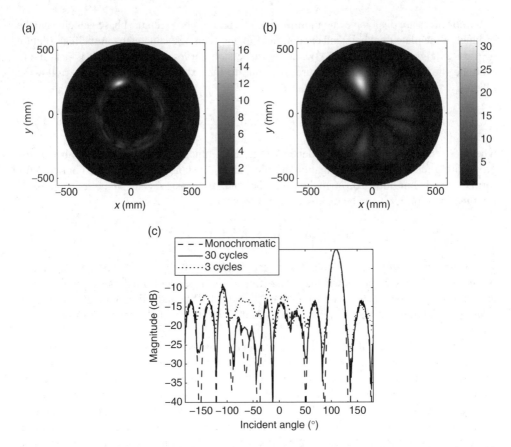

Figure 7.8 Target images obtained in the star-like STMR setup for the single A_0 mode excited with windowed tone burst consisting of 3 (a) and 40 sine cycles (b). Beam patterns obtained for the tone-burst excitation signals and for the monochromatic excitation (c)

function of the azimuth are presented in Figure 7.8(c) where they can be compared with the theoretical directivity characteristics of the same array obtained from Equation (7.6), assuming infinite, monochromatic exciting signal. Good agreement of the compared beam patterns can be observed, however, the response obtained for the 30-cycle excitation is closer to the monochromatic theoretical beam pattern, since it has deeper minima, due to the destructive interference.

7.3.2.2 Multistatic Imaging

In the next simulation the synthetic aperture concept was illustrated in the multistatic setup consisting of multiple transmitters and multiple receivers (MTMR) implemented in the form of a star-shaped array of transmitters and receivers, as presented in Figure 7.7(c). A point-like scatterer was localized at a distance of 250 mm and an angle of 110° from the array. The

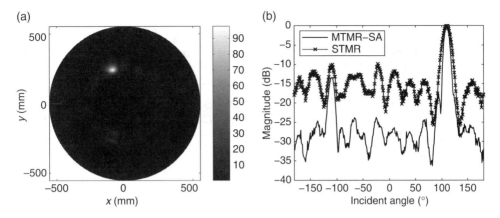

Figure 7.9 Target image obtained using the MTMR-SA concept in the star-like setup (a). Comparison of the resulting beam pattern with that obtained with STMR [Figure 7.8(a)] (b)

simulated responses from each emitter – receiver pair formed a matrix consisting of 8×24 time traces. The snapshots were then processed with the DAS algorithm and the resulting image obtained using Equation (7.15) is presented in Figure 7.9(a). If this image is compared with that in Figure 7.8(a) it can be easily seen that, although the same topology of the array and the same number of elements was used, a considerably improved result can be obtained this way. Both images were processed to find maxima occurring at the successive azimuths and their beam patterns are presented in Figure 7.9(b). A comparison of the results shows that the multistatic approach is superior to the monostatic one in terms of main lobe width and range resolution.

7.3.2.3 Imaging using Active Transmitting Array

In the next series of MTMR simulations the star-shaped array, presented in Figure 7.7(c), was also used in the sweeping PA mode. In this mode the elements of the horizontal transmitting array were fired using time-shifted signals to form a plane wave sweeping the ROI. The backscattered signals were received by the star arms and post-processed to obtain a high resolution image. In other words, focusing in the transmission was done in the material and focusing in the reception was performed offline by the DAS operation of the captured snapshots.

In the presented simulation the azimuth of a far-field point-like scatterer was, as before, $110°$. Since the transmitting array was linear and it produced a mirrored lobe, the transmission angle was limited to the range $0–180°$, but various angular sweeping steps were considered. In the first step the beam was transmitted at a set of azimuths with a step of $10°$, which led to the 19×24 matrix of time traces (19 azimuths and 24 receiving elements). The same resolution was used to process the captured snapshots; the result can be seen in Figure 7.9(a).

Next, the same data were processed with $1°$ angular resolution in the reception which yielded the result shown in Figure 7.10(b). Finally, the simulation of sweeping with $1°$ resolution both in the transmission and reception was performed resulting in a 181×24

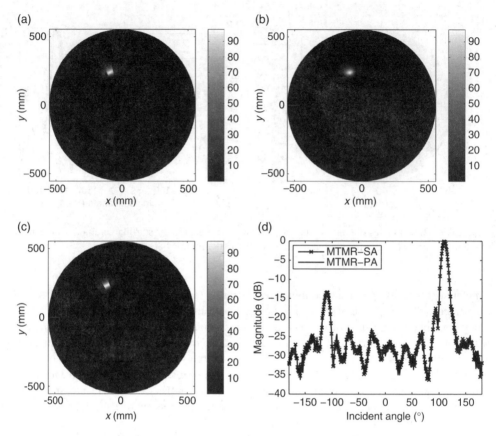

Figure 7.10 Target images obtained in the simulations of the MTMR-PA mode for the following emission/reception sweeping steps: 10°/10° (a); 10°/1° (b); and 1°/1° (c). The beam patterns obtained using MTMR-SA and MTMR-PA with azimuth sweeping step 1°/1° (d)

matrix of time traces (181 azimuths and 24 receiving elements); the results are presented in Figure 7.10(c). Comparison of the images shown in Figure 7.10(a), (b) and (c) shows that rather poor resolution obtained for the 10° step cannot be improved if the captured data are processed with a higher resolution in the reception and that a smaller step in the transmission is required for that.

The beam patterns obtained for the images from Figure 7.10(c) and Figure 7.8(a) are compared in Figure 7.10(d). From the plots it can be seen that both approaches, the MTMR-SA and the MTMR-PA, yield the same result.

7.3.3 Numerical Verification of the DORT-CWT Method

To examine the performance of the DORT and DORT-CWT methods, several numerical experiments were carried out and reported in Ambrozinski *et al.* (2012a). However, only those

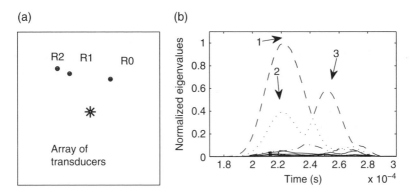

Figure 7.11 Damage location analysed in simulations and experiments (a). Normalized eigenvalues distribution obtained for the simulated data processed with the DORT-CWT algorithm. Arrows point to the peak eigenvalues used in the backpropagation (b)

related to the DORT-CWT method are presented below. In this section an introduction to the approach, used to simulate wave propagation, is followed by the description of the method used to calculate the backpropagated fields. In the final part of this section selected simulation results are presented.

7.3.3.1   Simulation Algorithm

The simulation of wave propagation phenomena was performed using the local interaction simulation approach with sharp interface model (LISA/SIM) described in Chapter 2. To perform the simulations a model of an aluminium plate with size $700 \times 700 \times 2$ mm was built. The plate was modelled using a high resolution grid with cell size $c = 0.5$ mm. Three artificial plate damages were introduced to the model: a notch of size 10×1 mm, denoted in Figure 7.11(a) as R2; a circular hole with diameter 12 mm, denoted as R0; and a square hole of size 5×5 mm, denoted as R1. The resultant model consisted of approximately 10 million cells. In order to capture backscattered reflections from the damages the analysis time was set to 0.5 ms. In all of the simulations a linear array placed in the centre of the plate was modelled. The number of array elements was 8 and the transducers spacing was equal to 5 mm. A broadband excitation signal was used −2 cycles of a sine with the centre frequency 100 kHz modulated with the Hanning window. It was observed that an enhanced asymmetric mode A_0, and a reduced, almost negligible symmetric mode S_0 were excited in the model. The wavelength of the dominant mode calculated from dispersion curves at the centre frequency of excitation was 12.9 mm. In each simulation one element of the array acted as the emitter, whereas the other elements acted as the receiver, therefore the number of analyses required to simulate all interelements responses, required to build the TRO matrix, was equal to the number of transducers used in the array. Both the dispersion effect and the geometrical spreading were taken into account.

7.3.3.2 Numerical Backpropagation

The last step of the DORT and DORT-CWT algorithms is backpropagation of the most significant eigenvectors, which can be performed in the real structure or numerically. In this chapter calculations of backpropagated fields were based on Equation (7.22), which defines the relation between the eigenvectors and signals that should be applied to the actuators to focus the wave on a target. If the backpropagated field is presented as a bitmap, the pixel value $S_{i,j}$ can be calculated using Equation (7.22) in the space domain as a sum of the signals emitted from all transducers as follows

$$S_{i,j} = \sum_p A_p \cos(k x_p^{i,j} - \phi_p),\tag{7.25}$$

where k denotes the wavenumber, and $x_p^{i,j}$ is the distance between pixel $S_{i,j}$ and transducer p. The value of the wavenumber k has to be calculated from the dispersion curves for the frequency at which the K^*K matrix was diagonalized. The backpropagated fields presented in the following sections were created using the pixel resolution 1×1 mm.

Only a coarse representation of the Lamb wave backpropagation can be obtained using this method since it assumes that a single frequency is excited. Practically, however, narrow-band signals can be used. Since the phase velocity of Lamb waves is a function of frequency, various frequency components travel with different velocities, which results in spreading the signal energy in space and time (Wilcox 2003b). However, as it will be shown below, the proposed representation can serve as a sufficient approximation to investigate the performance of the analysed methods.

7.3.3.3 Simulation Results

In the simulation scenario a model of the plate and defects was built according to the setup from Figure 7.11(a). The simulated interelement responses were processed with the DORT-CWT method and from the time distribution of the eigenvalues shown in Figure 7.11(b), it can be seen that the number of the significant peaks is equal to the number of damages. Additionally, the wave propagation distances were calculated for the time corresponding to the peaks of eigenvalues. The wave velocity was assumed as the group velocity of the A_0 mode for the diagonalization frequency. The calculated distances were 522, 525 and 595 mm, whereas the doubled distances from the centre of the array to the reflectors were 509, 510 and 602 mm, respectively; it is apparent that the calculated values are close to the expected ones.

The eigenvectors corresponding to the peak values, indicated in Figure 7.11(b) by arrows, were backpropagated numerically. The resultant backpropagated fields presented in Figure 7.12(a), (b) and (c), show that the beams were steered in the damages' directions, which is also confirmed by the polar plots shown in Figure 7.12(d), (e) and (f). This means that the damages were correctly resolved by the DORT-CWT method.

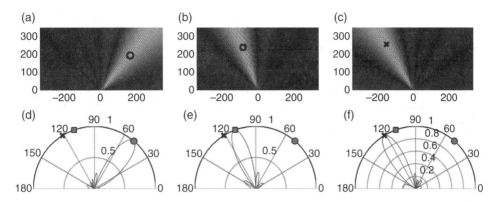

Figure 7.12 Numerical backpropgation of the eigenvectors obtained with the DORT-CWT method for the simulation. Wave focusing on the damages: (a) R0; (b) R1; (c) R2. Beampatterns (d), (e) and (f) correspond to the backpropagated fields (a), (b) and (c), respectively. Markers indicate damage locations. Dimensions of the backpropagated fields are in millimetre and beam azimuths are in degrees

7.4 Experimental Results

The experiment description in this section consists of two parts. First examples of imaging with the use of MTMR are presented. In these experiments a linear array of PZTs was used for multistatic and PA transmission, whereas the the measurement of the responses was conducted in a contactless manner by means of a laser scanning Doppler vibrometer (LSDV). In the second part of this section, the self-focusing DORT imagining technique is demonstrated using a star-shaped array of PZTs.

7.4.1 Experimental Setup

The experiments were conducted on an aluminium plate of size $1000 \times 1000 \times 2$ mm, presented in Figure 7.13(a). A uniform linear array of transmitters was mounted at the centre of the plate. The array consisted of 8 $2 \times 2 \times 2$ mm PZT elements, type CMAP 12 (Noliac, Denmark) spaced at a distance of 5 mm. The responses of the structure were captured on the other side of the plate using a laser scanning Doppler vibrometer. Tone-burst signals consisting of 3 cycles of sine modulated by Hanning window were used as an excitation. A PAS-8000 (EC Electronics, Poland), used as a signal generator, enabled both single-element excitation and simultaneous generation of the time-shifted signals in PA mode.

The laser scanning points were formed in a star-shaped array, presented in Figure 7.13, however, since the signals from the horizontal line were not used for imaging, the topology of the transmitting/receiving array corresponded to the one presented in Figure 7.7(c).

A great advantage of the contact-less measurement technique is that it facilitates modification of the measurement points grid which enables investigation of various topologies of the

(a)   (b)   (c)

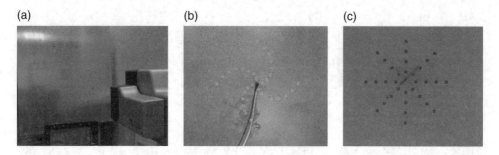

Figure 7.13 Experimental setup to investigate the MSI technique (a). Laser vibrometer measurement points in: spiral-shaped configuration used in STMR setup (b); and star-shaped configuration used in MTMR (c)

sensing subarray. This setup has, however, a serious limitation – effects of the inter-element scattering that may be encountered for a full 2D array of PZT elements cannot be taken into account.

7.4.2 Experimental Evaluation of Sensing Array

The method based on the LSDV measurements can be used to evaluate any shape of a 2D sensing array, however, only the topologies presented in Figure 7.14 will be discussed here. All arrays analysed in this section consisted of 32 sensors. The star-shaped array presented in Figure 7.14(a) consists of 4 linear subarrays intersecting at an angle of 45°. The elements of the subarrays are spaced at a distance $d = 5$ mm, therefore the outside diameter of the array was 40 mm. The circular-shaped topology shown in Figure 7.14(b) was made up of two concentric circles. The diameter of the bigger one was 25.62 mm and the distance denoted d in the figure was equal to 5 mm. Finally, the spiral-shaped array, presented in Figure 7.14(c), was created in the way described by Yoo *et al.* (2010). The transducers were placed on concentric circles with radii increasing with a step of 5 mm, 4 sensors were placed in each circle, which was rotated with an angle of 15° with respect to the former one. The spacing of the subsequent sensors varied from 5.11 to 8.87 mm and the outside diameter of the array was 80 mm.

The contact-less measurement technique made possible not only easy modification of the sensing arrays, but also rotation of measurement points. In this way, performance of the sensing array for a wave impinging from a set of angles could be evaluated. All of the investigated topologies have at least two axes of symmetry, therefore evaluation of the beam patterns was limited to the range of 0–90° and mirrored for the remaining directions.

The signals acquired in the setup presented in Figure 7.13(b) were processed with the DAS algorithm and an example of beam patterns obtained using this procedure can be seen in Figure 7.15(a), (b) and (c) for the wave with an arbitrary selected incident angle of 60°, for spiral, circular and star-shaped array, respectively.

Good agreement of the directivity characteristics, obtained in simulations, with those based on the structure's transfer function, and experiments could be observed for all of the topologies. However, higher noise level and no signal cancellation was observed in the measurement

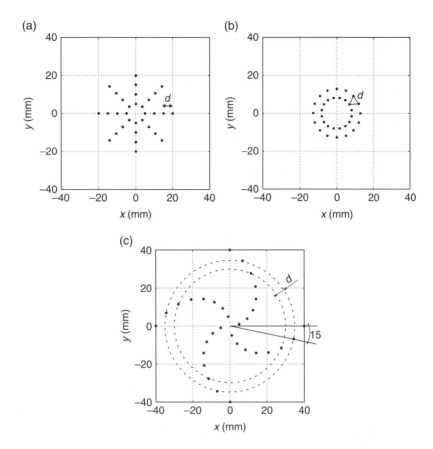

Figure 7.14 Investigated topologies of the array; (a) star-shaped; (b) circular; and (c) spiral

results compared with the simulated ones. The beam patterns obtained for the waves with incident angle in the range of 0–90° were analysed and it appeared that the main lobe width, estimated as its angular width at the level of −3 dB, was the narrowest for the spiral-shaped array, which was expected since this array has the largest aperture. The widest main lobe revealed the circular array, which has the smallest aperture. In the terms of the maximal side-lobe level, the lowest value was obtained for the circular array and the highest for the spiral-shaped array. Therefore, among the investigated topologies, the star-shaped array was found to be a compromise between beam width and side-lobe level. The details of the selected parameters estimated from the characteristics can be found in our previous study (Ambrozinski *et al.* 2012b).

7.4.3 *Experimental Evaluation of Effective Aperture*

The first experiment performed with this setup consisted of evaluating the effective aperture of the array for an exemplary azimuth. A scatterer was placed in the array's far field at an angle

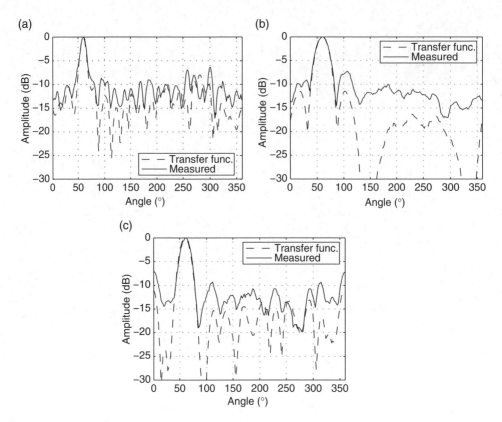

Figure 7.15 Beam patterns evaluated for the wave with incident angle 60° for: (a) spiral; (b) circular; and (c) star-shaped array

of 112°. In the first step, the MTMR-SA concept was used in which the subsequent elements were fired and the respective responses were captured. The snapshots were used for offline imaging that yielded the result shown in Figure 7.16(a).

In the next experiment (MTMR-PA) the elements of the transmitting array were excited simultaneously with time-shifted signals. The sweeping was performed for a set of angles in the range 0–180° with a step of 10°. The acquired data were used for imaging with a resolution of 1° and the results are presented in Figure 7.16(b).

Comparing the results it can be seen that MTMR-PA offers much worse resolution and lower image quality than MTMR-SA, which is not surprising since the resolution in the active transmission was low. Based on the image obtained with the use of SA, the beam pattern presented in Figure 7.16(c) was obtained; the plot of the result of simulation performed using the method described in Section 7.3.2 is also shown. Comparison of both beam patterns shows that the simulated and experimental results are in good agreement in terms of main lobe width, however, higher side-lobe levels and noise level can be observed for the experimental result.

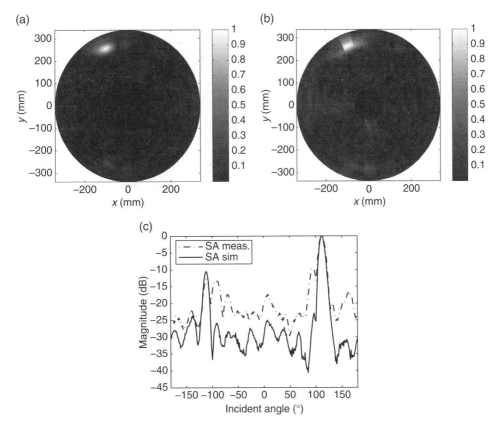

Figure 7.16 Experimental results of damage imaging with the use of MTMR (a) and MTMR-PA excitation (b). Comparison of the beam patterns obtained for the experimental and simulated data (c)

7.4.4 Damage Imaging Using Synthetic Aperture

In the next experiment, three scatterers were placed on the plate according to the setup presented in Figure 7.17(a). The result of imaging performed with the MTMR-SA approach is shown in Figure 7.17(b). From the image it can be observed that equivocal damage localization can be obtained with the use of SA and for each reflector a corresponding area of the image with high contrast exists.

7.4.5 Experimental Validation of the DORT-CWT Method

Since the performance of the methods was confirmed in simulations, two series of experiments were carried out with the use of an aluminium plate, and a linear array made up of 8 transducers. In the first step, which was the same for all conducted experiments, the interelement responses for the array elements were recorded. However, the backpropagation performed in the second step remained numerical in the first experiment series.

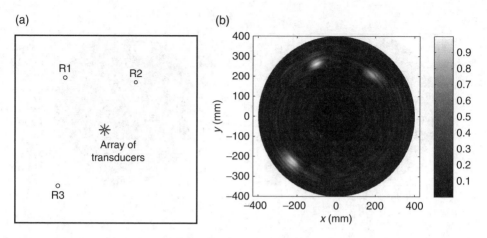

Figure 7.17   Multiple scattering experiment. Distribution of scatterers on the plate (a) and results of imaging using MTMR-SA (b)

The calculation of the backpropagated fields was performed, likewise in the simulations above, using Equation (7.25). As the numerical backpropagation can only yield coarse information about the wave propagation, the second experiment series was conducted with the array operating in transmission mode and generating signals defined by the investigated DORT algorithms. The wave propagation in the plate was investigated with the use of a laser scanning vibrometer.

7.4.5.1   Experimental Setup

The experiments were conducted using an aluminium plate with dimensions of $1000 \times 1000 \times 2\,mm$ and a $10 \times 1\,mm$ notch, denoted as R3 in Figure 7.11(a). Additionally, two small masses (R1, R0) were acoustically coupled to the investigated structure using wax. The masses attached to the plate simulated backscatters (damages) that could be shifted to various positions. A linear array consisting of 8 CMAP12 PZTs square multilayered transducers (Noliac, Denmark) was attached with wax in the middle of the plate. The transducers' size was $2 \times 2 \times 2\,mm$ and their spacing was $5\,mm$. The number of elements in the array was limited to 8 by the hardware. A basic version of PAQ 16000 with PAS 8000 (EC Electronics, Poland) was used for signal generation and data acquisition. Bursts of 2 cycles of $100\,kHz$ sinusoid modulated with Hanning window were used to obtain inter-element responses. Data acquisition was performed at sampling frequency of $2.5\,MHz$. The Noliac actuators used for the examined plate excited an enhanced asymmetric mode A_0, and a reduced, almost negligible symmetric mode S_0. The wavelength of the dominant mode at a frequency of $100\,kHz$ was $12,9\,mm$.

7.4.5.2   Experimental Results-Numerical Backpropagation

The signals obtained with the experimental setup described above were processed with the DORT-CWT method. The time distribution of eigenvalues is presented in Figure 7.18(a)

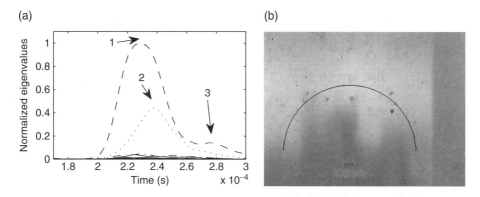

Figure 7.18 Eigenvalues distribution obtained for the data from the measurement processed with the DORT-CWT algorithm. Arrows point to the peak eigenvalues used in the numerical backpropagation (a). The arc-shaped measurement points used for vibrometer measurements. The linear array is located under the arc (b)

where the significant peaks are indicated by arrows. It can be observed that the number of peaks is equal to the number of damages in the investigated structure. Distances to the reflectors were calculated using the time corresponding to each peak value and the group velocity of the A_0 mode. The resultant distances, 488, 508 and 602 mm, are in a good agreement with the real distances from the centre of the array to the targets and back, which were 482, 509 and 603 mm, respectively. The eigenvectors corresponding to the significant peaks indicated in Figure 7.18(a) were numerically backpropagated and, as can be seen in Figure 7.19, all damages were correctly resolved. A high accuracy in the beam steering can be observed in the obtained beam patterns presented in Figure 7.19(d), (e) and (f).

7.4.5.3 Investigation of Backpropagation Using a Laser Vibrometer

The second series of experiments, carried out to verify the performance of the DORT-CWT method, consisted of monitoring the physical backpropagation in the plate. The setup presented in Figure 7.11 was considered and the distribution of eigenvalues presented in Figure 7.18 was used for the DORT-CWT.

In the experiment the array used previously to measure the inter-element responses, was employed, in a transmission mode, to generate the phase-shifted signals. The phase shifts were computed for each transducer according to Equation (7.22). Narrow-band pulse signals were used in order to perform the backpropagation. In these experiments bursts of 4 cycles of a sinusoid modulated with the Hanning window were generated. The centre frequency, 100 kHz, was equal to that used in the K^*K matrix diagonalization. A scanning laser vibrometer Polytec PSV-400 was used for monitoring the field generated by the array. The vibrometer enabled contactless measurements of the out-of-plane surface velocity in the plate. Time signals were acquired in points forming an arc-shaped grid, presented in Figure 7.19(b). The maximum values of the first arriving wave packets captured by the scanning vibrometer were used to create

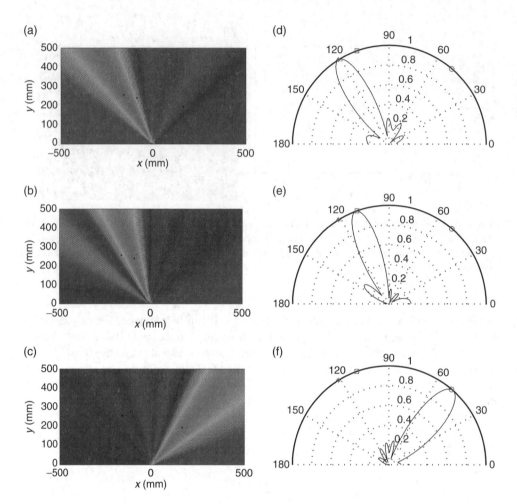

Figure 7.19 Numerical backpropgation of the eigenvectors obtained with the DORT-CWT method for the experiment. Wave focusing on the damage: (a) R2; (b) R1; and (c) R0. Beam patterns (d), (e) and (f) correspond to the backpropagated fields (a), (b) and (c), respectively. Markers indicate damage locations. Dimensions of the backpropagated fields are in millimetres and beam azimuths are in degrees

the beam patterns shown in Figure 7.20. High accuracy in the angle steering can be observed, which proves that the DORT-CWT algorithm correctly resolved the damages in the plate.

7.4.6   Damage Imaging Using Self-Focused Transmitting Array

Based on the configuration described in the previous sections and presented in Figure 7.7(c), a prototype of a star-shaped PA matrix was built. The array consists of 4 linear subarrays. The transmitting subarray can operate in transmission mode for excitation of time-shifted signals whereas the remaining subarrays operate in the reception mode to capture the response of

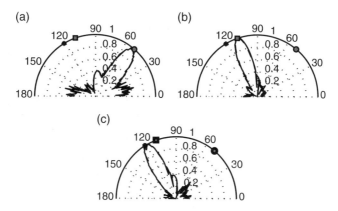

Figure 7.20 Beam patterns obtained using a scanning laser vibrometer for monitoring physical backpropagation. The beams steered in the direction of damage: (a) R0; (b) R1; and (c) R2

the structure. CMAP12 PZTs (Noliac, Denmark), were used to manufacture the array. Their size was $2 \times 2 \times 2$ mm and the spacing between the elements in the subarrays was 5 mm. As in the previous setup the data generation and acquisition system consisted of PAS800 with PAQ16000. All experiments were conducted using the aluminium plate, described above. The array was placed in the central point of the plate, using oil as a coupling agent. In the examined plate a notch, denoted in Figure 7.21(a) as R2, was introduced. Moreover, two additional masses, denoted in Figure 7.21 as R0 and R1, were added to simulate scatterers.

The aim of this experiment was imaging with the use of self-focused array, therefore the transmitting subarray was used to collect the inter-element impulse responses. The DORT-CWT was used to calculate the time – frequency distribution of the eigenvalues presented in Figure 7.21. The peak eigenvalues indicated in the distribution were backpropagated using the transmitting subarray with the time-shifts calculated according to Equation (7.22). The bandwidth and the excitation frequency of the generated signals were the same as in the previous experiment involving a vibrometer. The scatterers reflected signals were captured by the subarrays and the resulting damage images presented in Figure 7.21 (b), (c) and (d) were obtained in the case when the transmitting subarray was steered in the direction of the target denoted by R0, R1 and R2, respectively. Subsequently, the dispersion removal was applied to the snapshots and damage imaging was performed. From the results presented in Figure 7.21(b), (c) and (d), it can be seen that the damages were localized correctly. Moreover, in all images it can be observed that the illuminated targets are the strongest ones, which proves that the beams were steered correctly.

7.5 Discussion

In this chapter different approaches for the beamforming of Lamb waves have been outlined. In the first step monostatic and multistatic setups were compared for the simulated data. The images presented in Figure 7.8(a) and Figure 7.9(a) were obtained using arrays with almost

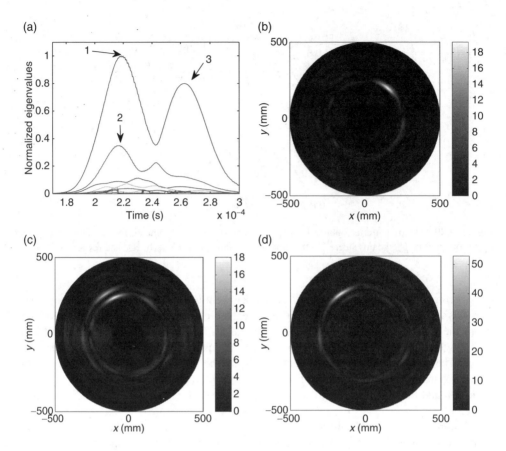

Figure 7.21 Distributtions of scatterers in the experimental setup (a). Damage imaging results obtained with the star-shaped array for the transmitting subarray steered in the direction of target denoted by R0 (b), R1 (c) and R2 (d)

identical topologies, the main difference was that in the monostatic case an additional transmitting element was used while in the multistatic scheme some of the elements were selected to operate as emitters. Note that those elements were not used for response measurement. Comparing the images obtained using both methods, it can be clearly seen that the multistatic scheme is superior to the monostatic one. It was confirmed with beam patterns presented in Figure 7.9(a), which showed that the MSI offers better results than STMR both in the terms of main beam width and side-lobe level.

The next series of simulations presented in the chapter was to discuss the possibility of MSI concept implementation with an active array operating in transmission mode (MTMR-PA) used for sweeping of the ROI. It appeared, however, that the implementation of the technique requires a large number of transmissions at a set of azimuths. The images, presented in Figure 7.10 (a) and (b), were obtained with a set of data consisting of 19×24 time traces (transmission at 19 azimuths, 24 receiving elements). Although a significant amount of data was used to create these images, they do not reproduce the quality of the image obtained with the use of MTMR-SA, presented in Figure 7.9(a). That was achieved only when the sweeping step was set to 1°, which was equal to the step used in the MTMR-SA. In order to produce the image presented in Figure 7.10(c) a set of 181 × 24 time traces was processed, and as it can be seen from Figure 7.10(d), the resulting beam pattern is identical to the one obtained using the MTMR-SA technique. Note, however, that in order to produce the image with the use of MTMR-SA 8 × 24 (8 transmitting elements, 24 sensors) snapshots were required. Moreover, offline imaging can be performed with arbitrary resolution, in both transmission and reception, limited only by the sampling frequency of the signals.

These above-mentioned simulation results were confirmed in experiments. For instance from Figures 7.16 (a) and (b) it can be seen that MSI offers better quality of the produced image than PA sweeping, although it requires less transmission – reception cycles. The beam patterns obtained with the use of experimental and simulated data processed with MSI can be compared in Figure 7.16(c). A good agreement of the main-lobe width can be seen in both characteristics, however, higher side – lobes and noise level can be observed for measurements.

The last experiment, performed for multistatic setup was conducted on a plate with 3 reflectors, as presented in Figure 7.17. Three reflectors, however, with a different layout, shown in Figure 7.21, were also considered in the self-focusing experiment. Comparing the results from both techniques, it can be seen that a single image, as presented in Figure 7.17(b), containing information on all of the reflectors is obtained with the use of MSI, whereas self-focusing technique makes selective illumination of the targets possible, which leads to a set of images as presented in Figure 7.21(b), (c) and (d).

7.6 Conclusions

In this chapter various approaches to beamforming of Lamb waves were presented; synthetic aperture was compared with monostatic and self-focusing approach.

Based on the results, it can be concluded that the multistatic approach takes advantage of multiple successive transmissions, which allows illuminating targets from a set of diverse positions and results in high quality of the produced image. Application of this technique enables narrower main lobe and a lower side-lobes level to be obtained compared with the monostatic setup using the same array topology.

It was shown in simulations that focusing can be performed partially in the inspected structure with the use of an active transmitting array operating in the PA mode. However, a large number of transmission reception cycles is required to achieve high image resolution. Moreover, the active approach requires a multiple channel system capable of simultaneous generation of time-shifted signals, while in the multistatic setup a single, multiplexed output channel is sufficient.

Therefore, the SA multistatic approach, in which the data can be acquired by means of relatively simple hardware and subsequently processed offline with an arbitrary resolution, is superior to the active scheme in SHM applications. However, if a self-focusing ability is available, the active array can be used to transmit a focused beam on scatterers. In this way separate images of the subsequent illuminated scatterers can be obtained.

References

Ambrozinski L, Stepinski T and Uhl T 2011 Self focusing of 2D arrays for SHM of plate-like structures using time reversal operator. In *Structural Health Monitoring Condition-Based Maintenance and Intelligent Structures* (ed. Chang FK). DEStech Publications, Inc., vol. 1, pp. 1119–1127.

Ambrozinski L, Stepinski T and Uhl T 2012b Experimental comparison of 2D arrays topologies for SHM of planar structures. *Proceedings of SPIE* **8347**, 834717.

Ambrozinski L, Stepinski T, Packo P and Uhl T 2012a Self-focusing Lamb waves based on the decomposition of the time-reversal operator using time – frequency representation. *Mechanical Systems and Signal Processing* **27**, 337–349.

Doctor S, Hall T and Reid L 1986 SAFT the evolution of a signal processing technology for ultrasonic testing. *NDT International* **19**(3), 163–167.

Drinkwater BW and Wilcox PD 2006 Ultrasonic arrays for non-destructive evaluation: A review. *NDT & E International* **39**(7), 525–541.

Engholm M and Stepinski T 2010a Direction of arrival estimation of Lamb waves using circular arrays. *Structural Health Monitoring* **10**(5), 467–480.

Engholm M and Stepinski T 2010b Using 2D arrays for sensing multimodal Lamb waves. *Proceedings of SPIE* **7649**, 764913.

Fink M 1992 Time reversal of ultrasonic fields. I. Basic principles. *IEEE Transactions on Ultrasonics, Ferroelectrics and Frequency Control* **39**(5), 555–566.

Fink M, Cassereau D, Derode A, Prada C, Roux P, Tanter M, Thomas JL and Wu F 2000 Time-reversed acoustics. *Reports on Progress in Physics* **63**(12), 19–33.

Giurgiutiu V 2008 *Structural Health Monitoring with Piezoelectric Wafer Active Sensors*. Academic Press.

Ing R and Fink M 1995 Surface and sub-surface flaws detection using Rayleigh wave time reversal mirrors. *IEEE Ultrasonics Symposium* **1**, 733–736.

Ing R and Fink M 1996 Time recompression of dispersive Lamb waves using a time reversal mirror-application to flaw detection in thin plates. *IEEE Ultrasonics Symposium* **1**, 659–663.

Ing R and Fink M 1998 Time-reversed Lamb waves. *IEEE Transactions on Ultrasonics, Ferroelectrics and Frequency Control* **45**(4), 1032–1043.

Karaman M, Wygant IO, Oralkan O and Khuri-Yakub BT 2009 Minimally redundant 2-D array designs for 3-D medical ultrasound imaging. *IEEE Transactions on Medical Imaging*.

Kerbrat E, Prada C, Cassereau D and Fink M 2002 Ultrasonic nondestructive testing of scattering media using the decomposition of the time-reversal operator. *IEEE Transactions on Ultrasonics, Ferroelectrics and Frequency Control* **49**(8), 1103–1113.

Liu L and Yuan FG 2010 A linear mapping technique for dispersion removal of Lamb waves. *Structural Health Monitoring* **9**, 75–86.

Lockwood GR, Li PC, O'Donnell M and Foster FS 1996 Optimizing the radiation pattern of sparse periodic linear arrays. *IEEE Transactions on Ultrasonics, Ferroelectrics and Frequency Control*.

Lockwood GR, Li PC, O'Donnell M and Foster FS 1998 Real-time 3-D ultrasound imaging using sparse synthetic aperture beamforming. *IEEE Transactions on Ultrasonics, Ferroelectrics and Frequency Control*/**45**(4), 980–988.

Michaels JE 2008 Detection, localization and characterization of damage in plates with an in situ array of spatially distributed ultrasonic sensors. *Smart Materials and Structures* **17**(3), 035035.

Mordant N, Prada C and Fink M 1999 Highly resolved detection and selective focusing in a waveguide using the DORT method. *The Journal of the Acoustical Society of America* **105**(5), 2634–2642.

Moreau L, Drinkwater B and Wilcox P 2009 Ultrasonic imaging algorithms with limited transmission cycles for rapid nondestructive evaluation. *IEEE Transactions on Ultrasonics, Ferroelectrics and Frequency Control* **56**(9), 1932–1944.

Prada C and Fink M 1998 Separation of interfering acoustic scattered signals using the invariants of the time-reversal operator. application to Lamb waves characterization. *The Journal of the Acoustical Society of America* **104**(2), 801–807.

Prada C, Manneville S, Spoliansky D and Fink M 1996 Decomposition of the time reversal operator: Detection and selective focusing on two scatterers. *The Journal of the Acoustical Society of America* **99**(4), 2067–2076.

Prada C, Wu F and Fink M 1991 The iterative time reversal mirror: A solution to self-focusing in the pulse echo mode. *The Journal of the Acoustical Society of America* **90**(2), 1119–1129.

Raghavan A and Cesnik CES 2007 Review of guided-wave structural health monitoring. *The Shock and Vibration Digest* **39**(2), 91–114.

Wilcox P 2003a Omni-directional guided wave transducer arrays for the rapid inspection of large areas of plate structures. *IEEE Transactions on Ultrasonics, Ferroelectrics and Frequency Control* **50**(6), 699–709.

Wilcox P 2003b A rapid signal processing technique to remove the effect of dispersion from guided wave signals. *IEEE Transactions on Ultrasonics, Ferroelectrics and Frequency Control* **50**(4), 419–427.

Xu B and Giurgiutiu V 2007 Single mode tuning effects on Lamb wave time reversal with piezoelectric wafer active sensors for structural health monitoring. *Journal of Nondestructive Evaluation* **26**, 123–134.

Yoo B, Purekar AS, Zhang Y and Pines DJ 2010 Piezoelectric-paint-based two-dimensional phased sensor arrays for structural health monitoring of thin panels. *Smart Materials and Structures* **19**(7), 075017.

8

Modal Filtering Techniques

Krzysztof Mendrok

Department of Mechatronics and Robotics, Faculty of Mechanical Engineering and Robotics, AGH University of Science and Technology, Poland

8.1 Introduction

The modal filter was first introduced by Meirovitch and Baruh (1982). They used it to overcome the spill-over problem in the control of distributed parameters systems. Spill-over is a phenomenon in which the energy addressed to the controlled mode is pumped into the uncontrolled modes. They decided to replace the control system of the $2n$th order system with n independent control systems to control each degree of freedom (DOF) of the systems separately. They used a representation of the modal filter for the distributed parameter system, which had the following drawbacks: the necessity to know the function of spatial distribution of mass (which is possible only for simple geometries) and the vibration response measurement performed at each point of the object. To avoid the second problem the authors proposed a method that uses an interpolated function of measurements from sensors located in a finite number of points. In such a case, however, the filtration accuracy, depends on the quality of interpolation. Therefore in the early 1990s the discrete modal filter was formulated (Zhang *et al.* 1989). The modal filter is a tool to extract the modal coordinates of each individual mode from the system outputs by mapping the response vector from the physical space to the modal space (Zhang *et al.* 1990). The construction of the rth discrete modal filter, which corresponds to the rth pole of the transfer function $H(\omega)$, starts with an assumption that the modal residue R_{rpp} is in the imaginary form:

$$R_{rpp} = j \cdot 1 \tag{8.1}$$

Advanced Structural Damage Detection: From Theory to Engineering Applications, First Edition.
Edited by Tadeusz Stepinski, Tadeusz Uhl and Wieslaw Staszewski.
© 2013 John Wiley & Sons, Ltd. Published 2013 by John Wiley & Sons, Ltd.

Next, the 1 DOF frequency response function $H_{pp}(\omega)$ is determined as follows:

$$H_{pp}(\omega) = \frac{R_{rpp}}{j\omega + \lambda_r} + \frac{R^*_{rpp}}{j\omega + \lambda^*_r} \qquad (8.2)$$

where λ_r is the rth pole of the system. For the given frequency range, the above FRF is determined by the k values:

$$H_{pp}(\omega) = [H_{pp}(\omega_1)H_{pp}(\omega_2) \cdots H_{pp}(\omega_k)]^T \qquad (8.3)$$

Assuming that the single excitation was used and response signals were measured in N points, the experimental FRFs matrix can be presented as the k x N matrix:

$$H_{kN}(\omega) = \begin{bmatrix} H_1(\omega_1) & H_2(\omega_1) & \cdots & H_N(\omega_1) \\ H_1(\omega_1) & H_2(\omega_1) & \cdots & H_N(\omega_1) \\ H_1(\omega_2) & H_2(\omega_2) & \cdots & H_N(\omega_2) \\ \vdots & \vdots & \ddots & \vdots \\ H_1(\omega_k) & H_2(\omega_k) & \cdots & H_N(\omega_k) \end{bmatrix} \qquad (8.4)$$

The FRF matrix formed in this way is used to determine the reciprocal modal vectors matrix Ψ_p:

$$\Psi_p = H^+_{kN} \cdot H_{pp} \qquad (8.5)$$

where $^+$ denotes a pseudo-inverse of the matrix. Reciprocal modal vectors should be orthogonal with respect to all the modal vectors except the one to which the filter is tuned, and thanks to that, they are applied to decomposition of the system responses to the modal coordinates η_r.

$$\eta_r(\omega) = \{\psi_r\}^T \cdot \{x(\omega)\} = \left(\frac{\{\phi_r\}^T}{j\omega - \lambda_r} + \{\psi_r\}^T \{\phi^*_r\} \frac{\{\phi^*_r\}^T}{j\omega - \lambda^*_r} \right) \cdot \{f(\omega)\} \qquad (8.6)$$

where ψ_r is the rth modal vector and $x(\omega)$ is the vector of system responses. Now, scaling the modal coordinates η_r by the known input, it is possible to determine the FRF with all peaks, except rth, filtered out.

Such a formulation of the modal filter was much more practical and soon it appeared that it is a very useful tool which can be used for other applications such as (Zhang *et al.* 1990): the vibration control of flexible structures, correlation analysis for experimental and analytical modal vectors and identification of operational forces from the system response. One more possible application of the modal filtration is damage detection. This will be further discussed in the next section.

8.2   State of the Art

One of the groups of damage detection algorithms used for damage detection is the group of the so-called vibration based methods, i.e. methods that use data in the form of accelerations of vibrations in the time or frequency domain. They are based on the observation of changes in the system vibration responses, which result from damage occurrence. The main idea of the method is model based diagnostics defined in the following way: the model of a particular system in an undamaged state is given, and this model is compared with the model identified from the data measured on the object in the current state. Differences between these two models indicate the object modification (e.g. stiffness or mass decrease), which may be caused by damage. The most convenient model, which can be applied in the described approach, is a modal model, due to the fact that it is relatively easy to identify, and by means of operational modal analysis, it may be identified only from response data. Application of the modal model based diagnostics within damage detection has, however, several limitations and faults. First of all, there is a serious problem with distinction of the parameters' change resulting from damage and that being the consequence of environmental changes, e.g. temperature or humidity. The changes in ambient temperature for civil engineering objects (bridges, viaducts, masts, tall buildings) may reach even dozens of degrees in a relatively short period of time. That results in stiffness changes and, finally, in modal parameter variation. This effect is multiplied when the object is unevenly heated, e.g. when one side is exposed to the sun and the other is kept at a constant temperature due to the proximity to the water surface. The influence of humidity variations is similar; concrete elements absorb moisture and this leads to their mass increase and variation in modal parameters. Naturally, there exist methods which enable the elimination of the influence of environmental changes on the diagnostic procedure correctness. In most cases of application, a lookup table is prepared in which modal parameters identified for different ambient temperatures and values of humidity are gathered together. Such a table is unique and independently prepared for every object as a result of a set of experiments. A more sophisticated method of elimination of the influence of weather conditions on the monitoring system's efficiency is application of the environmental filter (Peeters *et al.* 2000). It is generally an autoregressive model with moving average (ARMA) identified from a set of experimental data. When environmental changes are eliminated in this way, any modification of the object (e.g. lining another asphalt layer on a bridge deck) causes the necessity to repeat the entire set of measurements, to update the lookup table or environmental filter. A further limitation of the modal model based diagnostics involves difficulties with the automation of procedures. Despite the current development of autonomous modal analysis procedures in many scientific centres, total independence of an engineer's interference is still problematic in practice. It further results in the fact that diagnostic symptoms in the form of natural frequencies, modal damping coefficients and modal vectors are estimated periodically, and they depend on subjective assessment of a testing team. There is, however, a method which uses vibration data and a modal model of the object and, in addition, it overcomes all the problems listed above. This is the advantage of modal filtering applied to the data recorded on the object.

A modal filter was used for the first time for damage detection and SHM by Slater and Shelley (1993). They developed an approach to monitoring the health of a smart structure with sensing and actuation capability. The approach presented there was an integrated control and monitoring procedure whereby the sensors, which are assumed to be distributed spatially across the structure, are processed by a set of modal filters which automatically track the modal coordinates of specified modes, and similarly track changes in the modal parameters (Shelley *et al.* 1992). The adaptive modal filter is formulated and applied to track the time varying behaviour of the specified modes, thereby indicating the health of the structural

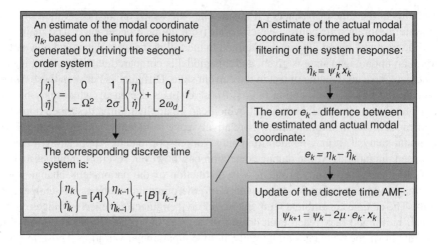

Figure 8.1   Scheme of the adaptive modal filter (AMF) – variant with known excitation force

system. The adaptive modal filter is insensitive to failures or calibration shifts in individual sensors and will automatically ignore failed sensors. It can also be used to detect disturbances entering the system as well as to identify failed actuator locations. The described method is very effective in damage detection and additionally it has some self-diagnosis abilities. However it can be used only with active vibration control systems, which greatly limits its practical applicability. The authors of the method developed two variants of the adaptive modal filter. The first of them operates in real time, but it requires knowledge of the excitation forces time history. The block diagram of this variant is presented in Figure 8.1.

The second of the modal filter updating procedures does not require loading forces measurements; it operates in the frequency domain and it cannot run in real time. The block diagram of the second variant of the adaptive modal filter is shown in Figure 8.2.

The method described above is quite effective in damage detection, and it has some self-diagnosis abilities, nevertheless, in the form presented by the authors it can be applied only together with an active vibration control system.

Gawronski and Sawicki (2000) proposed a different approach to applying modal filtration to damage detection. As a damage indicator they used the modal norm, calculated for each measuring sensor location and for each mode from the frequency band of interest. To calculate these norms the reciprocal modal vector matrix is required, that is the modal filter parameters. Next, the entire set of obtained modal norms is compared with an analogical set stored for the system in reference state. The method allows for damage detection and localization. Its scheme is presented in Figure 8.3.

The disadvantage of the method is mainly the large number of calculations which are required (modal norms are calculated for each mode and each measuring location).

Another application of the modal filter to damage detection can be found in El-Ouafi Bahlous *et al.* (2007). The suggested approach requires vibration data of the system in the undamaged and current stage along with FE model parametrized by means of specified damage parameters. With the use of modal filtration of the system response in the current

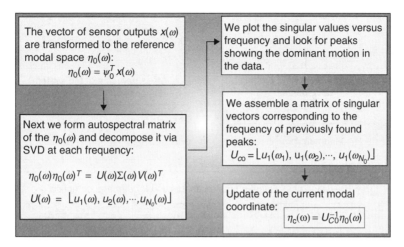

Figure 8.2   Scheme of the adaptive modal filter – variant with unknown excitation force. SVD, singular value decomposition

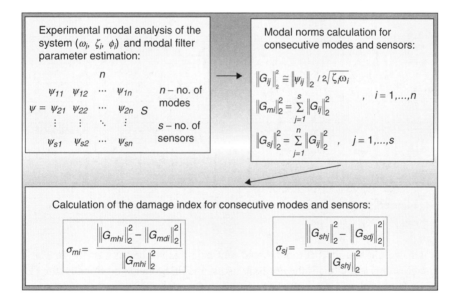

Figure 8.3   Flowchart of the damage detection procedure based on modal norms

stage the residuum function is calculated, which turns out to be normally distributed with a mean value equal to zero for undamaged system data and a mean value other than zero for the damaged case. To verify the statistical quantities of residua the generalized log-likelihood ratio test was proposed. This test allows for damage detection. Next, the procedure for damage localization and identification is started. It is based on multiple sensitivity and rejection tests (the number of tests equals the number of parameters θ). Steps of the damage diagnostic procedure are shown in Figure 8.4.

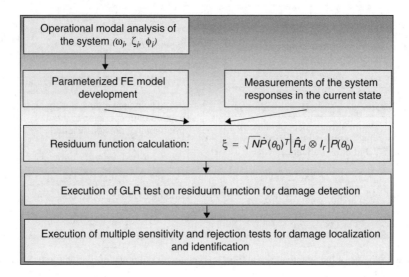

Figure 8.4   Block diagram of the diagnostic procedure. GLR, generalized log-likelihood ratio

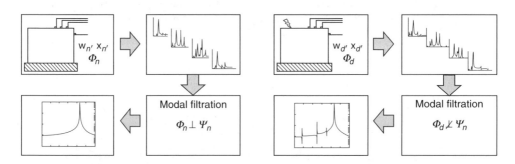

Figure 8.5   Procedure of damage detection and environmental effect filtration

What is more, the required computational power in this method is very high. The biggest disadvantage of this technique is the necessity to use the FE model. Additionally, the FE model has to be updated with respect to the large number of modes.

Another way of using modal filtering in SHM was presented by Deraemaeker and Preumont (2006). The frequency response function of an object filtered with a modal filter has only one peak corresponding to the natural frequency to which the filter is tuned. When a local change occurs in the object – in stiffness or in mass (this mainly happens when damage in the object arises), the filter stops working and on the output characteristic other peaks start to appear, corresponding to other, not perfectly filtered natural frequencies. On the other hand, global change of entire stiffness or mass matrix (due to changes in ambient temperature or humidity) does not corrupt the filter and the filtered characteristic has still one peak but slightly moved in the frequency domain. Graphical illustration of above statement can be seen in Figure 8.5.

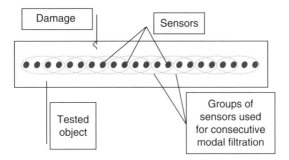

Figure 8.6   Procedure of damage detection and environmental effect filtration

The author found this approach very promising due to the following advantages:

- low computational cost (no modal analysis required at every diagnosis);
- possibility of autonomous operation (no engineer required to perform modal analysis);
- robustness for environmental changes (ambient temperature and humidity).

The only problem was that the modal filter did not work perfectly for the measured data, namely even for undamaged object data there were some small peaks in the area of natural frequencies. Due to this fact it is problematic to detect damage based only on the peak existence, as false alarms could appear. That is why another data interpretation was proposed by Mendrok and Uhl (2010).

The method described above was extended to damage localization by Mendrok and Uhl (2008), and further described in Mendrok and Uhl (2011). The idea for extension of the method by adding damage localization is based on the fact that damage, in most cases, disturbs the mode shapes only locally. That is why many methods of damage localization use mode shapes as input data. It is then possible to divide an object into areas measured with the use of several sensors and build separate modal filters for data coming from these sensors only. In the areas without damage, the shape of the modes does not change and a modal filter keeps working with no additional peaks on the filter output. When the group of sensors placed near the damage is considered, mode shape is disturbed locally due to damage and the modal filter does not filter perfectly characteristics measured by these sensors. The main idea of this extension is presented in Figure 8.6.

This technique is an extension of the previous one and possesses all its advantages, and moreover it allows to approximately localize damage.

8.3   Formulation of the Method

In this section the procedure of modal filter based damage detection will be formulated. First, let us look at the difference between modal filtration of the data recorded on a damaged object and of the data from a different temperature. Let us consider a dynamic system with proportional damping described by the equation of motion given in the matrix form:

$$M \cdot \ddot{x} + C \cdot \dot{x} + K \cdot x = f \qquad (8.7)$$

where M, C, K are the mass, damping and stiffness matrices, respectively, $C = c_1 \cdot M + c_2 \cdot K$ (c_1, c_2 are real constants) and f is the excitation forces vector.

The modal model of the system given in Equation (8.7) may be determined by solving the generalized eigenvalue problem formulated as follows:

$$\left(\left(\frac{s^2 + s \cdot c_1}{s \cdot c_2 + 1} \right) \cdot M + K \right) \cdot X(s) = 0 \tag{8.8}$$

As it was shown in the previous section, the modal filter uses reciprocal modal vectors to decompose the system's response on the components connected with consecutive modes. It occurs because the rth reciprocal modal vector ψ_r is orthogonal to all modal vectors except the rth. When damage occurs in the object, it results, in most cases, in a local drop of stiffness. The stiffness matrix is different and both eigenvalues (natural frequencies and modal damping coefficients) and eigenvectors (modal vectors) are changed. That further results in the fact that reciprocal modal vectors obtained for the undamaged object are not perfectly orthogonal to the modal vectors and consequently, the modal filter does not perfectly isolate other modes' influence from the system's response. So, why do environmental changes, such as temperature or humidity variations, not affect the model filter in this way? Let us now imagine that the temperature around the object increased and due to that, the Young's modulus of the object material decreased by a particular value. If we consider an ideal case, where the entire object has the same temperature and it is made of a homogeneous substance, the change in the stiffness matrix can be defined as follows:

$$K_t = \alpha \cdot K \tag{8.9}$$

where K_t is the stiffness matrix for higher temperature and α is the coefficient relating stiffness changes to temperature.

By solving Equation (8.9), it is possible to obtain the diagonal matrix Λ containing the eigenvalues and matrix Φ of the modal vectors. With such a solution the generalized eigenvalue problem for the undamaged system can be written in the following form:

$$-K \cdot \Phi = M \cdot \Phi \cdot \Lambda \tag{8.10}$$

Since the mass matrix M is square and invertible, the generalized eigenvalue problem can be presented as a standard eigenvalue problem:

$$M^{-1} \cdot (-K) \cdot \Phi = \Phi \cdot \Lambda \tag{8.11}$$

The same equation can be formed for the system at higher temperature:

$$M^{-1} \cdot (-\alpha \cdot K) \cdot \Phi_t = \Phi_t \cdot \Lambda_t \tag{8.12}$$

where the modal parameters for the system at higher temperature are denoted with subscript t.

Dividing Equation (8.11) by (8.12) and performing some simple mathematical operations, it is easy to derive the following:

$$\Lambda_t = \alpha \cdot \Lambda \tag{8.13}$$

Placing Equation (8.13) in (8.12) and comparing it with (8.11), it is immediately obvious that:

$$\Phi_t = \Phi \tag{8.14}$$

This guarantees further operation of the modal filter, i.e. there are no additional peaks on the filtered characteristic.

Basing on the assumptions presented in the previous section, one can formulate a diagnostic procedure in the following form:

1. Preliminary measurements and analysis: modal testing and analysis of the monitored object in the reference state.
2. Modal filter formulation: estimation of reciprocal modal vectors for the object in the reference state.
3. Measurements of selected characteristics of the object in the current state (FRFs, cross-power spectrum densities, power spectrum densities).
4. Filtration of recorded characteristics with the modal filter obtained for the reference model.
5. Calculation of the damage index from the filter characteristic in the region of the object natural frequencies.

The damage index mentioned in Point 5 of the procedure can be calculated in one of the following ways:

- Difference in amplitudes
- Difference in envelope:

$$DI_2 = \int_{\omega_s}^{\omega_f} |x_i(\omega)|d\omega - \int_{\omega_s}^{\omega_f} |x_{ref}(\omega)|d\omega \tag{8.15}$$

where ω_s is the starting frequency of the analysed band, ω_f is the closing frequency of the analysed band, x_i is the characteristic in the current state and, x_{ref} is the characteristic in the reference state.

- Envelope of scatter signal:

$$DI_3 = \int_{\omega_s}^{\omega_f} |x_i(\omega) - x_{ref}(\omega)|d\omega \tag{8.16}$$

- Scatter energy:

$$DI_4 = \frac{\int_{\omega_s}^{\omega_f} |x_i(\omega) - x_{ref}(\omega)|^2 d\omega}{\int_{\omega_s}^{\omega_f} |x_{ref}(\omega)|^2 d\omega} \tag{8.17}$$

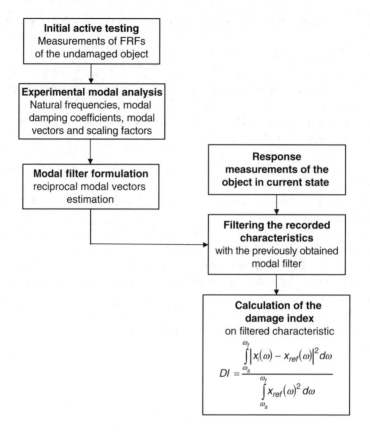

Figure 8.7   Block diagram of the proposed diagnostic procedure

The best results were obtained for DI_4 and this damage index is recommended for the implementation of the method.

The above procedure in the form of a block diagram is presented in Figure 8.7.

The damage index was calculated only for the frequency regions which are in the direct neighbourhood of the model's natural frequencies, except the one to which the modal filter was tuned. The bandwidth of the consecutive frequency intervals was established at 5 % of the considered natural frequency.

8.4   Numerical Verification of the Method

In the consecutive step the simulation verification of the method was performed. Extensive tests were conducted to verify its sensitivity to damage location, inaccuracy of the sensor's location in the consecutive experiments, noise of the measured characteristics, as well as changes in ambient conditions, such as temperature and humidity. This section presents the models used for simulation, results of these simulations and also the results of the model based probability detection (MPOD).

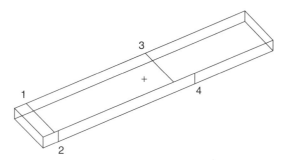

Figure 8.8 Cracks in the beam model

8.4.1 Models Used for Simulation

Three different models of the physical systems were created for the verification procedure. All the models were created using the finite element method (FEM). The first model was used to verify its sensitivity to damage location, inaccuracy of the sensors' location in the consecutive experiments and the noise of the measured characteristics. The FEM model was created directly using MSC.Patran 2008 R1 software. It represents a steel beam with dimensions of 200 x 40 x 10 mm supported at its free ends. It was modelled with 69 000 solid elements and 68 500 nodes. Such a dense mesh was required for tests on the inaccuracy of the sensor location. For the influence of damage location analysis, 4 cracks were modelled as discontinuity of a finite element mesh in different locations on the beam. For the purpose of further analyses, the following normal mode analysis scenarios were considered: beam without crack (reference state); and beam with cracks numbered 1, 2, 3 and 4, consecutively. In Figure 8.8 the locations of the cracks are presented.

The second model used for analysis was a metal frame that is presented in Figure 8.9. The model had a complex geometrical shape, was nonhomogeneous (steel – aluminium), had realistic boundary conditions and thermal expansion was included in it. The main goal of this simulation was to investigate the influence of different temperature loads on the modal filtration and to compare it with the effect of damage. It was also modelled in MSC.Patran and meshed with 10 000 hex-8 elements (the number of nodes was approximately 14 000). In order to investigate the behaviour of a damaged frame, a horizontal beam crack was modelled as node disconnectivity. There were two crack stage –5 % and 10 % of the cross-sectional area. Because the Young's modulus varies with temperature, temperature-dependent material properties were applied. The material properties of aluminium were applied only to the horizontal beam. To take into account temperature-dependent material properties, each design scenario was carried out in two substages: coupled mechanical – thermal analysis to obtain the temperature distribution, followed by normal mode analysis. The following cases were analysed: frame without crack at ambient temperature 20°C (reference state), frame with 10% crack on upper beam, frame without crack at ambient temperature 25, 30, 35, 40, 50, 0 and −50°C, frame without crack with upper beam heating (30 s) at 50°C, frame without crack with local heating of the right vertical bar (30 s) at 50°C.

The last model of a railway bridge used for the analysis of the damage detection procedure effectiveness was developed for two reasons. First, to examine how the ambient humidity can

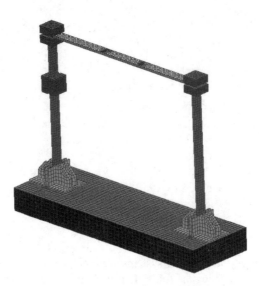

Figure 8.9   FE model of the frame

disturb the procedure, and secondly, to verify how it will work for such a complex structure. The CAD model was based on documentation of a real structure. The bridge is 27 m long and consists of steel (beams, barriers, rails, and reinforcement of main plate), concrete (main plate and pavements), soil and wooden (sleepers) elements. The CAD model is shown in Figure 8.10(a).

Afterwards, based on the CAD model, a FEM model was built in MSC.Patran. This consists of approximately 28 500 elements and 30 500 nodes. Solid, shell and beam elements were used in the model. The FEM model is shown in Figure 8.10(a). There were three cracks introduced in the model. The first and the second were vertical cracks in the web and the third was a flange crack. Crack localization was based on linear-static stress analysis. The crack was modelled as discontinuity of the FE mesh. To take into account the influence of moisture, different material densities were used. Depending on the analysis, material densities were appropriately adjusted. The following normal mode analysis scenarios were considered: dry bridge without crack (reference state); moist bridge without crack; wet bridge without crack; and bridge with cracks numbered 1, 2 and 3, consecutively.

8.4.2   Testing Procedure

To test the influence of the factors listed in previous sections on the modal filter based damage detection procedure, the following operations were conducted. First, the FRFs synthesis was performed for the models in reference states. The characteristics were synthesized for selected nodes of the models which simulated virtual measuring sensors.

Based on the reference modal model and synthesized FRFs of the reference systems, the reciprocal modal vectors were calculated − modal filter coefficients. With the use of these coefficients, sets of FRFs for the reference states and for consecutive damage states

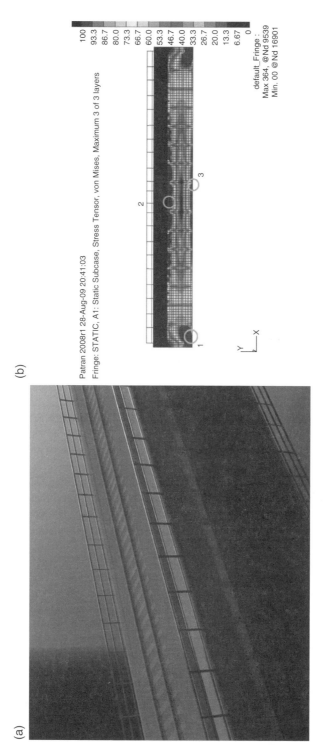

Figure 8.10 3D model of the bridge (a) and its FE mesh with crack localization (b)

Table 8.1 Damage index for different damage locations

No. of damage	Max. value of DI	No. of modal filter
1	3.5	7
2	13.5	5
3	12	1
4	810	1

were filtered. Damage index values were calculated for the obtained results according to Equation (8.17).

The damage index was calculated only for frequency regions, which are in the direct neighbourhood of the model's natural frequencies, except the one to which the modal filter was tuned. The bandwidth of the consecutive frequency intervals was established at 5 Hz.

8.4.3   Results of Analyses

In this section the results of analyses are presented. To save space only a few plots of modally filtered characteristics will be shown. The damage index values are presented only for the modal filter tuned to the mode shapes for which indication of the modelled crack was the best.

In the first stage of simulation the influence of the damage location on its detectability was examined. In the beam model four cracks were introduced at different locations (Figure 8.8). The size of each crack amounted to 5 % of the beam's cross-sectional area. In addition, in this section the author checked which natural frequency the modal filter should be tuned to in order to best detect the expected damage. In Table 8.1 the maximal values of the damage index calculated for the consecutive damages are presented, together with information about the natural frequency to which the modal filter was tuned.

The best detectable damage is Crack 4, in the middle of the beam, along the short side. The large difference in the damage index value between Crack 4 and the others results from the method of damage index calculation – the square of the characteristics' difference is considered [Equation (8.17)]. The far worse results obtained for Damage 3 arise from the fact that it is much shallower than Crack 4 (constant area with a much greater width). This is confirmed by the fact that for Cracks 1 and 2 the deeper one also gives higher damage index values. For Cracks 1 and 2, the highest damage index values were noted for filters tuned to Natural Frequencies 4, 7, 9 and 10, with a clear dominance of Natural Frequency 7. These frequencies correspond to torsional modes and higher bending modes. Cracks 3 and 4 are definitely best detected by Modal Filter 1 (tuned to Natural Frequency 1) – by far the highest amplitude of vibration in the fracture region. From the conducted analyses, it can be concluded that the detection of cracks should be performed with the use of a filter which is set to the mode in which the region of damage is subjected to the largest deformation. In Figure 8.11 the results of modal filtration for this simulation are presented.

To carry out the next verification test, it was assumed that in the consecutive measurements the sensors are slightly shifted against the reference position. The values of these shifts were defined as 1.5, 1 and 0.5 of the beam length. It was assumed that the sensor position error can

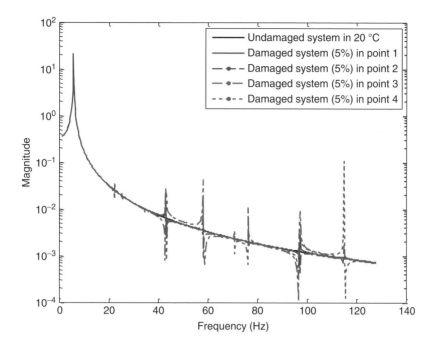

Figure 8.11 Results of modal filtration for different crack locations

Table 8.2 Damage index for incorrect sensors locations

Simulation scenario	Max. value of *DI*
Damage 4	810
Sensors shift 0.5 %	890
Sensors shift 1 %	2480
Sensors shift 1.5 %	9920

be done only along the *x*-axis – to the right or the left – and that, in a single measurement, sensors will be positioned with a constant error value (e.g. 1%). Taking 8 virtual sensors into account, there were 256 combinations to analyse for each considered value of inaccuracy. Modal filtering errors for all cases at the same assumed sensor shift value were at a similar level and the highest values were achieved by Modal Filter 1. For presentation in Table 8.2 the worst cases were chosen, that is, those with the worst accuracy of modal filtering. These results were compared with Damage 4.

The calculated values of the damage index confirm the significant impact of the sensor location in the subsequent measurements on the results of modal filtration, and thus the effectiveness of the method. In the case of the analysed beam, the smallest sensor shift gives a

Table 8.3 Damage index for noisy data

Simulation scenario	Max. value of DI
Damage 1	3.5
5 % noise on the data	0.39

comparable value of the index to 5% crack at the point where it is easiest to detect. The obtained results allow the following conclusions to be formulated:

- Since the method depends on the mode shape, it requires high repeatability for the location of sensors, and therefore, it is recommended for systems where a network of sensors is permanently attached to the object – a classic SHM system,
- if the method is used as an NDT technique, attention should be paid to the repeatable placement of sensors. Additionally, the level of damage that could be detected in this way should be raised to about 10% in order to avoid false alarms.

In the third stage of the simulation study the impact of noise on the modal filtration results was tested. The simulations tested the influence of interference generated by the measuring equipment on the modal filtering errors. Considering the characteristics of the sensors, their attachment (character – not location), as well as disruptions in wires and recording devices, the signal to noise ratio for the entire measurement path was set at 40 dB, which corresponded to the noise of 0.01% recorded signal amplitude. To ensure a large enough margin of error, it was decided that noise of a normal distribution, zero mean and amplitude of 5% of noise characteristics would be introduced. In Table 8.3, the values of the damage index for the noise characteristics and Crack 1 were compared, both for Modal Filter 7.

On the basis of this part of the simulations, it can be concluded that the noise associated with the measuring path does not affect the operation of the proposed method of damage detection.

One of the most important issues in the analysis of damage detection method properties is their sensitivity to changes in external conditions, especially ambient temperature. Therefore, it was decided that extensive simulation studies devoted to the influence of ambient temperature changes on the results of the modal filtration would be carried out. The tests were performed for various values of ambient temperature where the entire object had the same temperature as the environment, and for two special cases. The first one reflects the situation where the sun heats one of the parts of the object, while the others remain in the shade or are additionally cooled by their proximity to water. In this scenario, the simulated model was heated from above at the time, which prevented its total warming. The second special case is the situation in which the temperature changes only in one fragment of an object because of an artificial source of heat. Table 8.4 shows the damage index values collected for all changes in temperature as well as for the damage of 5% and 10%. In all cases the modal filter was tuned to Natural Frequency 2.

After the analysis of the damage index values, it can be seen that the impact of 5% damage is greater than a temperature change of $5\,^\circ$C. However, if one wants to use the method for a wider ambient temperature range, it is suggested that some kind of a modal filter bank should be built. In such a bank, one would have the reference model of the system identified for various ambient temperatures. In order to decide on the bank of filters designated for every $10\,^\circ$C, it is recommended to increase the minimum size of recognizable damage by

Table 8.4 Damage index for temperature changes

Simulation scenario	Max. value of DI
Damage 5%	5.8×10^6
Damage 10%	3.6×10^8
Ambient temp. 25 °C	2.7×10^5
Ambient temp. 30 °C	9.2×10^5
Ambient temp. 35 °C	2.5×10^7
Ambient temp. 40 °C	15.6×10^7
Ambient temp. 50 °C	1.9×10^8
Ambient temp. 0 °C	8.8×10^6
Ambient temp. −50 °C	2.3×10^9
Upper heating 50 °C	4.5×10^7
Local heating of the right bar 50 °C	5.7×10^9

Table 8.5 Damage index for bridge simulations

Simulation scenario	Max. value of DI
Damage 1	7210
Damage 2	2.7×10^4
Damage 3	7.3×10^7
Moist bridge	1670
Wet bridge	3.2×10^9

about 10%. The damage index calculated for such a damage is higher than the value of the difference in temperatures reaching up to 30 °C. In addition, the heating frame from the top gives a lower value of the damage index than the damage. The worst of the simulated cases were the temperature change of 70 °C (ambient temperature −50 °C) and heating only the vertical bar of the frame. Both of these cases, however, do not disqualify the method, since the 70 °C difference in temperature using even a small bank of filters should not occur, and a local high temperature change should be detected, because it can be regarded as failure.

To examine the operation of the method for a complex civil engineering object (real structure), and check the effect of humidity on the efficiency of the damage detection method again, the modal filter coefficients were calculated for the reference model, which is a dry bridge without any damage. Table 8.5 shows the damage index values for three cracks introduced to the bridge consecutively and for a moist (relative humidity 99%) and wet (after intense rainfall) bridge. All results are for Modal Filter 10.

The analysis of the damage index values confirmed the lowest detectability of Damage 1 – the one close to the support. To obtain higher values of the damage index for this crack, one should take into account the modal filters tuned to higher modes – the ones which deform the region of Damage 1 more. Regarding the impact of humidity changes, it is quite significant; the weight of soil and wood changes by over 10% and is a significant share of the object's total mass. On this basis, it was concluded that the method should not be applied to objects which change their mass greatly due to moisture. On the other hand, the method has shown its effectiveness in detecting small-scale damage for such a highly complex technical facility.

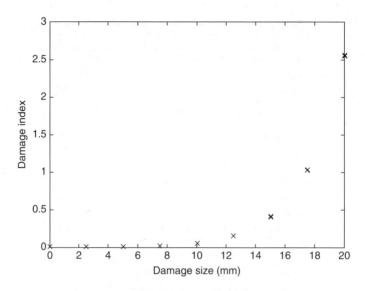

Figure 8.12   Damage index values for consecutive simulations

8.4.4   Model Based Probability of Detection

The research methodology is described in Chapter 3. Among the methods developed there, in the following studies the analysis of signal-response type is used.

In order to determine the model based probability of detecting damage the FE model of a steel – aluminium frame described above and shown in Figure 8.9 was applied. To prepare data for MPOD analysis at about 3/5 of the length of the frame upper beam the fracture was modelled as a discontinuity of nodes (double, unconnected nodes). The eight stages of damage were considered, which was a reduction of the beam cross section of 6.25% to 50% at regular intervals. Since the cross section of the frame upper beam has dimensions 10 x 400 mm (height x width), damage depth ranged 2.5–50 mm every 2.5 mm. For the nine consecutive models (undamaged and 8 stages of damage) the eigenvalue problem was solved and a synthesis of the FRFs for the 20 selected nodes of the model. These nodes, virtual sensors, were evenly spaced along the top (damaged) beam. Each transfer function has been disrupted by the addition of noise with a rectangular distribution, mean 0 and an amplitude equal to 5% of the amplitude of the noise characteristics. In addition, for each case of damage the noising process was carried out seven times to increase the amount of data to determine MPOD. One of the sets of characteristics for the undamaged model was used to estimate the reciprocal modal vectors and with their help further data sets corresponding to the consecutive stages of damage were filtered. In the last step of data preparation damage indexes were calculated for the simulations. In total there are 63 values presented as a function of the damage size in Figure 8.12.

Analysing the values of damage index one can observe that with the increase of damage, the scatter of damage index values also increases. For the greatest damage (20 mm) 5% noise of characteristics resulted in 2% scatter of damage index values. Figure 8.13 shows the damage index values for the undamaged model and with the smallest damage of 2.5 mm.

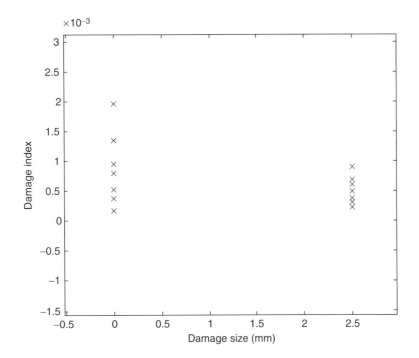

Figure 8.13   Damage index values for undamaged model and damage of 2.5 mm

As one can see, for the 2.5 mm damage the change of damage index is smaller than the scatter of its value due to the noisy characteristics.

The obtained values of damage indexes together with the corresponding sizes of damage were used in the procedure that creates MPOD curves. A signal – response procedure was launched with a threshold value of 0.0014, and generated the curve shown in Figure 8.14. Other thresholds gave very similar curves with the same confidence interval only shifted along the x-axis. The analysis of the obtained result allows the earlier conclusions of the testing method to be confirmed. The minimum amount of damage that can be detected without the risk of false alarms is a change in stiffness of approximately 10%. It is worth noting the very steep curve character and very narrow confidence intervals. Both of these features of the results obtained in the analysis of MPOD confirm the high efficiency of the method.

8.5 Monitoring System Based on Modal Filtration

In the next subsection the main assumptions on the design of the SHM system based on the described technique are given. Then, in subsequent subsections, the measuring diagnostic unit and dedicated software are presented.

8.5.1 Main Assumptions

The main assumption made for the designed system was that it should be completely independent. It means that the potential user should be able to perform a full diagnostic procedure

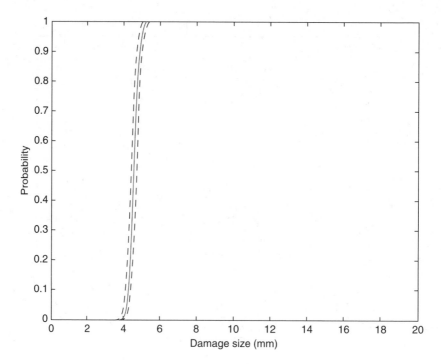

Figure 8.14   MPOD curve – signal–response

without the need to use any additional measuring devices or software. To fulfil the above requirement, the original 16 channel measuring diagnostic unit (MDU) was designed and the dedicated modal analysis and modal filtration software was written. Generally, the system composed of both hardware and software is supposed to work in one of three modes:

1. Operation in dynamic signal analyzer mode for the purposes of the modal testing. In this mode the modal filter coefficients are estimated for the reference structure.
2. Operation in diagnostic mode. In this mode the following actions are performed: acceleration/displacement of vibration measurements, estimation of selected characteristics (FRFs, PSDs), modal filtration of the above characteristics, damage index calculation, and finally, visualization of the filtered characteristics.
3. Operation in monitoring mode. In this mode the actions performed by the system are identical to in the diagnostic mode but they are done periodically and at the end of each diagnostic procedure the results are reported to the central unit.

8.5.2   Measuring Diagnostic Unit

From the technical point of view the diagnosis process is divided into a few basic steps: simultaneous synchronous acquisition of analogue signal (converted into digital domain) from 16 channels, digital signal processing applied to the measured signal and reporting processing results. The block diagram of the MDU is presented in Figure 8.15.

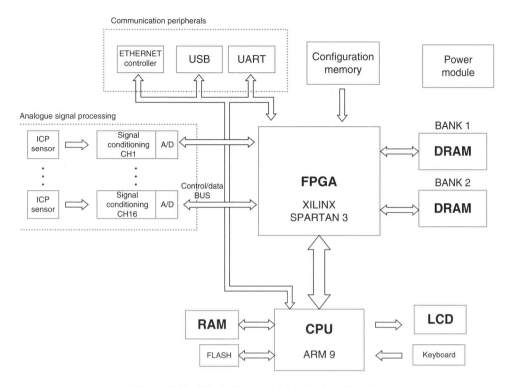

Figure 8.15   Block diagram of the developed device

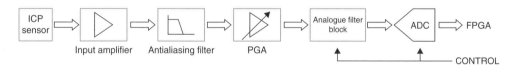

Figure 8.16   Analogue part of the measuring circuit

The diagnostic device contains two fully independent modules (CPU and FPGA) connected to each other. The CPU module is included for control purposes – it implements user interface using some peripheral devices such as a keyboard, LCD display and communication peripherals. Using this interface, it is possible i.e. to set gain or select the required analogue filter in each of 16 analogue signal processing modules, or to start a diagnostic process. The FPGA module contains all the logic modules needed for implementation of required digital signal processing. It is 'seen' by the CPU module as another peripheral device which can execute commands (such as the start data processing command) and send processing results. The MDU can be accessed via Ethernet or USB, which is needed in the system calibration phase or to read remotely processed results. The analogue signal processing module is shown in Figure 8.16.

Figure 8.17   Measuring diagnostic device MDFS 16000

The input analogue signal is delivered from ICP accelerometer sensors mounted on the examined object. The ICP signal standard is based on a 4–20 mA current signal transmission standard, the main advantage of which is the ability to transmit signal (with 1 kHz frequency band width) without any distortion at ranges of 100 m and more. A complete device can be seen in Figure 8.17.

The unit was named MDFS 16000, and was then tested under laboratory and industrial conditions the results of these tests will be presented later in this chapter.

8.5.3   Modal Analysis and Modal Filtration Software

The main goal of the software written for the described SHM system is the estimation of the modal filter coefficients. For this purpose, the application provides the following functionalities: geometrical model definition of the tested object, measurement points definition (namely, the assignment of specific points of a geometric model to the sensors placed on an object), execution of measurement and presentation of the results (time histories, PSD, FRF and coherence), data archiving and finally, execution of modal analysis by calculation of a stabilization diagram, estimation and visualization of mode shapes for selected poles, estimation of modal filter coefficients and visualization of filtration results. In Figure 8.18 the graphical user interface of the described software allowing for impulse modal testing and mode shape visualization control is presented.

All calculations related to the modal analysis are performed by the MATLAB® engine. The application provides the possibility to debug these functions from MATLAB® level. For this reason, at the user-specified location, mat-files are stored that contain input parameters for the appropriate MATLAB® functions.

It was assumed that in order to fluently visualize the mode shapes it is necessary to refresh with a minimum speed of 30 fps. Sufficiently effective controls that allow visualization and animation of 3D models with the assumed speed are not available on the market. Therefore, implementation of such control was done using the XNA environment. The control uses a graphics accelerator which allows refreshing at 60 fps at 10 000 points of a geometrical model.

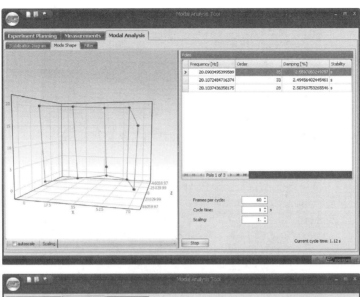

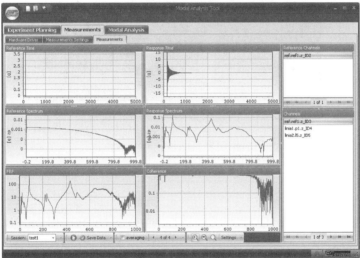

Figure 8.18   Graphical user interface of the described software

8.6   Laboratory Tests

This section contains the program, description and results of laboratory tests of the developed SHM system. The results of the POD are also shown and discussed.

8.6.1   Programme of Tests

The tested object was a cantilever beam made of aluminium alloy. Its dimensions were 50 x 4 x 1000 mm. The object was divided into nine measuring points equally spaced every

Figure 8.19   The laboratory stand

100 mm, placed along the longitudinal axis of the beam. In Figure 8.19 the measuring stand is presented. The sampling frequency was set at 1024 Hz and the length of recorded time histories was 200 s.

The following measurements were performed:

- undamaged beam, ambient temperature $20\,^\circ$C;
- undamaged beam, ambient temperature $20\,^\circ$C;
- undamaged beam, ambient temperature $20\,^\circ$C, sensors reassembled;
- undamaged beam, ambient temperature $31\,^\circ$C;
- damaged beam (added mass 4 g in point blk:5), ambient temperature $20\,^\circ$C;
- damaged beam (added mass 4 g in point blk:9), ambient temperature $20\,^\circ$C;
- damaged beam (added mass 15 g in point blk:5), ambient temperature $20\,^\circ$C;
- damaged beam (added mass 15 g in point blk:9), ambient temperature $20\,^\circ$C,

All the measurements were performed with the use of the tested MDU. Additionally, the visual assessment of modal filtration results was performed.

Table 8.6 Comparison of beam's modal parameters

Mode shape no.	Natural frequency (Hz)	Modal damping coefficient (%)
1	24.7	0.69
2	68.5	0.39
3	133.8	0.27
4	222.8	0.28
5	331.9	0.24
6	461.2	0.28

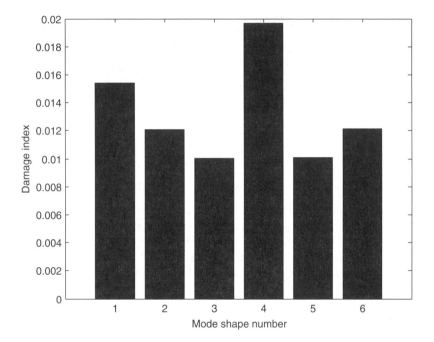

Figure 8.20 Damage index values for the undamaged object

8.6.2 Results of Experiments

In the first step the measurements were performed in the reference state. Then, on their basis the modal analysis was done and eigenfrequencies, the corresponding modal damping ratios and mode shapes were estimated. Table 8.6 shows the identified modal parameters.

All the measured scenarios listed in the previous subsection were evaluated for two cases. First, two measurements performed under identical conditions were considered to verify the stability of the method and robustness for excitation differences. Figure 8.20 presents the damage index values calculated with the use of Equation (8.17) for two measurements taken under the same conditions.

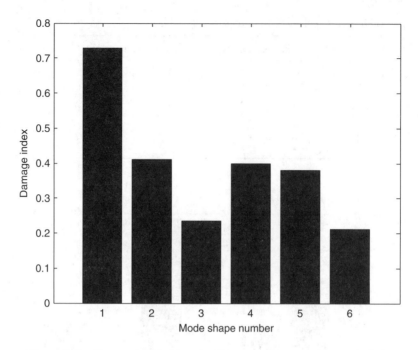

Figure 8.21   Damage index values for different ambient temperature(31 °C)

From the results shown in Figure 8.20 it is clear that the method acts properly — the damage index remained at a very low level. Furthermore, the temperature in the laboratory was increased by 11 °C and kept at this level for 1 h. In Figure 8.21 the damage index values calculated for the beam at different temperature are presented. This level of temperature difference has little effect on the method. Of course the ambient temperature for a civil engineering object can change more than 11 °C but one of the assumptions made during the SHM system design was that it should be equipped with a modal filter bank, namely, the reference data should be collected for a set of ambient temperatures changing every 10 °C. The results obtained during this analyses together with the mentioned assumption allow to state that the system is robust for ambient temperature changes.

The next stage of tests included the measurements also on the undamaged beam, but this time the sensors were taken off and placed again possibly in the same position. In Figure 8.22 the damage index values calculated for this case are presented.

For this case the damage index values rose slightly but they were still at an acceptable level. On the other hand, one has to keep it in mind that the sensors were remounted immediately after removal. In real nondestructive testing the period between measurements is usually much longer, and this increases the likelihood of bigger errors in sensors' placement. It is then recommended to apply the permanent or embedded sensor network.

The last group of tests concerned the damage detection by the proposed system. Before the beginning of the tests, the laboratory was cooled down to the stable temperature of 20 °C, and reference modal filter coefficients were calculated for the data obtained after sensor re-installation. The damage was introduced in a form of added mass to allow further tests

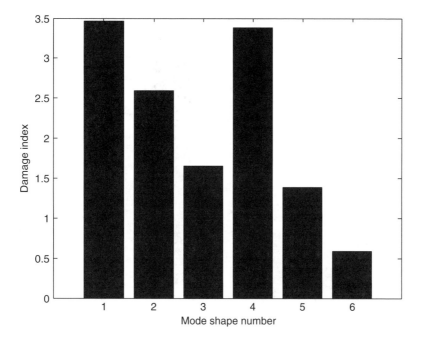

Figure 8.22   Damage index values for the sensors remounted

on the beam. The total mass of the beam was 0.544 kg; the added masses were 3.7 g and 15.5 g, which represented 0.68 % and 2.85 % of the beam mass and 6.8 % and 28.5 % of the 100 mm surrounding (distance between consecutive measuring points), respectively. The first added mass location was chosen in the middle of the beam (point blk:5) and the second location was near the beam fastening (blk:9). The damage index values for all damage cases are shown in *Figure 8.23, Figure 8.24, Figure 8.25 and Figure 8.26.*

A general conclusion that can be drawn from the presented results is that all the damage cases were properly detected. Damage index values grew significantly and they were much bigger than for all undamaged beam cases. It can be observed that different damage location affects different modal filters, and as it was concluded in simulation tests, the best detectability occurs when the modal filter is tuned to the mode shape that has biggest deflection in damage location. Furthermore, in Figure 8.27 the modally filtered characteristics for the modal filter tuned to Mode Shape 5 for larger damage are shown.

In Figure 8.27 the damage is clearly visible in the form of peaks near the 6th and 4th natural frequencies.

8.6.3   Probability of Detection Analysis

The object of this study was the same steel – aluminium frame, the model of which was used to analyse MPOD. The upper beam was divided into 10 equal sections, each with a length of 70 mm. The measuring points were marked, respectively, blk:1, ..., 10, starting from the right.

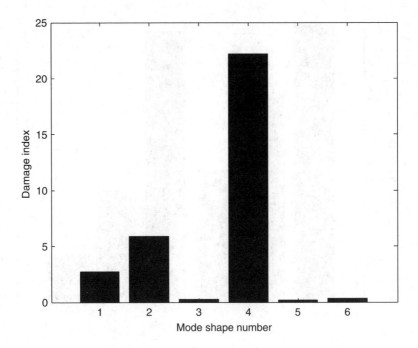

Figure 8.23   Damage index values for the 3.7 g added mass in point blk:5

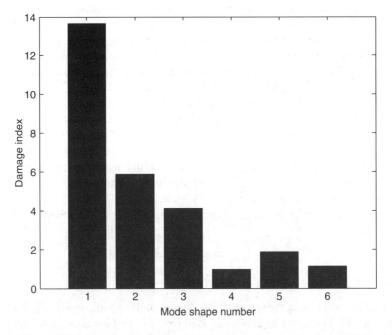

Figure 8.24   Damage index values for the 3.7 g added mass in point blk:9

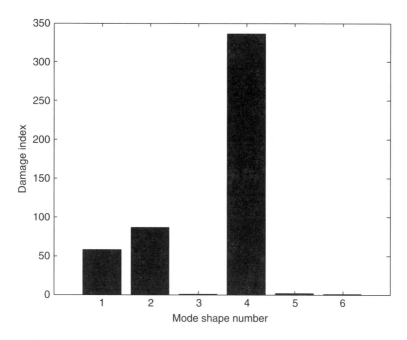

Figure 8.25 Damage index values for the 15.5 g added mass in point blk:5

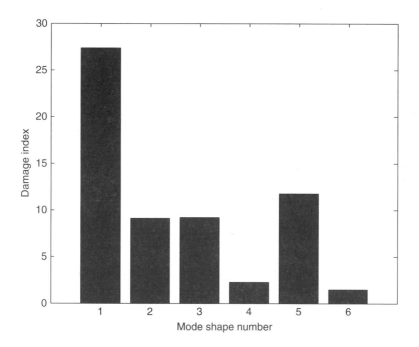

Figure 8.26 Damage index values for the 15.5 g added mass in point blk:9

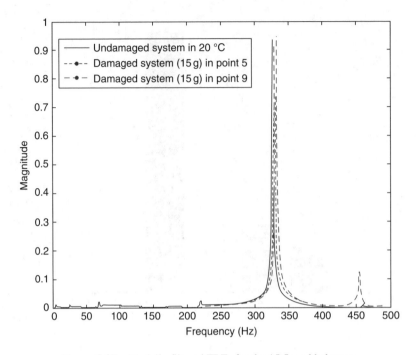

Figure 8.27   Modally filtered FRFs for the 15.5 g added mass

At every point an accelerometer was placed. The frame with sensors is shown in Figure 8.28. This was followed by an impulse test that was used for modal parameters and reciprocal modal vectors – modal filter coefficients estimation.

In the next step an electrodynamic shaker was connected to the test object and it was excited with bursts of random signal. This method of excitation was used to simulate the operating conditions. Under these conditions, a number of measurements was performed to determine the POD. The damage was introduced by cutting the top beam in the place between points blk:3 and 4, with the following depths: 0, 2, 3, 4, 6, 7, 9, 10, 12, 13, 15, 16, 19 and 20 mm. Therefore damage varied from 5 to 50% of the beam cross section. For each degree of damage, as well as for the undamaged object 7 measurements were performed, which gave 96 measurements in total. Then the damage index was calculated for the subsequent experiments. A graphical comparison of damage index values is shown in Figure 8.29.

A greater spread of damage index values is visible than in the artificial noise simulation data, especially for large defects.

The obtained damage index values together with the corresponding damage sizes were the input data for the procedures established to determine POD curves. The procedure signal–response was triggered by a threshold value of 0.165 and produced the curve shown in Figure 8.30. This was the minimum threshold for which it was possible to determine damage without the risk of a false alarm.

Next, the hit – miss analysis was performed. The convergence of the algorithm was achieved for the damage index threshold value of 0.19. The obtained POD curve is presented in Figure 8.31. The resulting POD curve of the signal – response method is very similar in

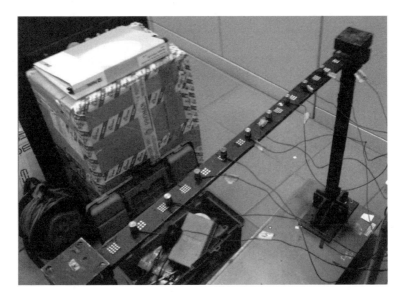

Figure 8.28   Set of the accelerometers on the upper beam

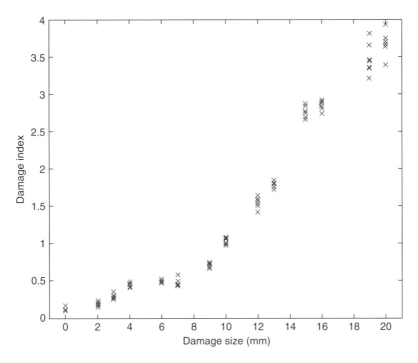

Figure 8.29   Damage index values for consecutive measurements

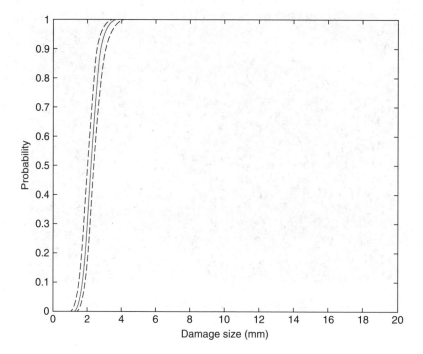

Figure 8.30   POD curve – signal–response

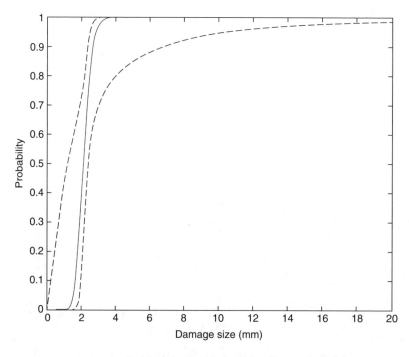

Figure 8.31   POD curve – hit–miss

Figure 8.32   Tested wing

shape to that from the numerical analyses. Interestingly it shows some damage detection even at a smaller size. Also, confidence intervals are similar to those obtained using simulation, although in this case a little wider. This follows from the fact that the scatter of damage index values was greater than for the MPOD. POD analysis showed that, with probability 1 for the test object, damage can be detected from about 3.5 mm, that is 8.75% loss in cross – sectional area. It is noteworthy that the experimental data also gave a steep curve and narrow confidence intervals. Both of these features confirm the high reliability of the method. With regard to the POD curve obtained by the hit – miss method it has almost the same shape as that of the signal – response method. However, the confidence interval shows a significant difference.

8.7   Operational Tests

The last stage of the developed SHM system testing was its application to the damage detection of the real technical object under operational conditions. The study was conducted in the State Higher Vocational School in Chelm, Poland. The object of the study was a wing of the training jet fighter PZL I 22 Iryda. Possible structural changes in the wing of the aeroplane were examined. During the tests the aircraft was set on wheels outside the hangar, the wing was tested in the closed state, the tail was in the neutral position and the fuel tanks were empty. (Figure 8.32).

In order to build a reference model and to estimate modal filter coefficients, the modal impulse test was performed, which can be characterized by the following features: the impulse excitation in point wing:ref (connecting point of the front girder and the second frame from the tip of the wing), simultaneously the accelerations of vibrations were measured in 5 points evenly placed along the front girder of the wing (every 1200 mm), the frequency

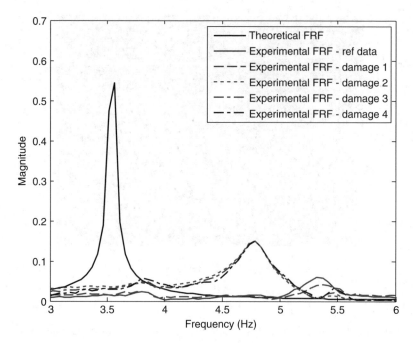

Figure 8.33 FRFs of the tested wing after modal filtration with the filter tuned to 1st mode (region of 1st, 2nd and 3rd modes)

band was set up to 25 Hz with resolution 0.125 Hz. The following measurements were performed:

- Three measurements in the same configuration as the reference one, performed every 0.5 h in order to check the stability of the procedure and the impact of environmental conditions on the results of its operations.
- Measurement after strong deformation of the wing – the wing was loaded with approximately 1 kN and then it was released – simulating measurement after flight.
- Measurement with added mass of 3 kg in measuring point 5 (the end of the wing).
- Measurement with added mass of 3 kg in measuring point 4.
- Measurement with added mass of 3 kg in measuring point 3.
- Measurement after removing the extra mass.

First, the modal analysis was performed and on the basis of its results the modal filter coefficients were calculated for the reference measurements. Diagnostic testing was then performed using the test procedures implemented in the MPOD. In the same configuration (without removing the sensors) consecutive impulse tests were performed in accordance with the scenario shown above.

During the testing the FRFs together with response spectra were estimated and stored. These characteristics were next filtered, and the results of modal filtration were used to calculate the damage index. In Figure 8.33 and Figure 8.34 the exemplary results of the modal

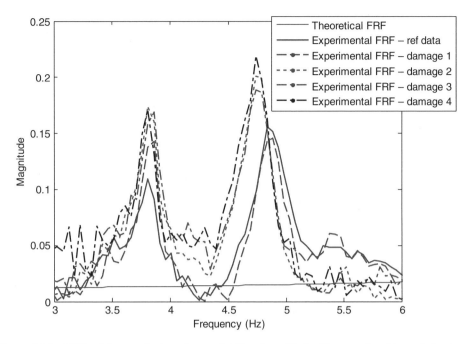

Figure 8.34 FRFs of the tested wing after modal filtration with the filter tuned to 5th mode (region of 1st, 2nd and 3rd modes)

Table 8.7 Damage index values

Test scenario	Summed damage index
Undamaged wing no. 1	1.4767
Undamaged wing no. 2	1.2110
Undamaged wing no. 3	0.8259
Applied and released 1 kN load	1.0681
Added mass in wing:5	5.5794
Added mass in wing:4	4.2769
Added mass in wing:3	4.3029
Undamaged after all measurements	1.2806

filtration for selected tests are presented. As it can be seen, in the region of the 2nd and 3rd mode, deterioration of filtration quality has arisen as a result of the structural change. This effect is small, as the used mass was only a small change in the weight of the entire wing. Since the effect was observed in the region of the 1st mode the damage (added mass) did not affect the filtration accuracy. In the case of filters tuned to higher modes, the closer the added mass was to the anti-node of the mode, the stronger filtration deterioration was observed, i.e. more detectable damage. Table 8.7 summarizes the damage indexes calculated for the subsequent tests by the MDU.

Values presented in Table 8.7 confirm the applicability of the developed SHM system in real technical structures. The system detected properly all the damage scenarios, although the applied structural changes were rather small.

8.8 Summary

In this chapter, the application of modal filtration in SHM was shown. Based on the work presented above, the following conclusions and comments can be made:

- A modal filter is an effective damage detection tool with the ability to detect structural change in objects on levels no lower than other vibration based methods.
- The use of modal filtration as a method based on mode shape is insensitive to changes in environmental conditions, such as temperature and humidity.
- The conducted simulations showed that insensitivity to changes in external conditions designated theoretically is, in actual fact, disturbed by such factors as inhomogeneous material of the object, complex geometric shape, boundary conditions, and uneven heating of the object. All these factors result in the inhomogeneous change in stiffness and mass matrices which refute the considered theoretical assumptions.
- Other factors which may cause false alarms are: inaccurate positioning of sensors in successive measurements; and large changes in the structural mass caused by, for example, heavy rainfall,
- The performed laboratory tests confirmed the advantages of the method and the SHM system.
- MPOD and POD analyses allowed to determine the damage level that is possible to be detected without a risk of false alarm as a 8–10% change of structural parameter,
- Both the system and the method were successfully applied to structural change detection in a real technical object.

Further work being conducted on the development of the described methods is foreseen. It will be directed at achieving the following aims:

- The use of principal component analysis to create methods which are independent of structural change unconnected with the damage, e.g. a change in weight caused by vehicles travelling on a bridge.
- The development of new scaling procedures for operational mode shapes with the aim of constructing a modal filter for exploitation data.
- Extension of the capabilities of the system for the location of damage according to the procedure presented by Mendrok and Uhl (2011).

References

Deraemaeker A and Preumont A 2006 Vibration-based damage detection using large array sensors and spatial filters. *Mechanical Systems and Signal Processing* 20(7), 1615–1630.

El-Ouafi Bahlous S, Abdelghani M, Smaoui H and El-Borgi S 2007 Modal filtering and statistical approach to damage detection and diagnosis in structures using ambient vibrations measurements. *Journal of Vibration and Control* 13(3), 281–308.

Gawronski W and Sawicki J 2000 Structural damage detection using modal norms. *Journal of Sound and Vibration* 229(1), 194–198.

Meirovitch L and Baruh H 1982 Control of self-adjoint distributed parameter system. *Journal of Guidance Control and Dynamics* 8(6), 60–66.

Mendrok K and Uhl T 2008 Modal filtration for damage detection and localization. *Proceedings of the Fourth European Workshop on Structural Health Monitoring. DEStech Publications, Inc.*, pp. 929–936.

Mendrok K and Uhl T 2010 The application of modal filters for damage detection. *Smart Structures and Systems* 6(2), 115–134.

Mendrok K and Uhl T 2011 Experimental verification of the damage localization procedure based on modal filtering. *Structural Health Monitoring* 10(2), 157–171.

Peeters B, Maeck J and De Roeck G 2000 Dynamic monitoring of the z24-bridge: Separating temperature effects from damage. *Proceedings of the European COST F3 Conference on System Identification and Structural Health Monitoring.* Madrid, Spain, pp. 377–386.

Shelley S, Freudinger L and R.J.A 1992 Development of an on-line parameter estimation system using the discrete modal filter. *Proceedings of the 10th IMAC.* San Diego, USA, pp. 173–183.

Slater GL and Shelley SJ 1993 Health monitoring of flexible structures using modal filter concepts. *Proceeding of SPIE* 1917, 997–1008.

Zhang Q, Allemang RJ and Brown DL 1990 Modal filter: Concept and applications. *Proceedings of the 8th IMAC.* Orlando, USA, pp. 487–496.

Zhang Q, Shih CY and Allemang RJ 1989 Orthogonality criterion for experimental modal vectors. *Proceedings of the 12th Biennial ASME Conference on Mechanical Vibration and Noise.* Montreal, Canada, pp. 251–258.

9

Vibrothermography

Łukasz Pieczonka and Mariusz Szwedo
Department of Mechatronics and Robotics, Faculty of Mechanical Engineering and Robotics, AGH University of Science and Technology, Poland

9.1 Introduction

This chapter introduces the use of thermographic measurements in NDT and SHM. The physical principles of measurements and the classification of thermographic techniques are presented with special focus on vibrothermography (VT). An overview of the measurement equipment, including thermographic cameras and excitation sources, is given. In addition, the perspectives for performing virtual testing are discussed. In particular, numerical simulations of coupled thermomechanical phenomena with the use of explicit finite elements (FEs) are presented. On the basis of the literature sources and on the authors' experience it is concluded that presently the good qualitative agreement with measurements that can be obtained is a starting point for performing virtual testing, test preparation and optimization. Hardware and software components of the VT measurement system developed by the authors at AGH University of Science and Technology are described in the second part of this chapter. Functionality of the system is illustrated by the description of laboratory measurements performed on composite and metallic samples. A parametric study of the influence of measurement parameters on the resultant thermal response are also discussed. Finally, measurements performed with the developed test system on a military aircraft fuselage and wing panels are reported. Practical issues related to field measurements are also discussed. The chapter is concluded with considerations on practical applications of VT in SHM.

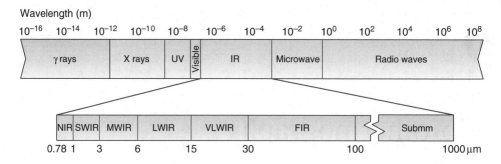

Figure 9.1   The electromagnetic spectrum

9.2   State of the Art in Thermographic Nondestructive Testing

This section introduces the basics of thermographic techniques and their use in the area of NDT. Thermography, as the name implies, relates to the process of mapping temperature (prefix *thermo-*) distribution (sufix *-graphy*) of an object's surface. Typically, the process of detecting surface temperatures is carried out by capturing the emitted infrared (IR) radiation of an object. Thermography is therefore a non contact measurement. Originally, the term *thermography* referred to a contact measurement whereas *infrared thermography* referred to a noncontact measurement. Nowadays, however, contact techniques are obsolete and the term thermography has become equivalent to infrared thermography. This transition was mainly due to the technological progress in IR detectors technology, which made them affordable for most typical applications.

Before discussing the applications of thermographic techniques in NDT let us briefly recall the basic theory behind these measurements techniques. IR radiation covers a portion of the electromagnetic spectrum from 0.78 to 1000 µm as shown in Figure 9.1. It is the part of the electromagnetic spectrum that is not visible to the human eye. It was discovered in 1800 by Friedrich Wilhelm Herschel, who was testing optical filters to observe the sun. In a series of experiments Herschel passed the sunlight through a prism and discovered that the temperature measured beyond the red end of the visible spectrum was higher than the temperature measured in the visible range. Herschel termed this region of the spectrum 'calorific rays' which was later renamed as IR radiation. The IR spectrum is typically partitioned on the basis of the different source and detector technologies used in each region of the spectrum (Rogalski 2011). The commonly used partitioning of the IR spectrum is shown in Table 9.1.

IR is emitted by all objects at temperatures above the absolute zero, and the amount of radiation increases with temperature. This is a consequence of the atomic nature of matter. The vibration of all charged particles, including atoms, generates electromagnetic waves. The higher the temperature of an object, the faster the vibration, and thus the higher the spectral radiant energy (Rogalski 2011).

An object reacts to incident IR in three ways: it absorbs, reflects and transmits the radiation. A blackbody is an object that absorbs all incident radiation. Conversely, according to Kirchhoff's law, a blackbody is also a perfect radiator. The radiative properties of a blackbody as a function of its temperature and the wavelength of the emitted radiation are described by

Table 9.1 Partitioning of the infrared spectrum

Infrared region	Wavelength range (μm)
Near infrared (NIR)	0.78–1
Short wavelength IR (SWIR)	1–3
Medium wavelength IR (MWIR)	3–6
Long wavelength IR (LWIR)	6–15
Very long wavelength IR (VLWIR)	15–30
Far infrared (FIR)	30–100
Submillimetre (Submm)	100–1000

Table 9.2 Emissivity values of common materials. After Fluke Corporation (2009)

Material name	Emissivity value
Aluminium, polished	0.05
Aluminium, strongly oxidized	0.25
Cast iron, polished	0.21
Cast iron, rough casting	0.81
Concrete	0.92
Iron, hot rolled	0.77
Iron, shiny, etched	0.16
Lacquer, black, dull	0.97
Rubber	0.93
Steel, galvanized	0.28
Steel, rough surface	0.96
Steel, sheet, rolled	0.56

Planck's law (Rogalski 2011). *Radiant exitance* is the density of radiant flux leaving a surface at a point. It has units of [W m^{-2}]. The ratio between the radiant exitance of the actual source and the exitance of a blackbody at the same temperature is defined as *emissivity*. The emissivity is therefore a dimensionless parameter that takes values between 0 and 1, where 1 is the emissivity of a blackbody. Emissivity values of some common engineering materials are summarized in Table 9.2. Emissivity is an important factor in practical applications of thermography which will be discussed later. Another factor that influences measurements is the medium in which the radiation propagates from the emitter to the detector. In most cases it is the atmosphere that attenuates radiation due to absorption and scattering by gases and particles present in the atmosphere. The phenomena are similar to the ones that we all know from the visible part of the electromagnetic spectrum, where rain, fog or clouds can significantly reduce visibility. The attenuation is strongly dependent on the radiation wavelength. There are, however, regions of the IR spectrum for which the attenuation is small. These regions correspond to the MWIR and LWIR ranges, that are typically used in practical applications of thermography. An interested reader can find more theoretical considerations related to IR radiation in (Maldague 2001; Minkina and Dudzik 2009; Rogalski 2011).

In engineering, active thermography refers to a family of NDT methods based on temperature measurements aimed at revealing structural damage. This family of testing methods is also commonly referred to as thermographic nondestructive testing (TNDT). There are several factors that determine feasibility and performance of thermographic testing. The basic factors that should be considered in practical applications of TNDT of structures are: (1) emissivity of an object's surface; (2) thermal diffusivity; and (3) thermal effusivity, among others.

The **emissivity** of an object's surface determines the amount of IR radiation that can be detected by an IR camera. In certain cases (e.g. polished steel or aluminium components) the emissivity of a surface may be too low for TNDT applications. Low emissivity results in poor thermal contrast measured by an IR camera and hence in very low damage detection sensitivity. Additionally, emissivity may not be uniform over the measured area. This causes problems in the analysis of measurement data and extraction of damage features, which is based on the detection of anomalies in surface temperature distribution. The remedy in both cases is coating the surface with a thin layer of graphite or washable black paint. It greatly improves the emissivity and makes it uniform over the coated area.

Thermal diffusivity is the ratio of thermal conductivity to mass density and specific heat capacity of a material.

$$\alpha = \frac{k}{\rho c_p} \tag{9.1}$$

where α is thermal diffusivity in $m^2\ s^{-1}$, k is thermal conductivity in $W\ m^{-2}\ °C^{-1}$, ρ is mass density in $kg\ m^{-3}$ and c_p is specific heat capacity in $J\ kg^{-1}\ °C^{-1}$.

Thermal diffusivity is a measure of how rapidly heat is conducted in a material. Materials with high thermal diffusivity (e.g. aluminium) conduct heat quickly and are therefore more demanding for TNDT applications because a high frame rate thermographic camera should be used to follow surface temperature evolution. Materials with low thermal diffusivity, on the other hand (e.g. epoxy), conduct heat slowly which results in longer TNDT inspection times. Most composite materials [e.g. carbon fibre reinforced polymers (CFRPs)] have intermediate thermal diffusivities which makes them very well suited for TNDT inspections.

Thermal effusivity is the square root of the product of thermal conductivity, mass density and specific heat heat capacity of a material.

$$\beta = \sqrt{k\rho c_p} \tag{9.2}$$

where β is thermal effusivity in $J\ m^{-2}\ s^{-\frac{1}{2}}°C^{-1}$.

Thermal effusivity is a measure of thermal inertia of a material. Materials with high thermal effusivity (e.g. aluminium) exhibit low temperature changes in response to thermal excitation, and are therefore more problematic for TNDT inspection. The reason is that low temperature changes are harder to measure, especially in the presence of measurement noise, and a high sensitivity thermographic camera should be used for inspection. Materials with low thermal effusivity (e.g. epoxy) exhibit high temperature changes in response to thermal excitation, which is easier to detect and makes the TNDT process easier and more effective.

Thermographic cameras (also referred to as thermal imagers or IR cameras) are used to measure the amount of IR radiation emitted by an object. The amount of IR energy is then converted to temperature values by means of calibration curves formulated on the basis of the physical principles discussed above. The measurements are used to create a thermal image, which can be easily interpreted by a human, by assigning colours corresponding to certain temperatures. The construction of a thermographic camera is in principle similar to a digital video camera. The main components are: (1) an IR detector; (2) a lens that focuses IR radiation onto a detector; and (3) the necessary electronics and software for processing the signals (FLIR Systems 2012). An IR detector is typically a focal plane array (FPA) of micrometre size pixels made of various materials sensitive to IR wavelengths. The FPA detectors fall into two main categories: thermal detectors and quantum detectors (Rogalski 2011). Thermal detectors are typically semiconductor microbolometers that do not require cooling, but they have slower response and lower sensitivity than photon detectors. Quantum detectors, on the other hand, offer very good signal-to-noise performance and a very fast response, but require cryogenic cooling for proper operation. The cooling is typically realized by Stirling cycle refrigerator units or by liquid nitrogen cooling systems. This makes the quantum detector technology much more expensive than thermal detector technology. Quantum detector based thermographic cameras are more bulky, heavier and therefore less convenient than thermal detector based cameras. Quantum detectors are typically made from materials such as: Si for the NIR range; PbS, InGaAs for the SWIR range; InSb, PbSe, PtSi for the MWIR range; and HgCdTe (MCT) for the LWIR range (FLIR Systems 2012; Rogalski 2011). Optical components dedicated to IR applications also need special attention. Ordinary glass does not transmit radiation beyond 2.5 μm in the IR region, therefore IR camera lenses are typically made of silicon (Si) and germanium (Ge) materials. Si lenses are used for the MWIR range and Ge lenses are used in the LWIR range. Additionally, both materials have favourable mechanical properties, which makes them suitable for technical applications (FLIR Systems 2012).

In recent years a number of different damage detection methods have been developed and applied in various technical applications (Balageas *et al.* 2006; Inman *et al.* 2005; Staszewski *et al.* 2004). The success of these methods often depends on three major factors, namely: (1) on the simplicity of interpretation of the results that they provide; (2) on the necessity to use the baseline reference data (measured in undamaged state); and (3) on the cost and complexity of their implementation. TNDT is a group of methods particularly advantageous in all three aspects. The thermographic methods can be divided into two major groups, i.e. passive and active approaches, as shown in Figure 9.2.

Passive methods rely on temperature measurements without introducing any external excitation. These methods are typically qualitative and provide global information on potential anomalies. They are mainly used in applications such as leakage detection, examination of heat losses in buildings, monitoring of power stations or surveillance, among other possible applications. Apart from engineering applications, passive thermography is widely used in medicine, e.g. for skin temperature mapping. Passive methods have found only a limited application in NDT.

Active methods require the delivery of energy from an external source to a measured object. The energy can be delivered to an object in two ways: by external or internal excitation. Measurement configurations for both excitation scenarios are depicted in Figure 9.3. Active thermographic methods with external excitation [Figure 9.3(a)] can be applied in the reflection mode [for angle $\alpha \in (0°, 180°)$] and in the transmission mode [for angle $\alpha \in (180°, 360°)$].

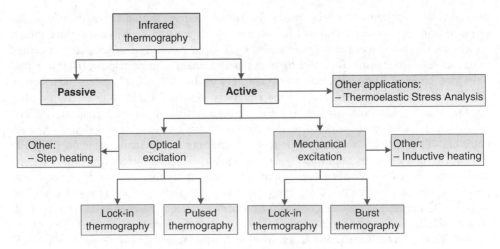

Figure 9.2   Classification of thermographic techniques

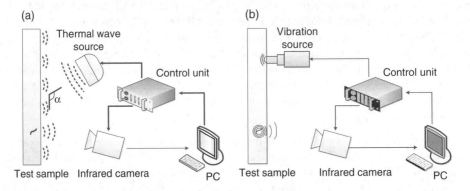

Figure 9.3   Schematics of two basic arrangements of active thermography: (a) variant with external excitation; and (b) variant with internal excitation

The energy can be delivered to an object by means of halogen, IR or flash lamps that heat up its surface. Three different stimulation configurations are commonly used:

- Pulsed thermography is the most commonly applied stimulation type. Thermal stimulation is delivered to a measured object as an impulse, as shown in Figure 9.4(a). The pulse duration may vary from a few milliseconds to a few seconds, depending on the thermal properties of an inspected structure. The method is simple to apply and offers short inspection time. Diagnostic information is extracted in the cooling phase that follows the rapid heating phase caused by the applied thermal pulse. The presence of structural defects is inferred from the anomalies in surface temperature distribution that are due to differences in thermal diffusion caused by defects.
- Lock-in thermography is based on delivering thermal energy to the structure in a periodic manner, as shown in Figure 9.4(b). The data acquired by a thermographic camera

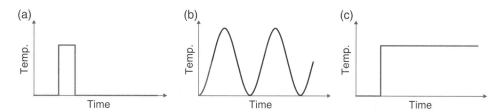

Figure 9.4 Stimulation types in active thermography with external excitation: (a) pulse heating; (b) periodic (lock-in) heating; and (c) step heating

are analysed at the excitation frequency to extract both phase and amplitude information. By changing the modulation frequency it is possible to investigate the material at different depths (Breitenstein *et al.* 2010; Maldague 2001; Wu and Busse 1998). The main advantage of lock-in thermography is the fact that it is able to measure very weak temperature signals, due to its averaging nature. This, however, comes at a cost of longer measurement time.

- Step thermography is an approach in which the stimulation is applied as a step function [Figure 9.4(c)], typically with the use of an IR lamp. The method can be applied for coating thickness evaluation or inspection of multilayered structures (Maldague 2001).

Active thermographic methods with internal excitation utilize vibration or inductive heating for internal heat generation. Vibrothermography is of special interest in this group of methods due to its efficiency. The method was originally proposed by Hennecke *et al.* in the 1980s (Henneke *et al.* 1986; Reifsnider *et al.* 1980), but was popularized by Favro *et al.* (2001) almost 20 years later. In scientific literature, vibrothermography is also known as ultrasonic infrared thermography, acoustic thermography, thermosonics, sonic IR, elastic-wave-activated thermography, thermal vibration method or vibroIR. The gap between the original work of Hennecke *et al.* and the widespread interest in the method, after the publication of Favro *et al.*, was due to several factors, one of the most important being the availability of more efficient and affordable IR cameras that are used in measurements. Vibrothermography is a special deployment of active thermography that uses mechanical vibration excitation. A simplified principle of operation of the method can be described as follows: periodic stress waves introduced to a test structure cause energy dissipation at discontinuities (e.g. delaminations, fatigue cracks) by conversion of mechanical energy into thermal energy generating heat. Thus, the heat source in vibrothermography, unlike in other thermographic techniques, is the discontinuity itself, which makes the identification of defects simpler. Heat generated at discontinuities propagates to the surface where the temperature change is measured by a sensitive IR camera. Typical IR cameras that are used for this type of tests measure electromagnetic radiation in the MWIR range. The principle of operation of vibrothermography is illustrated in Figure 9.3(b).

Two different configurations of mechanical stimulation are commonly used in vibrothermography:

- Burst vibrothermography in analogy to pulsed thermography is the most commonly applied stimulation type. Mechanical stimulation is delivered to the measured object as a short vibration burst in the ultrasonic frequency range as shown in Figure 9.5(a). The burst

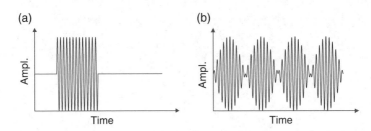

Figure 9.5 Stimulation types in active thermography with internal excitation: (a) burst signal; and (b) modulated signal

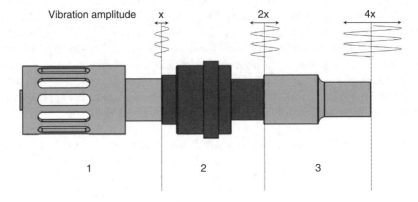

Figure 9.6 Typical ultrasonic excitation source: (1) converter; (2) booster; and (3) sonotrode

duration is typically a fraction of a second. This setup offers the same advantages as the pulsed thermography described above.

• Lock-in vibrothermography is based on delivering vibrational energy to the structure in a periodic manner, as shown in Figure 9.5(b). Again, the advantages of this measurement configuration are analogous to the lock-in thermography described above.

Excitation in vibrothermography is typically applied by an ultrasonic device that comprises three elements: (1) converter – a bolt clamped Langevin type transducer consisting of a piezoceramic stack bolted between two metal pieces (this configuration allows high power and narrowband frequency operation); (2) booster – a metal piece (typically aluminium or titanium alloy) that is used to clamp the entire ultrasonic assembly and to amplify the vibration amplitude; and (3) sonotrode – an element that comes into contact with the inspected structure and further amplifies the vibration amplitude. A typical excitation device is presented in Figure 9.6. For proper operation, it is important that specific natural frequencies of these elements are perfectly matched. This configuration is typical for ultrasonic welding devices. Mechanical stimulation can be also applied by means of an electromagnetic shaker or a magnetostrictive transducer, but these options are less used.

The original research on vibrothermography done by Hennecke *et al.* considered excitation of a tested object at its resonant frequencies. This is, however, impossible with the use of an

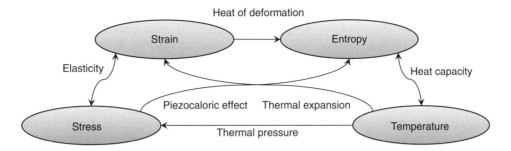

Figure 9.7 Physical phenomena relating different domains. After Tinder (2008)

ultrasonic device such as the one depicted in Figure 9.6, which is dedicated, by its design, to narrowband frequency operation. Han *et al.* (Han 2004; Han *et al.* 2002) proposed the use of a nonlinear coupling, between the sonotrode tip and the tested object, that causes a hammering effect and enriches the frequency spectrum of the excitation. The problem with this solution is, however, the lack of repeatability of excitation, which in this case is nonlinear and potentially chaotic. Holland (2007) showed the feasibility of using a piezoelectric stack driven by a chirp signal to obtain a broadband excitation spectrum for vibrothermographic measurements. The main issue of concern in this case is, however, the amount of energy that can be delivered to the measured object. The ultrasonic welder design is meant for high amplitude operation, which is hard to achieve with a piezoelectric stack without the additional wave guide. There is still room for improvement in the design of vibration excitation sources for vibrothermography.

Despite many research efforts over the last few years, the mechanisms of vibration energy dissipation on damage are not yet fully understood. The interaction between the strain, stress, entropy and temperature fields results in a complex thermo mechanical response of the material that lies behind the measured temperature field. The relationships between the physical domains involved are depicted in Figure 9.7. The most probable mechanisms of heat generation during a vibrothermographic test include the following:

- **Frictional heating** seems to be the dominant mechanism of heat generation in vibrother-mography (Homma *et al.* 2006; Mabrouki *et al.* 2009; Mian *et al.* 2004; Renshaw 2009; Renshaw *et al.* 2008, 2011; Rothenfusser and Homma 2005). Mechanical vibration deliv-ered to a sample causes relative movement of the asperities on the opposing crack (or delamination) faces. The contact and friction processes, present between these asperities, are invariably accompanied by heat generation, which can be used for diagnostic purposes. The power dissipated during the process, i.e. the frictional heat generation, is proportional to the tangential shear stress and to the slip rate.

$$P_{fr} = \tau \cdot \dot{\gamma} \qquad (9.3)$$

where τ is tangential shear stress between the rubbing surfaces and $\dot{\gamma}$ is slip rate.

Frictional rubbing may also induce plastic deformation of rubbing asperities at the micro level, due to their roughness (Mabrouki *et al.* 2010; Renshaw *et al.* 2011).

However, since the plastic zones are formed on crack faces, rather than on crack tip, the test remains nondestructive (Mabrouki *et al.* 2010). For the sake of completeness it should be noted, however, that some authors claim that damping in the elasto plastic zone, rather than friction, is the main cause of energy dissipation observed on damage (Bovsunovsky 2004).

- **Plastic heat generation** appears if a yield stress of the material is exceeded. The irreversibility of plastic flow causes an increase in the amount of entropy in the body and the inelastic work is dissipated into heat. Vibrothermographic testing is believed to be nondestructive as the vibration-induced stresses are unlikely to reach the yield stress of the material during the application of mechanical stimulation. Crack tips are, however, specific locations where much higher stress concentrations exist. According to the classical linear fracture mechanics theory, the stress field diverges at the crack tip. In reality, the stress field at the crack tip is much more complex and depends on the residual stresses in the material and crack shape on the macro- and microscale (Homma *et al.* 2006; Renshaw 2009; Renshaw *et al.* 2011). To support the claim of a nondestructive nature of vibrothermographic testing Tsoi and Rajic (2004) have shown that, in aluminum samples, there is no evidence of crack growth caused by the vibration applied during vibrothermographic tests.

- **Material damping** causes bulk heating of the structure in the stress concentration locations. The energy is dissipated into heat in the hysteresis loop of the stress–strain relation of the material. The effect is negligible in most metals but can have a significant influence in polymers and polymer based composites.

- **Thermoelastic effect** is the relationship between the temperature and the applied stress in an elastic material. It has been shown that there exists proportionality between the change of stress and temperature in a material exposed to elastic deformation (Harwood and Cummings 1991). The basic theoretical relation employed is:

$$\Delta \varepsilon = \frac{(1 - 2v)\Delta \sigma}{E} + 3\alpha \Delta T \tag{9.4}$$

where $\Delta \varepsilon$ is the change of main strains, $\Delta \sigma$ is the change of the first stress invariant, v is the Poisson's ratio, ΔT is the change of temperature, α is the coefficient of thermal expansion and E is the Young's modulus. Thermoelastic effect can be significant in some circumstances and it is used in engineering applications for Thermoelastic Stress Analysis (Harwood and Cummings 1991). In vibrothermography, however, it is typically averaged out of the measured data. This is due to the use of a high frequency vibration excitation that has a much shorter vibration period than the integration time of a thermographic camera (Renshaw *et al.* 2011).

Vibrothermography is a promising and rapidly evolving TNDT method that has a number of advantages over the traditional TNDT methods. It will be shown in the following sections that the method is a viable complement to the existing NDT methods which offers good performance in different applications. However, certain problems with repeatability and reliability of measurements need to be solved, prior to the industrial application of the method.

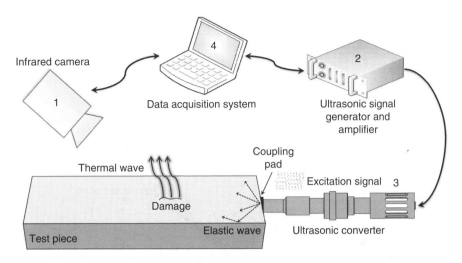

Figure 9.8 Schematics of a vibrothermographic test system

9.3   Developed Vibrothermographic Test System

The research group at the Department of Robotics and Mechatronics at AGH University of Science and Technology in Kraków is conducting research on vibrothermography within the scope of the research project Monitoring of Technical State of Construction and Evaluation of its Lifespan, acronym MONIT (AGH-UST 2012; MONIT Project 2012). This section discusses technical details of the vibrothermographic test system that has been developed within the scope of the project and used in subsequent analyses.

The basic components of the developed vibrothermographic test system are depicted in Figure 9.8. The system is composed of: (1) thermographic camera; (2) ultrasonic signal generator and amplifier; (3) ultrasonic excitation source; and (4) personal computer for measurement control, data acquisition and signal processing.

The measurement system was developed in two variants, as shown in Figure 9.9. Both variants share the four aforementioned basic components but differ in the design of the holder for the ultrasonic excitation device. The first variant, depicted in Figure 9.9(a), is a stationary system designed for laboratory use. The bearing structure is composed of a light aluminium frame, pneumatic press system and fixture for the ultrasonic excitation assembly. The second variant, depicted in Figure 9.9(b), is a mobile system for field measurements. The bearing structure in this case has been reduced to a fixture for the ultrasonic excitation assembly with an ergonomic hand grip for easy operation.

For the sake of completeness, the main technical parameters of the constituent components of the measurement system are briefly summarized below. The parameters have been specified after an extensive literature study and preliminary numerical and experimental investigations with prototype devices.

1. The thermographic camera is a high performance series, cooled detector MWIR camera. The main parameters of the camera are summarized in Table 9.3.

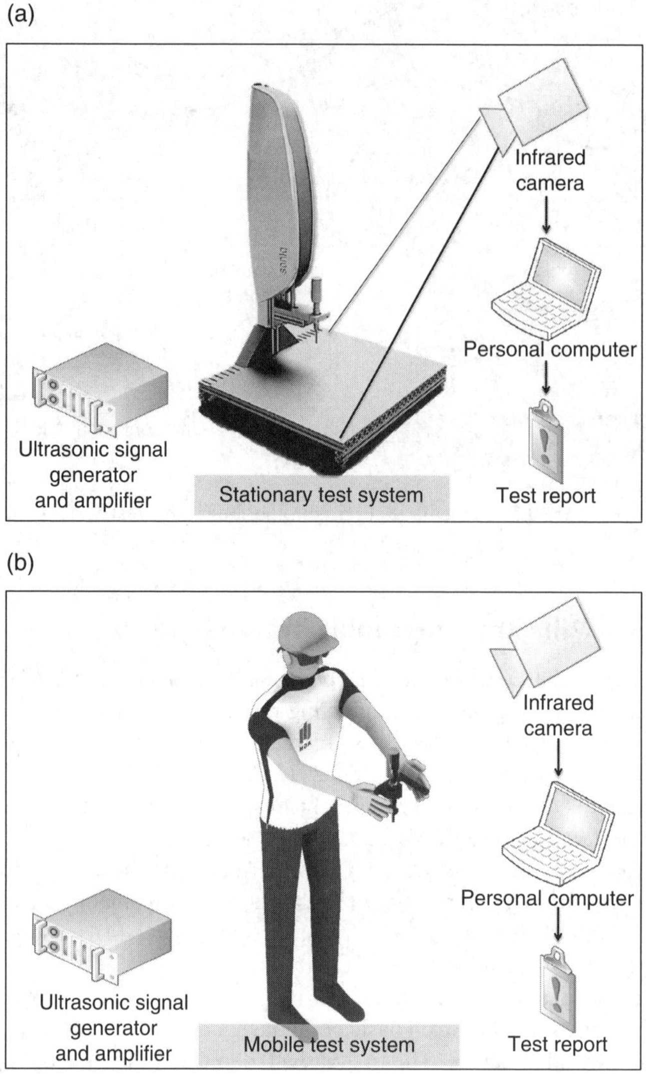

(a)

(b)

Figure 9.9　Developed diagnostic system in two variants: stationary (a); and mobile (b)

Table 9.3　Specification of the infrared camera

Sensor type	InSb
Sensor resolution	320 × 256
Spectral range	2.5–5 μm
NETD	<25 mK
Interface	USB, CameraLink, GigE

Table 9.4 Specification of the ultrasonic amplifier

Frequency range	20–50 kHz
Output power	1.5 kW
Trigger in/out	TTL
Communication interface	USB (Matlab, C/C++ libraries)
Output signal	Continuous, Burst, Modulated

2. The ultrasonic signal generator and amplifier was designed by the research team in close cooperation with a local electronic systems manufacturer, to meet the specific needs of vibrothermographic measurements. High voltage and high power output were required for proper operation of the ultrasonic excitation system. The signal generator had to be equipped with a frequency tuning circuit to match the working resonant frequency of the transducer assembly. Moreover, appropriate trigger outputs had to be designed to allow seamless operation with the IR camera. Trigger outputs were also necessary for lock-in measurement configurations. All parameters of the signal generator and amplifier are controllable from the software layer. The main parameters of the prototype design are summarized in Table 9.4.
3. The ultrasonic excitation source is a typical ultrasonic welding setup discussed in the previous section and depicted in Figure 9.6. Two frequency variants are used in the measurement setup – 20 kHz and 35 kHz. The excitation source can be mounted on a pneumatic press system or in a hand grip, for field measurements.
4. An important part of the measurement system is the software for measurement control, data acquisition and signal processing. The software allows the parameters of both the thermographic camera and the ultrasonic generator to be controlled. This task can be done either manually or following a measurement wizard, customized for typical measurement configurations. A measurement can be set up and triggered from the software. A thermographic image sequence can be acquired and post-processed with the use of typical tools such as point and line plots, image subtraction and differentiation, among others. The identified anomalies in surface temperature distribution can be visualized as a 3D pixel intensity plot or overlaid on the thermal image of the measured object. The software also has a reporting feature, useful in repetitive measurement tasks. An exemplary screenshot of a program window is depicted in Figure 9.10.

9.4 Virtual Testing

This section introduces the aspects of numerical modelling of vibrothermographic tests. The basic information regarding the problem is given together with a computational example of a composite plate with impact damage.

Virtual vibrothermographic testing is important for better understanding of the method and may lead to improvements in its efficiency. The possibility of numerical simulation of a vibrothermographic test would allow a designer to estimate the capability of the method to detect certain defect types and sizes in a designed structure. As a result, safety factors can be chosen and maintenance schedules planned accordingly. Virtual damage detection would

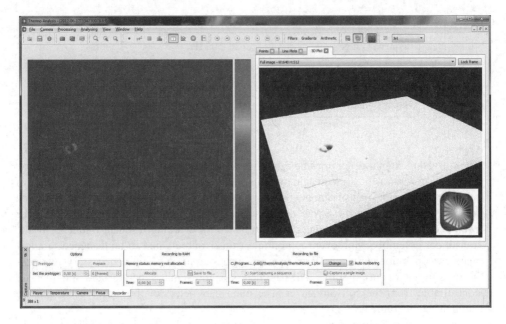

Figure 9.10   The main window of the ThermoAnalysis software (MONIT SHM LLC 2012)

allow a maintenance engineer to customize the test setup and parameters before performing the actual physical test. This can save time and money, especially if tests have to be done in a remote or poorly accessible location. The ultimate benefit of a credible virtual testing procedure would be associated with increased efficiency and reliability of the method as well as with reduced cost of its implementation.

Virtual damage detection is inevitably related to the fields of computational mechanics and experimental testing. The tools of computational mechanics allow to reproduce the physics behind the measurement while experimental testing delivers the necessary parameters for the computational model. The numerical simulation of the problem is complex due to the presence of strong nonlinearities (contact and friction), high frequency excitation (typically in the range from 20 to 40 kHz) and coupling between the mechanical and thermal fields. All these factors result in large computational times and difficulties in formulating the computational model (e.g. thermal and friction parameters, residual stresses at cracks, etc.).

The previous work in the area of numerical modelling of vibrothermography includes studies on the efficiency of monoharmonic and chaotic ultrasonic excitations in generating heat around fatigue cracks (Han *et al.* 2005, 2006), detection of cracks in metallic plates (Han *et al.* 2009; Mabrouki *et al.* 2009, 2010; Plum and Ummenhofer 2010; Saboktakin *et al.* 2010) and detection of fatigue damage in composite materials (Mian *et al.* 2004). Different models assume different heat generation mechanisms and different model parameters to replicate the experimental data. The most common assumption is that the frictional energy dissipation has a dominant effect on the observed temperature field. The computational studies performed to date consider only material samples of a limited size and simplified geometry. One of the problems in performing a full scale virtual vibrothermographic test is the amount of computational

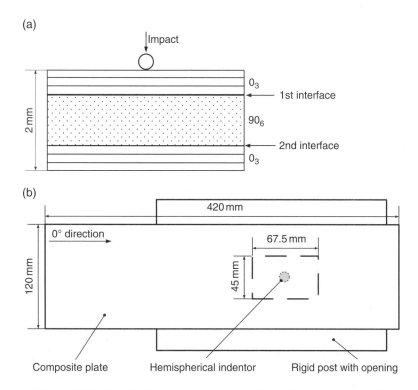

Figure 9.11 Details of the laminate structure (a) and impact test setup (b)

power required to solve a coupled thermo mechanical problem at high frequencies. Another, even more important, factor is the understanding of the physical mechanisms behind the heat generation observed during the vibrothermographic test. A credible computational model can be formulated only if the mechanisms are identified and understood. At this stage of our understanding of the problem, we are able to obtain a fair qualitative agreement with the measured data. There is, however, a lot of ongoing research in this area so, the situation may change in the near future.

As an example of a virtual vibrothermographic test let us discuss the problem of detecting barely visible impact damage (BVID) in a CFRP plate (Pieczonka *et al.* 2012).

The specimen was a rectangular composite plate with $[0_3/90_3]_s$ ply stacking sequence, as shown in Figure 9.11(a). The dimensions of the plate are $120 \times 420 \times 2$ mm as shown in Figure 9.11(b). The plate was a laminate made from Seal Texipreg HS160/REM carbon/epoxy prepreg with 61.5% fibre weight fraction. The specimen was ultrasonically C-scanned prior to testing to assess the quality of the laminate and to exclude the presence of manufacturing defects.

The damage was introduced in the composite plate with the use of a drop-weight impact testing tower. The plate was subjected to two impacts at adjacent locations, close to the main symmetry axis of the plate. The distance between the impact points was approximately 12 mm. During the test, the composite specimen was simply supported on a steel plate with

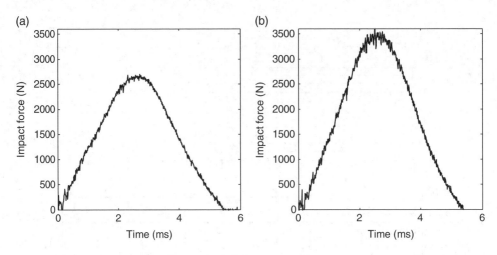

Figure 9.12 Experimental impact forces: (a) 1st impact (3.9 J); and (b) 2nd impact (6 J)

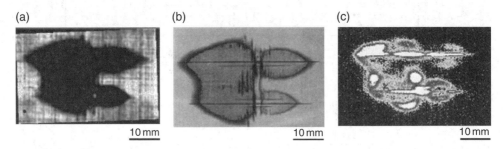

Figure 9.13 Damage identified in the plate by means of: ultrasonic C-scan (a); penetrant-enhanced X-radiography (b); and vibrothermography (c)

a rectangular opening 45 mm × 67.5 mm in size (Figure 9.11). The energy of the first impact was equal to 3.9 J and the energy of the second impact was equal to 6 J, as shown in Figure 9.12.

The damage induced in the plate was characterized with the use of three different methods: vibrothermography, ultrasonic C-scan testing and penetrant-enhanced X-radiography (PEXR). The extent of impact damage revealed by all three methods was in very good agreement, as shown in Figure 9.13. The vibrothermographic test was performed with the use of the stationary version of the developed test system described in the previous section. The ultrasonic excitation device was operating at 35 kHz and at 30% power. The excitation was applied outside the field of view of the camera. No coupling was used between the sonotrode and the test plate, which was motivated by the low operating power of the ultrasonic device and therefore there was no risk of damaging the surface of the plate. The IR camera was recording temperature evolution at the bottom surface of the plate, i.e. the side opposite to the impacted surface. No special treatment of the surface of the analysed plate was necessary and the whole measurement took less than 1 min. The differential IR image showing temperature increase,

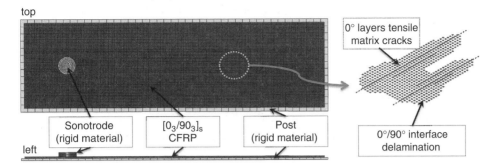

Figure 9.14 Virtual model and damage details

taken 0.5 s after the ultrasonic excitation was introduced to the plate, is presented in Figure 9.13c. Extensive delamination at the interface between the 90° and the 0° plies farthest from the impact side is, as expected, the major damage mode induced by the impact. Other fracture phenomena include tensile matrix cracks in the 0° plies and shear matrix cracks in the central 90° plies.

A coupled thermo mechanical simulation was performed at high frequencies to reproduce the vibrothermographic test. An explicit time integration FEM has been chosen for this investigation. The method is well suited for simulations with large nonlinearities and complex dynamics, as in the analysed case. A commercial LS-Dyna FE solver was used [Livermore Software Technology Corporation (LSTC) 2012a]. Frictional energy dissipation was assumed to be the dominant heat generation mechanism in the model. As mentioned earlier, the rate of frictional energy dissipation is given by the product of the frictional stress and the slip rate. The amount of this energy released as heat on each of the contacting surfaces is assumed to be the same. On account of the short duration of the measurement and a low temperature difference between the sample and the environment, the heat exchange by convection and radiation between the test specimen and the environment has not been modelled.

The FE model of the measurement setup was prepared with the use of the LS-PrePost FE pre-processor [Livermore Software Technology Corporation (LSTC) 2012b]. The FE model consisted of three main components: (1) rigid post; (2) composite plate with $[0_3/90_3]_s$ ply stacking sequence; and (3) rigid cylinder simulating the sonotrode, as shown in Figure 9.14. The entire model comprised approximately 260 000 nodes and 202 000 hexahedral solid elements, of which 900 were rigid elements. There were four elements across the thickness of the plate – one solid element for every three plies – allowing reproduction of the desired ply stacking sequence.

The extent of delamination at the 0°/90° interface and tensile matrix cracks in the 0° layers were modelled as surface to surface thermal contact areas. The shape and size of the delamination was based on the NDT investigations described above. It was assumed, in the numerical model, that the frictional work is entirely converted to heat and enters the contacting faces (master and slave) in equal portions. Friction in LS-Dyna is based on the Coulomb model. In the present analysis the coefficient of static friction had the value of 0.3 and the coefficient of dynamic friction had the value of 0.25 for all contact segments in the model.

Table 9.5 Elastic and thermal properties of the composite material used

Mass density	$\rho = 1.6 \text{ g cm}^{-3}$
Elastic constants	$E_1 = 93.7 \text{ GPa}, E_2 = 7.45 \text{ GPa}$
	$G_{12} = G_{23} = G_{31} = 3.97 \text{ GPa}$
	$v_{12} = 0.261, v_{21} = 0.0208$
Thermal expansion coefficients	$\alpha_1 = -0.3 \times 10^{-6} \text{ K}^{-1}, \alpha_2 = 28.3 \times 10^{-6} \text{ K}^{-1}$
Thermal conductivity	$k_1 = 25 \text{ W m}^{-1} \text{ K}^{-1}, k_2 = 6 \text{ W m}^{-1} \text{ K}^{-1}$
Specific heat	$c_p = 700 \text{ J kg}^{-1} \text{ K}^{-1}$

Friction coefficients were assigned on the basis of the limited literature data (Chand and Fahim 2000; Gay et al. 2002).

Sliding of contact surfaces occurs when the tangential force is greater than the friction force. Frictional behaviour depends on the characteristics of the contact surfaces, such as surface roughness or temperature and loading conditions, such as normal force and relative velocity. Transversely isotropic elastic and thermal properties of the composite plate were applied in the numerical model as reported in Table 9.5. There was no information from the manufacturer about thermal properties of the composite, therefore, thermal parameters of the investigated composite plate were estimated on the basis of a literature review (Gay et al. 2002).

Boundary conditions in the numerical model were specified to represent the experimental setup in the best possible way. A rigid post was fixed in all degrees of freedom. The contact boundary condition was specified between the composite plate and the post. A rigid cylinder, modelling the sonotrode, was fixed in all degrees of freedom except translation in the z direction (i.e. normal to the composite plate). Translation of the cylinder in the z direction was enforced by a sinusoidal displacement boundary condition. The amplitude of vibration was set to 5 μm and frequency was set to 35 kHz. A coupled field thermo mechanical analysis has been performed for a duration of 0.2 s. The time step size for the mechanical part was set to 50 ns and the time step size for the thermal part was 100 μs. The thermal time step could be substantially greater than the mechanical time step due to the much slower dynamics of thermal phenomena with respect to ultrasonic wave propagation in the structure. The Crank–Nicolson time integration method was used in solving the heat transfer equation.

Figure 9.15 presents the comparison between experimental and numerical temperature distributions at three different time steps. Images in the top row in Figure 9.15 present images from a thermographic camera and images in the bottom row represent simulation. Thermographic images from the experiment were obtained by subtracting the first recorded frame from all other frames in the recorded sequence. This allowed the analysis of temperature increase in the plate that was due to the energy dissipated on damage. Uniform temperature distribution on the image, at time equal to 0 s, is the result of the data processing. At time equal to 0.1 s it can be seen that only the matrix cracks are visible in the thermal image for both experimental data and numerical prediction. For time equal to 0.2 s the whole delaminated area is visible in the measured data and in the numerical prediction. The time delay between the presence of crack and the presence of delamination in the thermal image is defined by the time necessary for the thermal wave originating from the sub surface delamination to reach the surface. The simulated results are very well matched with the experimental trend. However, the maximum temperature increase in the damaged area after

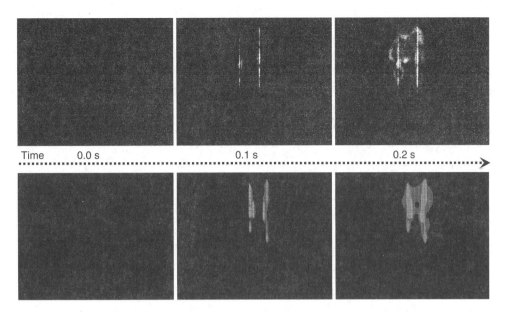

Figure 9.15 Comparison of surface temperature distribution obtained experimentally (top row) and numerically (bottom row)

0.2 s was equal to 0.1 K in experimental measurements, whereas the equivalent temperature increase in numerical simulations was equal to 0.06 K.

The differences between measured response and numerical prediction in the damaged area can be attributed to several factors. The only mechanism of heat generation was frictional heating, while viscoelastic material damping, that is present in CFRP, was not included. Thermal properties of the composite plate were estimated from a literature review and there is a large degree of uncertainty about their corectness. Friction coefficients were not known precisely for the material used in the current investigations. Delamination and cracks were modelled in a simplified manner as FE mesh discontinuities (double nodes) and thermal frictional contact was assumed between the opposite faces. Delamination and crack faces were therefore ideally planar. This assumption does not match reality and certainly influences the frictional behaviour. Despite all these discrepancies, the simulation results are very encouraging.

9.5 Laboratory Testing

This section describes the experimental testing campaign that has been undertaken in order to validate the developed vibrothermographic test system and to perform a parameter influence study to reveal the most important factors affecting vibrothermographic measurements.

Numerous laboratory tests have been performed on metallic and composite test specimens, using both stationary and mobile versions of the developed test system in order to validate system components and explore measurement parameters. Hardware platform and software

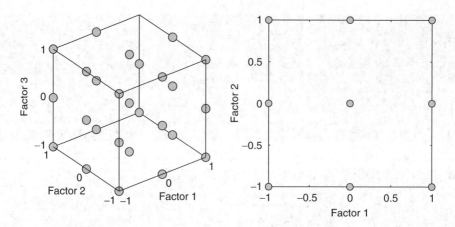

Figure 9.16 Three-level full factorial design for three factors

Table 9.6 Parameter values used in the DOE

Parameter	Level		
	Low	Medium	High
Time (s)	0.1	0.5	1
Force [N]	50	250	500
Power [%]	30	60	100

procedures were verified on laboratory test specimens, with known damage features, before the procedures could be applied for damage detection in real working conditions. The following main parameters of the system were investigated in order to verify the dependence of the amount of heat dissipated on damage in relation to:

- the time of excitation;
- the contact force between the sonotrode and the specimen;
- the power of the ultrasonic converter.

Design of Experiments (DOE) was performed in order to find the influence of the three above-mentioned parameters on the measured thermal signal. The three-level full factorial design configuration was chosen for the analysis. The design is typically used in factor screening analyses. The number of sampling points equals k^n where k is the number of parameters (factors) and n is the number of parameter levels. In the analysed case, the design resulted in $3^3 = 27$ parameter configurations that had to be tested. The design in depicted in Figure 9.16. The values of parameters, that were used in the experiments, are summarized in Table 9.6.

The test specimen was a carbon epoxy prepreg plate, as shown in Figure 9.17(a). The plate was damaged in a low velocity impact event that resulted in matrix and fibre cracking in the vicinity of the impact location. The vibrothermographic tests were performed on the stationary

(a)   (b)

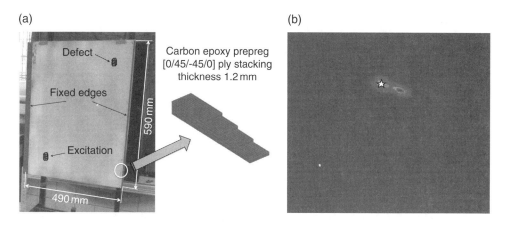

Figure 9.17   Analysed test specimen (a) and exemplary test result (b)

test system, using 35 kHz excitation source. The field of view of the IR camera was set to the upper right corner of the plate where the defect was located. The plate was excited in a lower left corner, i.e. on the diagonal from damage. The excitation location was outside the field of view of the thermographic camera. The measurement configuration is shown in Figure 9.17(a).

A sample thermal image of the damaged area of the plate, acquired during the vibrothermographic test, is shown in Figure 9.17(b). The star-shaped marker indicates the location for which the subsequent comparisons were performed. Image processing techniques implemented in the developed software package were used for subsequent analyses. The implemented thermal images processing can be used to estimate sizes and localizations of defects in the tested component.

The experiments were performed on the test specimen, according to the prepared DOE plan. The performance metric was a maximum temperature gain for a chosen point on the damaged area [marked with a star in Figure 9.17(b)]. The quality of the vibrothermographic test is dependent on the amount of heat generated at structural defects. The more vibration energy is dissipated on defects, the better is the contrast of the thermal image. Thus, evaluation of damage parameters is easier.

All experiments in the DOE plan were performed according to the same measurement plan composed of the four main steps:

1. Stabilization of the temperature of the test piece to a reference ambient temperature (21 °C).
2. Synchronized start of the ultrasonic excitation and data acquisition with thermographic camera.
3. Acquisition of thermal images for 10 s.
4. Post-processing of the acquired image sequence.

The influence analysis was performed on the results obtained from experimentation, in order to identify the most important factors that influence the thermal response. The first-order effects were computed for the three process parameters. The results are shown in Figure 9.18. It can be seen, that all considered parameters have an influence on the measured thermal

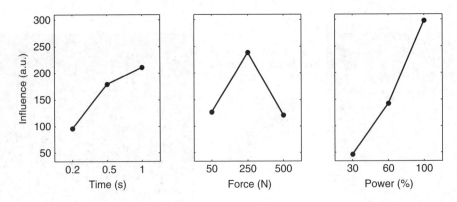

Figure 9.18   Main effects analysis of measurement parameters

signal. The most important factor is the power of the ultrasonic converter. This is intuitive, as the amount of heat dissipated on damage should be proportional to the amount of mechanical energy delivered to the structure. The duration of excitation is influential for the same reason. The longer the ultrasonic excitation, the more energy is delivered to the structure, thus more energy dissipates on damage. The contact pressure between the sonotrode and the test piece is an interesting factor. This parameter is representative for the transfer of ultrasonic energy generated by the converter to the test specimen. As can be seen in Figure 9.18, the measured thermal signal increases when the force is increased from 50 to 250 N and then diminishes when the force is further increased to 500 N. If the contact pressure between the sonotrode and the specimen is too low, the hammering effect may occur. Ultrasonic energy is not transmitted efficiently because the sonotrode tip separates from the surface of the specimen and it may become damaged. If the contact pressure is too high, the surface of the specimen may also become damaged and the vibration of the sonotrode is limited, resulting in poor excitation of the sample. For a very high force level the blocking force of the piezoceramic stack inside the ultrasonic converter may be exceeded, and no vibration occurs. This is important especially for low power operation. There should be, therefore, an optimal force level resulting in the best coupling between the sonotrode and the specimen that results in the best thermal response. In the analysed case, the optimal contact force is around 250 N.

According to the literature survey, proper coupling between the ultrasonic assembly and the test specimen is indeed an important measurement parameter (Lick *et al.* 2007; Shepard *et al.* 2004). Certain configurations of coupling pads and coupling pressure can produce the acoustic chaos effect mentioned before (Han 2004; Han *et al.* 2002).

In conclusion, it can be said that the designed measurement system for vibrothermographic testing was verified as fully functional. Preliminary testing was performed on a composite specimen in order to reveal measurement parameters that have the largest influence on the measured thermal response. Operating power of the ultrasonic converter, duration of mechanical excitation and contact force were all found influential for the measured thermal response. A few aspects of the measurements need further investigations. In particular, there exists a maximal operating power level above which the test sample may be locally damaged. The same configuration of the excitation device, used in vibrothermography, is used for ultrasonic

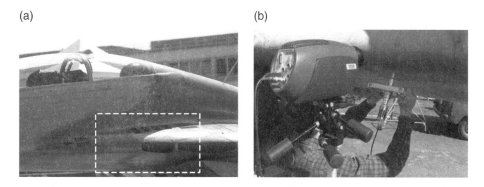

Figure 9.19 Measured section of the MiG-29 fuselage (a) and experimental setup (b)

welding of plastics where the material undergoes plasticization to form a durable joint. This is, clearly, not desired in NDT applications. It is, however, difficult to assess the maximum power level that can be used for a specific test configuration. Another aspect is the identification of an optimal contact force between the sample and the sonotrode, as discussed above. At this stage, in development of the method, these questions remain open.

9.6 Field Measurements

This section describes some results from the field testing campaign performed using the developed vibrothermographic test system, and discusses some practical aspects of vibrothermographic testing.

Field measurements have been performed on the fuselage and wing panels of a Mikoyan MiG-29 jet fighter aircraft, as depicted in Figure 9.19(a). Tests were performed using the mobile version of the developed vibrothermographic test system. An ultrasonic device working at 35 kHz was used to excite the structure and a Cedip Silver 420M IR camera was used to acquire thermal image sequences. The experimental setup is shown in Figure 9.19(b).

One of the main goals of the test was the identification of loose rivets that may exist in the fuselage panels. The measurements were performed directly on the aircraft. Ultrasonic excitation was applied for 500 ms and a sequence of thermal images with a total duration of 5 s was recorded. Figure 9.20(a) presents an exemplary measurement result, obtained from the developed software. The image is an overlay picture of the baseline thermal signature of a panel with identified area of thermal change. The area is located around one of the rivets. Figure 9.20(b) shows temperature evolution in the identified area (solid line) compared with background temperature change (dotted line). The ultrasonic excitation was applied at time equal to 1.2 s. It can be clearly seen that a rapid temperature change takes place in the damaged area, while the temperature in the healthy area remains unaltered.

An illustration of a problem related to the proper choice of the excitation location is presented in Figure 9.21. Figure 9.21(a) shows a location on the aircraft fuselage where three different panels meet. The excitation has been applied in the upper right corner just outside of the field of view of the IR camera. It can be seen, in Figure 9.21(b), that the energy delivered

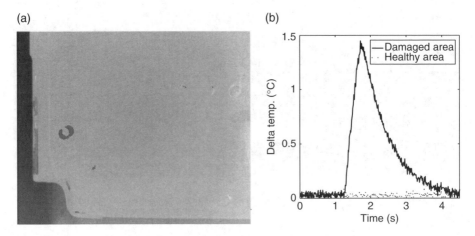

Figure 9.20 Experimental results: (a) identified loose rivet; and (b) temperature evolution on the loose rivet (solid line) and on a healthy part of the panel (dotted line)

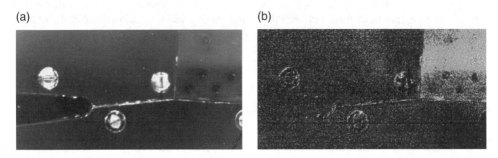

Figure 9.21 Experimental results: (a) baseline image; and (b) temperature change in the area of the panel's connection

to the structure was constrained inside the panel that was excited, and was unable to cross the connection point. In such cases it is necessary to excite more than one location in order to obtain the desired result.

The presented illustrative examples are only an excerpt from an extensive experimental campaign that was performed in order to assess the feasibility of performing vibrothermographic testing in real working conditions. The method was proven useful for a number of applications such as the detection of closed cracks, delaminations or loose rivets (Aymerich *et al.* 2010; Pieczonka *et al.* 2012; Szwedo *et al.* 2012). It was verified to be applicable to composite, metallic and ceramic components. The problems mentioned in conclusion of the previous section are, however, equally important for the field tests. In particular, the amount of ultrasonic energy, that has to be delivered to the tested structure, is a tradeoff between the amount of heat generated on damage and the danger of locally damaging the structure. The location of excitation on a build-up structure has to be carefully chosen in order to deliver ultrasonic energy to all structural components of interest. Alternatively, the excitation has to be applied at several locations and measurement results combined.

9.7 Summary and Conclusions

Vibrothermography is a very promising and rapidly developing TNDT method. It has a number of advantages over the traditional TNDT approaches: it is a dark field method (i.e. the source of heat is the defect itself); it offers short measurement time and requires only minimal effort for measurement preparation; and it can detect closed cracks and delaminations not detectable by thermography with optical excitation. The main issues of concern in industrial application of vibrothermography are still repeatability and reliability of measurements. Repeatability of measurements is mostly influenced by repeatability of vibrational stimulation. While it is possible to control excitation parameters in a laboratory environment, it is not so easy to do it during field tests. Reliability of measurements is also related to the excitation. The frequency and the location of an excitation source influence the amount of heat dissipated at defects. The exact amount of heat that is expected to dissipate on certain defects is not known *a priori* due to the lack of understanding of the physical phenomena behind heat generation in different cases.

Vibrothermography has a great potential in many industrial and research applications, which has been experimentally verified by the authors. The configuration of the measurement system is flexible and can be fairly easily adapted to a specific application. The main fields of application of the method, as identified from the performed experiments and from a literature survey, are the detection of:

- cracks in metallic and composite materials;
- delaminations in composite materials;
- defects in welded joints;
- loose rivets and bolted connections.

The main practical advantages of vibrothermography are:

- nondestructive and noncontact testing procedure;
- short measurement time (typically only a few seconds are enough to obtain satisfactory results);
- detection, localization and size of the defect can be evaluated from thermal image processing;
- ease of interpretation of the results.

The authors see vibrothermography as a valuable complement, rather than a replacement, of the existing NDT methods, which can help to improve the safety of engineering structures.

References

AGH-UST 2012 http://krim.agh.edu.pl/ (last accessed 5 October 2012).

Aymerich F, Pieczonka L, Jenal RB, Staszewski WJ, Uhl T and Szwedo M 2010 Comparative study of image-based impact damage detection in composite materials. *Proceedings of the 5th European Workshop on Structural Health Monitoring* (eds. Casciati F and Giordano M). Sorrento, Italy. DEStech Publications, Inc., pp. 547–553.

Balageas D, Fritzen CP and Güemes A 2006 *Structural Health Monitoring*. Wiley-ISTE.

Bovsunovsky A 2004 The mechanisms of energy dissipation in the non-propagating fatigue cracks in metallic materials. *Engineering Fracture Mechanics* **71**(16-17), 2271–2281.

Breitenstein O, Warta W and Langenkamp M 2010 *Lock-in Thermography*, vol. 10 of *Springer Series in Advanced Microelectronics*. Springer.

Chand N and Fahim M 2000 *An Introduction to Tribology of FRP Materials*. Allied Publishers Pvt. Ltd.

Favro LD, Han X, Ouyang Z, Sun G and Thomas RL 2001 Sonic IR imaging of cracks and delaminations. *Analytical Sciences* **17**, 451–453.

FLIR Systems 2012 *The Ultimate Infrared Handbook for R&D Professionals*. FLIR AB.

Fluke Corporation 2009 Emissivity values of common materials. Technical report 905, Fluke Corporation.

Gay D, Hoa SV and Tsai SW 2002 *Composite Materials: Design and Applications*. CRC Press.

Han X 2004 Acoustic chaos for enhanced detectability of cracks by sonic infrared imaging. *Journal of Applied Physics* **95**(7), 3792.

Han X, Ajanahalli A, Newaz G and Favro LD 2009 Modeling to predict sonic IR detectability of defects in metals. *Review of Quantitative Nondestructive Evaluation* **28**, 467–472.

Han X, Islam S, Newaz G, Favro LD and Thomas RL 2005 Finite-element modeling of acoustic chaos to sonic infrared imaging. *Journal of Applied Physics* **98**(1), 014907.

Han X, Islam S, Newaz G, Favro LD and Thomas RL 2006 Finite element modeling of the heating of cracks during sonic infrared imaging. *Journal of Applied Physics* **99**(7), 074905.

Han X, Li W, Zeng Z, Favro LD and Thomas RL 2002 Acoustic chaos and sonic infrared imaging. *Applied Physics Letters* **81**(17), 3188.

Harwood N and Cummings W 1991 *Thermoelastic Stress Analysis*. Taylor & Francis.

Henneke EG, Reifsnider KL and Stinchcomb WW 1986 Vibrothermography: Investigation and development of a new nondestructive evaluation technique. Technical report, U. S. Army Research Office, ARO 18787.5-MS.

Holland SD 2007 First measurements from a new broadband vibrothermography measurement system. *Review of Progress in Quantitative Nondestructive Evaluation* **26**, 478–483.

Homma C, Rothenfusser MJ, Baumann J and Shannon R 2006 Study of the heat generation mechanism in acoustic thermography. *Review of Quantitative Nondestructive Evaluation* **25**, 566–574.

Inman DJ, Farrar CR, Lopes Junior V and Steffen Junior V 2005 *Damage Prognosis: For Aerospace, Civil and Mechanical Systems*, 1st edn. John Wiley & Sons, Ltd.

Lick KE, Wong CH and Chen JC 2007 Determination of the minimum energy required for sonic-IR detection. *Review of Quantitative Nondestructive Evaluation* **26**, 515–523.

Livermore Software Technology Corporation (LSTC) 2012a http://www.lstc.com/lspp/ (last accessed 5 October 2012).

Livermore Software Technology Corporation (LSTC) 2012b http://www.lstc.com/products/ls-dyna (last accessed 5 October 2012).

Mabrouki F, Thomas M, Genest M and Fahr a 2009 Frictional heating model for efficient use of vibrothermography. *NDT & E International* **42**(5), 345–352.

Mabrouki F, Thomas M, Genest M and Fahr A 2010 Numerical modeling of vibrothermography based on plastic deformation. *NDT & E International* **43**(6), 476–483.

Maldague XPV 2001 *Theory and Practice of Infrared Technology for Nondestructive Testing*. John Wiley & Sons, Ltd.

Mian A, Newaz G, Han X, Mahmood T and Saha C 2004 Response of sub-surface fatigue damage under sonic load a computational study. *Composites Science and Technology* **64**(9), 1115–1122.

Minkina W and Dudzik S 2009 *Infrared Thermography Errors and Uncertainties*. John Wiley & Sons, Ltd.

MONIT Project 2012 http://www.monit.pw.edu.pl/index.php/eng/ (last accessed 5 October 2012).

MONIT SHM LLC. 2012 http://www.monitshm.pl/ (last accessed 5 October 2012).

Pieczonka L, Aymerich F, Brozek G, Szwedo M, Staszewski WJ and Uhl T 2012 Modelling and numerical simulations of vibrothermography for impact damage detection in composites structures. *Structural Control and Health Monitoring* DOI: 10.1002/stc.1483.

Plum R and Ummenhofer T 2010 Structural-thermal FE simulation of vibration and heat generation of cracked steel plates due to ultrasound excitation used for vibrothermography. *10th International Conference on Quantitative Infrared Thermography*. Québec, Canada.

Reifsnider KL, Henneke EG and Stinchcomb WW 1980 The mechanics of vibrothermography. In *Mechanics of Nondestructive Testing* (ed. Stinchcomb W). Plenum Press, pp. 249–276.

Renshaw J 2009 *The mechanics of defect detection in vibrothermography*. PhD thesis, Iowa State University.

Renshaw J, Chen JC, Holland SD and Thompson RB 2011 The sources of heat generation in vibrothermography. *NDT & E International* **44**(8), 736–739.

Renshaw J, Holland SD and Thompson RB 2008 Measurement of crack opening stresses and crack closure stress profiles from heat generation in vibrating cracks. *Applied Physics Letters* **93**(8), 081914.

Rogalski A 2011 *Infrared Detectors*, 2nd edn. CRC Press.

Rothenfusser MJ and Homma C 2005 acoustic thermography: vibrational modes of cracks and the mechanism of heat generation. *Review of Quantitative Nondestructive Evaluation* **24**, 624–632.

Saboktakin A, Ibarra-Castanedo C, Bendada A and Maldague XPV 2010 Finite element analysis of heat generation in ultrasonic thermography. *10th International Conference on Quantitative Infrared Thermography*. Québec, Canada.

Shepard SM, Ahmed T and Lhota JR 2004 Experimental considerations in vibrothermography. *Proceedings of SPIE* **5405**, 332–335.

Staszewski WJ, Boller C and Tomlinson GR 2004 *Health Monitoring of Aerospace Structures: Smart Sensor Technologies and Signal Processing*. John Wiley & Sons, Ltd.

Szwedo M, Pieczonka L and Uhl T 2012 Application of vibrothermography in nondestructive testing of structures. *Proceedings of the 6th European Workshop on Structural Health Monitoring*. Deutsche Gesellschaft für Zerstörungsfreie Prüfung e.V.

Tinder RF 2008 *Tensor Properties of Solids*. Morgan & Claypool.

Tsoi KA and Rajic N 2004 Effect of sonic thermographic inspection on fatigue crack growth in an Al alloy. Technical report, Australian Government Department of Defence, Defence Science and Technology Organisation.

Wu D and Busse G 1998 Lock-in thermography for nondestructive evaluation of materials. *Revue Générale de Thermique* **37**(8), 693–703.

10

Vision-Based Monitoring System

Piotr Kohut and Krzysztof Holak
*Department of Mechatronics and Robotics, Faculty of Mechanical Engineering
and Robotics, AGH University of Science and Technology, Poland*

10.1 Introduction

SHM (Uhl *et al.* 2011) involves integration of sensors, data transmission, processing and analysis in order to detect, localize and assess damages within a structure, which can lead to its failure at the present time or in the future. SHM methods can be divided into two groups: local and global methods. The latter are applied if a global change in the geometry of a structure can be observed. Damages in SHM can be defined as changes in the material properties or the geometry of a structure which can affect its overall performance. A damage developed in the structure decreases its rigidity and alters the dynamic and static properties to a certain extent.

The most frequently used methods of damage detection are based on variation of dynamic properties caused by a damage (Kohut and Kurowski 2005, 2006). The vibration based damage identification methods can be classified as model based and non model based. The model based methods identify damage by comparing a theoretical model, which is usually based on the FEs, with test data obtained in measurements of the damaged structure (Perera and Ruiz 2008; Jaishi and Rex 2006). Correlating the updated model with the original one provides an indication of damage and information on the damage location and its severity. Non model based damage detection methods apply signal processing algorithms for the analysis of structures' response signals in time and frequency, as well as time frequency, domains. Changes in the natural frequencies (Cawley and Adams 1979), changes in the modal assurance criteria (MAC) across substructures (West 1984), changes in the coordinate modal assurance criterion (COMAC) (Lieven and Ewins 1988), changes in modal strain energy (Guan and Karrbhari 2008; Salehi *et al.* 2009) and changes in mode shape curvature (Pandy *et al.* 1991) are formulated as indicators to localize damage. However, obtaining the dynamic data

Advanced Structural Damage Detection: From Theory to Engineering Applications, First Edition.
Edited by Tadeusz Stepinski, Tadeusz Uhl and Wieslaw Staszewski.
© 2013 John Wiley & Sons, Ltd. Published 2013 by John Wiley & Sons, Ltd.

requires devices, such as a laser vibrometer, which are expensive to set up and maintain as well as difficult to be automated. Moreover, excitation of large structures can be prohibitively costly and difficult. The collection of static profiles requires much less effort, which makes the damage detection methods based on changes in the deflection shape more attractive.

In most of these methods, the densely sampled deflection curve obtained from many measurement points is necessary for the correct damage detection and localization. As a structure becomes more complex, more sensors are needed and the cost increase may be significant. Sometimes it is extremely challenging to attach sensors to a structure because of the environmental conditions or the geometrical constraint. Non contact experimental techniques have been developed as alternatives. For example, Jang et al. (2007) presented an example of the use of computer vision technology to capture the static deformation profile of a structure, and then employ profile analysis methods to detect the locations of damages. The vision systems can be a good alternative to contact type transducers. They can be characterized as easy to use, accurate, low cost and universal tools which can be used for vibration and deformation measurements.

Nowadays, there is increased availability of vision systems using digital image correlation for the measurement of displacements, strains and stresses of objects. The most important companies offering vision measurement systems existing at present on the market are: Correlated Solution, GOM, LIMESS Messtechnik Software, Dantec Dynamics and Metris (Krypton 9000) (Schmidt et al. 2003; Tyson 2000).

In this chapter the developed vision based method dedicated for in-plane measurement of a civil engineering structure's displacement fields is presented. The deflection curve is obtained from two images of the construction: the reference one and the one acquired after application of the load.

The principle of the method is calculation of the object points' displacement by means of a normalized cross correlation coefficient. The analysis of the deflection curve course of the beam under the load makes it possible to apply various damage detection, localization and assessment methods based on the strain energy or the deflection's shape curvature. Image registration techniques were introduced in order to increase the flexibility and accuracy of the method. Perspective distortions of the construction's image are removed by means of homography mapping, which allows two photographs of the object to be taken from two distinct points in space. In order to calculate the correspondences between matching features on both images, new techniques of marker detection and shape filtering, as well as subpixel corner detection are introduced.

The developed and presented vision system is characterized by many novel features with respect to the existing systems. First, it is dedicated to large-scale civil engineering structures. It has two basic modes of operation. It can be used by a skilled user as a manual measurement device, like the other commercially available systems. On the other hand, it can be a part of a fully automatic measurement system used for long-term monitoring of the structure. The method makes it possible to acquire the images of the construction from two distinct points in space, which makes the use of costly positioning devices unnecessary. Moreover, if the analysed structure has a highly random texture, the method can be applied even without a special set of markers. The system can operate with non expensive digital SLR cameras, which reduces even further the cost of the system's installation on the object.

10.2 State of the Art

Optical measurement techniques making use of digital image correlation for calculation of displacements, deflections and strain were introduced to experimental mechanics in the late 1980s. The basic principle of the method is calculation of the similarity degree between two patterns in images: before and after deformation. The displacement field of the analysed object is obtained as a result. In the paper by Chu *et al.* (1985), the authors described a specimen's deformation measurement method by means of correlation coefficient calculation. The first laboratory tests' results were presented there, in which rigid displacements and deformations of the specimen were examined. The authors have also developed a method of 3D motion and deformation measurement from the images obtained by a stereovision system of two cameras.

In the paper by Tiwari *et al.* (2007), the application of an ultra fast camera system was presented. Noncontact measurement techniques using digital image correlation have found broad application in many areas of experimental mechanics, material science, quality control of manufacturing process and diagnostics (Budzik *et al.* 2010; Wieczorkowski *et al.* 2010) or even in fine art (Gancarczyk and Gancarczyk 2010). Nowadays, the method is used for static and dynamic measurements. In recent years many applications have been developed, for example: IC packages' reliability analysis by measurement of mechanical and thermal induced strain and deformation (Teo and Xue 2007), examination of anisotropy of metal sheets' mechanical properties (Rossi *et al.* 2008), structural analysis of welded joints (Sierra *et al.* 2008), measurement of low level strain fields with high resolution necessary in analysis of mechanical properties of refractory castables (Robert *et al.* 2007), and examination of mechanical properties of biological tissues (Cheng *et al.* 2007; Schmidt *et al.* 2002). The scientists developed methods of experimental analysis of nano- and micro-scale deformation processes induced by mechanical (Berfield *et al.* 2007) and thermal factors (Li *et al.* 2008).

Correlation methods are used for testing specimens during the fracture process (Helm 2008; Rethore *et al.* 2005; Roux and Hild 2006). The development of fast image sensors enabled the correlation coefficient method to be used for dynamical phenomena. In the existing applications there are systems that analyse the processes induced during shock tests of the objects and measure the surface vibrations using digital speckle correlation (DSP) (Fujun and Xiaoyuan 2006) in which an image of the pattern is projected on the examined surface. The other application areas involve: a real-time print-defect detection system for web offset printing (Shankar *et al.* 2009), and image analysis based displacement-measurement system for geotechnical centrifuge modelling tests (Zhang *et al.* 2009) and roughness measurement by electronic speckle correlation and mechanical profilometry (Spagnolo *et al.* 1997).

Homography transformation (Hartley and Zisserman 2004; Kosetska *et al.* 2004) has broad applications in computer image processing and analysis methods and other areas of science and technology. One of the first applications of homography was developed for stereo-vision systems. The transformation (Loop and Zhengyou 1999) was used for rectification of the images registered by the cameras in order to map epipolar points of each of the images to the ideal points at infinity. The mapping reduced complexity of the correspondence search problem. A number of applications have been developed, examples of which are: correction of projective distortions of text documents and graphical symbols in order to prepare them for an analysis by OCR systems (Zhu *et al.* 2008) or control of the shape of image projected by the projector depending on the orientation of the plane on which the image is projected in systems of augmented reality (Borkowski *et al.* 2003).

The next broad application of homography is mosaicing (Behrens *et al.* 2011; Brown and Lowe 2003; Scoleri *et al.* 2005), which is a method of connecting many registered photographs into one panoramic image. The mosaicing technique is used in processing of medical images (Behrens *et al.* 2011), hand writing acquisition systems (Zhang and He 2007) and analysis of video sequence in order to divide it into the background and foreground of the scene (Mei *et al.* 2005). Homography mapping is used in the process of making models of objects, vehicles, buildings and other architectonic scenes from their images (Infantino and Chella 2001; Liebowitz and Zisserman 1998). There exist methods of multi-camera tracking systems (Benhimane and Malis 2007) based on homography. The methods are implemented in: buildings surveillance systems (Eshel and Moses 2008), tracking of players' movement during sport games (Iwase and Saito 2002) and methods of objects tracking (e.g. cars, people) independent of scale and camera orientation (Park and Trivedi 2007). Fang *et al.* (2002) developed a method of control and navigation of a mobile robot based on homography mapping. Transformation is also used in problems of 3D motion and structure reconstruction (Kosetska *et al.* 2004).

10.3 Deflection Measurement by Means of Digital Image Correlation

The developed vision based method of the in-plane constructions' deflection measurement consists of the following steps (Kohut *et al.* 2008, 2011; Uhl *et al.* 2009, 2011): system calibration, rectification of the chosen plane and deflection curve measurement.

In the first step, the scale coefficient is computed. It is used for rescaling the calculated displacements to metric units. The calibration is performed by means of specially manufactured calibration patterns or certified length standards. The full calibration of the intrinsic parameters of the camera is carried out as well in order to remove the optical distortions introduced to the images by lens imperfections.

In the next step, two registered images of the construction are analysed: the first one is the reference image and the second one captured after the application of a load. It is assumed that the consecutive photographs of the object can be acquired from distinct points in space. The reference image is captured from the point at which the optical axis of the camera is perpendicular to the construction's plane, but the position and orientation of the camera used for image acquisition can be arbitrary. The captured images have projective distortions and therefore have to be transformed before the deflection calculation. The rectification of the construction's plane is carried out. A set of rectangular markers is placed on the surface of a structure in such a way that it is coplanar with the construction's plane and does not move as the structure deforms under the applied loads. The markers are detected and matched automatically by means of the developed image analysis algorithms. The homography matrix **H** is computed from the set of corresponding markers and the image is transformed by mapping. It is overlaid with the reference image and the deflection curve course can be calculated by means of digital image correlation. The construction's image is divided into parts, namely into intensity patterns. The position of each pattern on both images is obtained by normalized cross-correlation (NCC) coefficient analysis and the point's displacement is computed as a difference between the pattern's position on the reference and analysed images of the

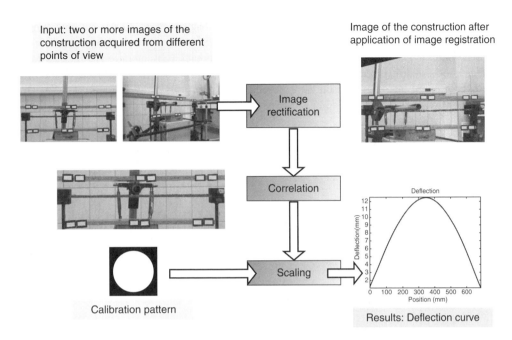

Input: two or more images of the construction acquired from different points of view

Image of the construction after application of image registration

Image rectification

Correlation

Calibration pattern

Scaling

Results: Deflection curve

Figure 10.1   The main steps of the developed measurement method

structure. The algorithm carried out for all points results in a full field deflection curve. Additionally, the system checks if the maximum deflection exceeds the threshold value and sends information to the user.

After the measurement, the deflection curves can be analysed for detection and localization of the damage. The wavelet based damage detection and localization method has been implemented in the system. The overview of the developed measurement technique is shown in Figure 10.1.

In general, the 2D digital image correlation method consists of the following three steps: object preparation; acquisition of the object's image before and after loading; and processing of the acquired images to obtain the displacement field. The construction's surface should have a random grey intensity distribution which deforms together with the surface. The speckle pattern can be a natural texture or an artificially made special random intensity pattern attached to the structure or painted on it. The object's surface should remain in the same plane during loading and out-of-plane motion of the object during loading should be negligible. The basic principle of digital image correlation is tracking or matching the same image patch between the two images acquired before and after deformation (Uhl *et al.* 2009, 2011). In order to compute the displacement of a construction's point, a square or rectangular reference subset of pixels centred at a point is chosen from the reference image and used to match its corresponding location in the deformed image. Assuming that the deformations of the object are small, only a local search is performed in the search window centred on the position of a centre of a pixel subset on the reference image. To evaluate the similarity degree between the reference image patch and the search region on the deformed image, a cross-correlation (CC) criterion or sum-squared difference (SSD) criterion is applied (Kosetska *et al.* 2004). The matching is carried out by searching for the peak position of the correlation coefficient function over the search window. The difference in the position of the reference subset centre and the deformed image subset centre gives the displacement of a given point. In the developed method the NCC coefficient has been chosen as a correlation criterion given by Equation (10.1). This coefficient is more robust and is not sensitive to changes in the scene illumination.

$$NCC(u, v) = \frac{\sum_{xy}(f_n(x, y) - \overline{f_n})(f_d(x - u, y - v) - \overline{f_d})}{\sqrt{\sum_{xy}(f_n(x, y) - \overline{f_n})^2 \sum_{xy}(f_d(x - u, y - v) - \overline{f_d})^2}}, \qquad (10.1)$$

where

$f_n(x, y)$ is an intensity value for a pixel with coordinates (x, y) on the reference image;
$\overline{f_n}$ is the mean value of intensities of the pattern on the image before deformation;
$f_d(x - u, y - v)$ is the intensity value for a pixel with coordinates (x, y) on the image after deformation;
$\overline{f_d}$ is the mean value of intensities of the pattern after deformation;
x, y is the position of the pattern on the reference image;
u, v is the displacement of the pattern between the images.

In the process of deflection measurement, the reference image of the unloaded construction is divided into random speckle intensity patterns. Each of the patterns is matched with corresponding pixel subsets on the image of loaded structure by means of the NCC coefficient. The displacement vector is computed as a difference between positions of the pattern on the two images. The method performed on each of the points of interest gives a complete course of deflection of the analysed object. The correlation coefficient computes the position of each pixel as an integer value on the pixel grid. When the subpixel methods are introduced, the measurement's accuracy increases up to $0.01 - 0.1$ parts of pixel. The developed algorithm takes an integer value of the pixel's position given by the correlation coefficient and fits a second degree polynomial into the 4-element neighbourhood of the pixel. The new subpixel location of the correlation coefficient's maximum is the point for which the fitted parabola reaches its maximum value. The overview of the deflection measurement is presented in Figure 10.2.

10.4   Image Registration and Plane Rectification

Image registration (Zitova and Flusser 2003) is a method applied in order to overlay two or more images. The technique is frequently used in areas like remote sensing, medicine or cartography. Image registration techniques are divided into four main subgroups: multiview analysis, multitemporal analysis, multimodal analysis and scene-to-model registration. All image registration techniques consist of four steps: feature detection, feature matching, homography matrix estimation and image transformation. In this work, homography mapping (Hartley and Zisserman 2004) was introduced to reduce the perspective distortions which enabled the deflection's course to be measured from images of a construction taken from distinct locations. Homography is the mapping between two sets of points on a plane (Figure 10.3). If coplanar points' positions are given in homogenous coordinates, homography

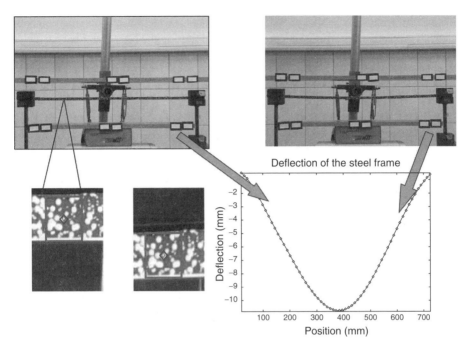

Figure 10.2 Digital image correlation applied to the deflection measurement. The displacement of the intensity pattern is found by the NCC coefficient

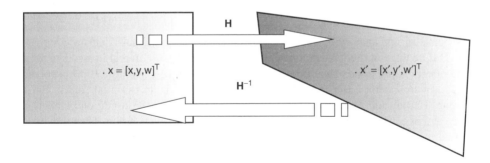

Figure 10.3 Homography mapping transforms a set of points on the plane into a different set of points on the plane

can be represented by a 3×3 matrix denoted as **H**. In general, the matrix **H** represents a full projective transformation of the plane and consists of similarity (translation, rotation and scaling), affine and projective transformations. Homography preserves collinearity, order and incidence of points and lines.

Homogenous coordinates of the point on the construction's reference image are denoted as $\mathbf{x} = [\mathbf{x}, \mathbf{y}, \mathbf{w}]^T$ and homogenous coordinates of the corresponding point on image with projective distortions as $\mathbf{x}' = [\mathbf{x}', \mathbf{y}', \mathbf{w}']^{\mathrm{T}}$. Transformation which maps coplanar points on the image with projective distortions on corresponding points on the reference image is given by:

$$\mathbf{x} = \mathbf{Hx'}, \tag{10.2}$$

where the homography transformation matrix is given by:

$$\mathbf{H} = \begin{bmatrix} h_{11} & h_{12} & h_{13} \\ h_{21} & h_{22} & h_{23} \\ h_{31} & h_{32} & h_{33} \end{bmatrix}. \tag{10.3}$$

In Equation (10.3) the matrix elements h_{13} and h_{23} represent translation, while the elements of the upper left 2×2 sub matrix are associated with rotation, shear transformation and scaling. The third row of the matrix $\mathbf{H}$ carries information about the perspective and non-uniform foreshortenings. The last element of the normalized form of $\mathbf{H}$ is always equal to 1. The homography matrix $\mathbf{H}$ is computed from a set of corresponding points by the linear least squares algorithm (DLT) (Hartley and Zisserman 2004). Equation (10.2) can be converted to a different form:

$$\begin{bmatrix} \mathbf{0}^T & -w'\mathbf{x}^T & y'\mathbf{x}^T \\ w'\mathbf{x}^T & \mathbf{0}^T & -x'\mathbf{x}^T \\ -y'\mathbf{x}^T & x'\mathbf{x}^T & \mathbf{0}^T \end{bmatrix} \begin{bmatrix} \mathbf{h}^1 \\ \mathbf{h}^2 \\ \mathbf{h}^3 \end{bmatrix} = \mathbf{0}. \tag{10.4}$$

in which h^{jT} denotes the jth row of the matrix $\mathbf{H}$. Two equations necessary for computation of matrix $\mathbf{H}$ are obtained from two corresponding points. From n pairs of correspondences, a $n \times 9$ matrix $\mathbf{A}$ is constructed, which satisfies the system of Equation (10.5):

$$\mathbf{Ah} = \mathbf{0}. \tag{10.5}$$

Four pairs of coplanar corresponding points are necessary and sufficient to compute the matrix $\mathbf{H}$ if no three of them are collinear. In the presence of noise in real image data, a larger set of corresponding points should be used and the system of Equation (10.5) is solved by the least squares method (DLT). Normalization of the data is an essential step in the DLT algorithm (Hartley and Zisserman 2004). It is needed to improve the condition number of the DLT system of equations and reduce the divergence from the true solution. The normalization proposed in Hartley and Zisserman (2004) is applied. The normalization matrices for data points are given as:

$$\mathbf{T} = \mathbf{T}_s \mathbf{T}_t \tag{10.6}$$

$$\mathbf{T}_t = \begin{bmatrix} 1 & 0 & -\bar{x} \\ 0 & -\bar{y} & 0 \\ 0 & 0 & 1 \end{bmatrix} \tag{10.7}$$

$$\mathbf{T}_s = \begin{bmatrix} \frac{\sqrt{2}}{d(x)} & 0 & 0 \\ 0 & \frac{\sqrt{2}}{d(x)} & 0 \\ 0 & 0 & 1 \end{bmatrix}. \tag{10.8}$$

(a)   (b)

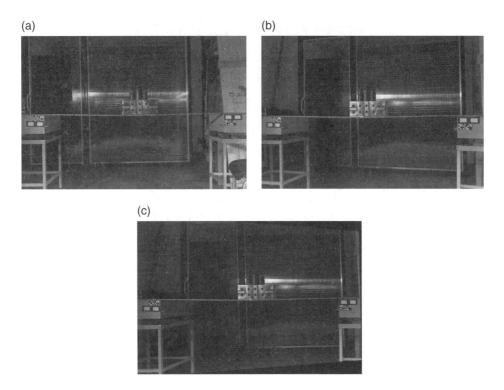

(c)

Figure 10.4   The image rectification: (a) the reference image; (b) the image of the deflected frame captured with projective distortions; and (c) the image rectified by homography matrix computed from positions of coplanar rectangular markers

Matrices are set up as follows: (1) points are translated so that their centroid moves to the origin; and (2) points are scaled so that their average distance $d(x)$ from the origin is equal to the square root of two. Two sets of points obtained from a pair of images have different normalization matrices. The homography matrix has to be denormalized after computation. The plane of the construction is rectified when all image points are transformed by homography mapping. The bicubic interpolation of pixels' intensity values is applied. Results of the rectification performed on the image of the laboratory setup are presented in Figure 10.4.

10.5   Automatic Feature Detection and Matching

The marker detection and matching algorithm was developed in order to increase the level of automation of homography mapping calculation. Sets of rectangles are identified on two images by means of binary image processing, contour detection and filtering. The corresponding points for homography computation, vertices of the rectangles, are extracted from the image by the Harris corner detector with a sub pixel accuracy (Harris and Stephens 1988). The markers' positions are expressed in a polar frame of reference and sorted with respect to its centre of mass. The binary image I of resolution M by N is an image which consists of

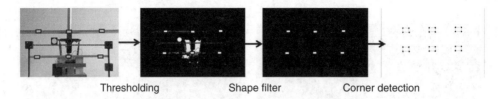

| Thresholding | Shape filter | Corner detection |

Figure 10.5 Rectangle detection algorithm carried out on the laboratory setup image

two kinds of pixel areas (Gonzales and Woods 2002): foreground A and background B, where A and B are defined as:

$$A = \{(x, y) : 0 \leq x \leq M, 0 \leq y \leq N \wedge I(x, y) = 1\}, \quad (10.9)$$

$$B = \{(x, y) : 0 \leq x \leq M, 0 \leq y \leq N \wedge I(x, y) = 0\}. \quad (10.10)$$

Let D_8 be the 8-neighborhood of the pixel $p_i = (x, y)$. The closed contour (or boundary) of the foreground region A on the binary image is a set of pixels defined as follows:

$$\text{contour} = \left\{ p_{i=1,2...N} \in A : (\forall_{pi} \exists_{p_i \in (B \wedge D_8(p_i))} \wedge p_i = p_N) \right\}. \quad (10.11)$$

The first step of the algorithm is conversion of a grey scale image to binary image. The threshold value is obtained by an interactive analysis of the intensity histogram's peak values and can be adapted to variable illumination. In the next step, the boundaries enclosing all foreground objects are detected. The contours are transformed to the chain polygon representation in which only endpoints of the line elements approximating the contour are stored. The boundaries are filtered by a shape filter whose response is the strongest for convex, rectangular contours with user defined range of area, width to height ratio and angle between adjoining sides of the quadrilaterals. The method is illustrated in Figure 10.5.

It is assumed that there is no rotation around the optical axis of the camera with respect to the camera coordinate frame from which the reference image was captured. In the first step, the centre of the mass of the set of markers' vertices is computed. The calculated point becomes the origin of the new polar coordinate frame. All of the points have to be expressed in this coordinate frame. Next, sorting of points is performed. The points' polar angles and radial distances from the origin are an input to the comparison function passed to the sorting algorithm. The sorting is carried out on sets of markers on both images of the construction.

The scale coefficient $\alpha_{mm/pix}$ is computed from an object on the scene with known geometric dimensions. The value of $\alpha_{mm/pix}$ can be obtained from a planar circular or rectangular marker or from the certified length standard. Two different methods of scale calibration were implemented in the system. The first one makes use of the binary image of the calibration pattern, the second one introduces the Hough transform for circle detection and high accuracy ellipse fitting technique. In the first method, the circular intensity pattern with a known diameter, (D_{mm}) is used. The diameter of the pattern on the image (D_{pix}) is computed by means of a binary image region analysis technique (Gonzales and Woods 2002). Scale coefficient $\alpha_{mm/pix}$ carries information about the number of pixels in 1 mm and can be derived from:

$$\alpha_{mm/pix} = \frac{D_{mm}}{D_{pix}}. \quad (10.12)$$

The scale calibration introduces the Hough transform for circle detection and high accuracy ellipse fitting technique. The Hough transform is a method for the specific geometric objects' detection on the image. In the developed method, the Hough transform for circular shape detection was applied. Circles with a user specified radius are detected in an iterative process. The ellipse is fitted to the points of the contour recognized as a circle in the previous step of the algorithm by the least square method. The new values of the circle's centre and radius are obtained from the ellipse fitting step.

10.5.1 Deflection-Shaped Based Damage Detection and Localization

Recently, a lot of damage detection methods based on analysis of changes in the deflection shape have been developed. Jang *et al.* (2007) presented a strain damage locating vector (DLV) method, i.e. a method combining DLV and static strain measurements. Guo and Li (2011) applied strain energy and evidence theory for damage detection. The evidence theory method was proposed to identify structural damage locations. Then, structural modal strain energy was utilized to quantify the extent of structural damage. Chen *et al.* (2005) developed a method of damage identification that uses limited test static displacement based on the grey system theory. In their work the grey relation coefficient of deflection curvature was defined and used to locate a damage in the structure. Patsias and Staszewski (2002) presented an application of the wavelet transform for damage detection based on optical measurements. The continuous wavelet transform (CWT) coefficients of both the damaged mode and the approximation function were computed, and thus, a reliable damage index could be obtained by taking their difference. On the other hand, Li *et al.* (2011) proposed a new damage identification method based on fractal theory and wavelet packet transform. The location of damage in the beam was determined by a large fluctuation appearing on the contour of the estimated fractional dimension (FD) and the extent of the damage was estimated by the FD based damage index. Rahmatalla and Eun (2010) presented a method of damage detection based on the distribution of flexural curvatures and constraint forces along a structural beam member.

The deflection curve course obtained from the vision method can be used for damage detection and localization. The crack detection method applied in the system is based on the change in the deflection's curvature analysis. In order to detect the curvature's discontinuities the CWT is applied. The CWT of a given signal $y(x)$ is given as:

$$W_y(u, s) = \frac{1}{\sqrt{s}} \int_{-\infty}^{\infty} y(x) \psi^* \left(\frac{x - u}{s} \right) dx, \qquad (10.13)$$

where function $\psi(x)$ is called a mother wavelet. The function can have real or complex values and is used to create the wavelet family $\psi_{u,s}(x)$ which is applied in CWT for signal decomposition. Any arbitrary function belonging to the family is given as:

$$\psi_{u,s}(x) = \frac{1}{\sqrt{s}} \psi \left(\frac{x - u}{s} \right), \qquad (10.14)$$

where s and u are the scaling and shift of the wavelet, respectively. The family consists of stretched and shifted copies of the mother wavelets. $W_y(u, s)$ is the wavelet coefficient which is a measure of frequency given by a scale s content in a signal in the neighbourhood of the space point which is given by a shift u. The scaling factor in Equation (10.14) is applied to normalize the energy of each wavelet.

The wavelet transform belongs to the methods of non-stationary signals' analysis and can be used to obtain the spatiotemporal information of the signal. It can be applied to damage detection and localization by change and discontinuities detection in the transducers' signals. The deflection curve course measured by the vision system is transformed by CWT. The maximum values of the 2D $W_y(u, s)$ function or the largest change of the values carry information about the damage position. Additionally, the value of the CWT amplitude for the point corresponding to crack localization is proportional to the decrease of the stiffness in this point.

The most important property of the wavelet for the damage detection is the vanishing moment. The wavelet has n vanishing moments if it is orthogonal to polynomial of degree k (Rucka and Wilde 2006):

$$\int_{-\infty}^{\infty} x^k \psi(x)dx, k = 0, 1, 2 \ldots n - 1,$$

(10.15)

Mallat proved that for the wavelet which has n vanishing moments, there exists a function defined as:

$$\psi(x) = (-1)^n \frac{d^n \theta(x)}{dx^n}.$$

(10.16)

The wavelet with n vanishing moments is an nth derivative of a function $\theta(x)$. Therefore, CWT can be understood as a multiscale differential operator:

$$W_y(u, s) = s^n \frac{d^n}{du^n}(y * \theta_s)(u),$$

(10.17)

$$\theta_s(x) = \frac{1}{\sqrt{s}}\theta\left(\frac{-x}{s}\right).$$

(10.18)

The convolution in Equation (10.17) can be interpreted as averaging of the signal y. The computation of CWT of the signal y is therefore equivalent to calculation of the n-derivative of the averaged version of the signal. The method is less susceptible to noise than the straightforward numerical computation of the curvature as the second derivative of the displacement signal. The type of wavelet which can be applied to crack detection depends on the number of its vanishing moments. The chosen wavelet should generate a CWT signal in which most of the coefficients are as close to zero as possible and only coefficients corresponding to damage localization have larger values. Any wavelet with the number of vanishing moments greater or equal to 2 can be applied for damage localization (Rucka and Wilde 2006). In the literature, Gauss, Haar, Gabor, Symlet or Coiflet are reported to be used in SHM systems. In the system, the Gaussian family of wavelet has been applied. The literature study and numerical experiments proved that the application of this wavelet resulted in easy to detect maxima of the CWT signal in positions corresponding to cracks.

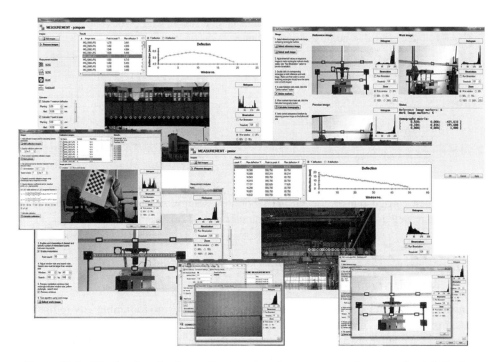

Figure 10.6   The developed software for vision based measurement of constructions' deflections

10.6   Developed Software Tool

The developed software (Figure 10.6) enables constructions' deflection measurement with the use of digital SLR cameras for remote image acquisition and provides image processing and vision algorithms to calculate the deflection curve course. Two modes of operation are available: the online and the offline. In the first case, the user specifies the date and the number of measurements and then the system works fully automatically, carrying out the image acquisition and deflection measurement. The offline mode provides analysis of the images registered by other devices or in previous sessions. The live view makes it possible to change the camera parameters in real time. The software has embedded tools for camera calibration and scale coefficient calculation from special calibration patterns. The result browser module carries out the visualization of the calculated curves of deflection, storing of the data and automatic reporting in a popular format, such as PDF, Excel spreadsheet or HTML. An additional feature of the software is detection when the allowed level of maximum deflection is exceeded and sending alerts to a client by e-mail.

10.7   Numerical Investigation of the Method

The main goal of the numerical simulation of the developed method was homography matrix elements' and deflection curve's uncertainty analysis as well as the deflection measurement's uncertainty evaluation in the case in which the camera with imaging sensor is parallel to the

Table 10.1 The focal lengths used in simulations with associated reference distance to the construction

Focal length (mm)	Reference distance (mm)
12	500
35	1200
50	1400
80	2500

construction's plane and for images after application of the rectification. A series of numerical tests have been carried out on the virtual laboratory setup consisting of the model of the construction as well as the camera system (Kohut *et al.* 2012a).

10.7.1 Numerical Modelling of the Developed Vision Measurement System

The model of the scene was implemented in the MATLAB® programming environment. It consisted of a pinhole camera and a virtual construction. The acquisition of the image by a camera was modelled as a central projection with a centre in the origin of the camera's coordinate system. The camera had a known set of internal and external parameters $(\mathbf{R},\mathbf{T})$ and the projection of the homogenous 3D points of the scene on the image plane was given by Equation (10.19). The virtual camera with a 21.1 mega pixel resolution sensor and lens with four different focal lengths ($f = 12$, 35, 50 and 80 mm) was modelled. Each of the focal lengths was associated with a reference distance between the camera and the construction, as summarized in Table 10.1.

$$\lambda \begin{bmatrix} u \\ v \\ 1 \end{bmatrix} = \begin{bmatrix} s_x & s_\theta & u_0 \\ 0 & s_y & v_0 \\ 0 & 0 & 1 \end{bmatrix} \begin{bmatrix} f & 0 & 0 \\ 0 & f & 0 \\ 0 & 0 & 1 \end{bmatrix} \begin{bmatrix} 1 & 0 & 0 & 0 \\ 0 & 1 & 0 & 0 \\ 0 & 0 & 1 & 0 \end{bmatrix} \begin{bmatrix} \mathbf{R} & \mathbf{T} \\ \mathbf{0} & 1 \end{bmatrix} \begin{bmatrix} X \\ Y \\ Z \\ 1 \end{bmatrix}. \tag{10.19}$$

The model of the construction consisted of a simply supported beam loaded with a point force acting centrally and a set of markers coplanar with the beam's plane, necessary for rectification. The FEM model of the beam was constructed in MATLAB® from 1D Euler beam elements. A steel beam of length 800 mm, cross-sectional area 500 mm^2 and Young's modulus equal to 2.1×10^5 MPa was modelled. The construction was loaded with a point force of magnitude 1000 N. The set consisting of 6 markers was positioned symmetrically with respect to the structure (Figure 10.7).

10.7.2 Uncertainty Investigation of the Method

The following cases were investigated in the analysis of the variability of the measured deflection of the beam: (1) the change of the camera's distance to the structure and the angle of the

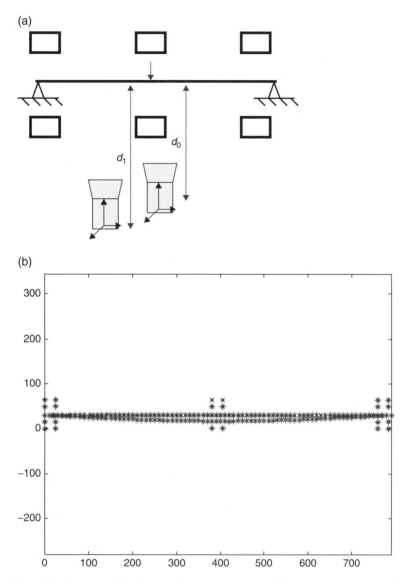

Figure 10.7 The virtual model of the laboratory setup. (a) The camera setup modelled in the simulations: two pinhole projection cameras at a distance d_0 (reference) and d_1 from the construction. (b) The virtual construction's plane

optical axis with respect to the normal vector of the construction's plane; (2) the change of the camera's focal length; and (3) the camera undergoing 'orbiting' motion about the construction, e.g. the camera moving on an elliptical path with its optical axis directed at the construction all the time (cf. Figure 10.8).

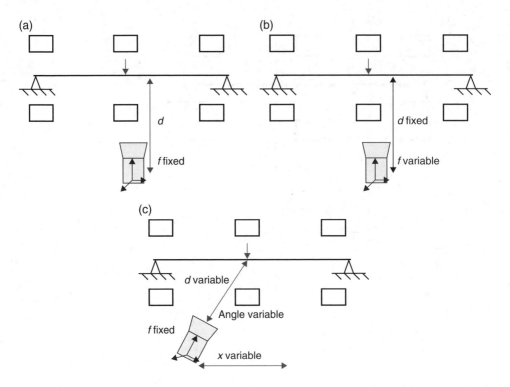

Figure 10.8   The virtual scene: (a) the first simulation – the variation of distance of the camera; (b) the second simulation – the variation of focal length; and (c) the third simulation – the camera orbiting around the construction

The uncertainty was considered for all the corner coordinates of the 6 markers used in the simulations. A random noise described by a normal probability distribution function was added to the nominal values of both vertical and horizontal coordinates of corner points. For each of the investigated cases, four standard deviations attached to the localizations of marker corners were assumed, i.e. 0.05, 0.5, 1, 1.5, all defined with pixels as units.

In the next discussion, the following notation concerning the simulation experiment will be used: each of the measurement series will be designated as $fXdY$, where X is the focal length value (in mm) and Y is a reference distance for a given focal length (in mm), for example $f50d1800$ means the images obtained by a virtual camera with a focal length of 50 mm and a reference distance of 1800 mm.

The first analysis [Figure 10.8(a)] deals with the change of the camera's distance with respect to the construction. Figure 10.9(a) presents the maximal standard deviation of static deflection (found in the endpoints) of the studied beam computed for the focal length equal to 50 mm. Irrespective of small fluctuations observed for the plotted curves, linear relationships between input and output standard deviations were found. Figure 10.9(b) shows an exemplary group of deflection curves calculated for all measurement points localized along the beam and defines the relationship between the standard deviation of the static deformation and the input

(a)

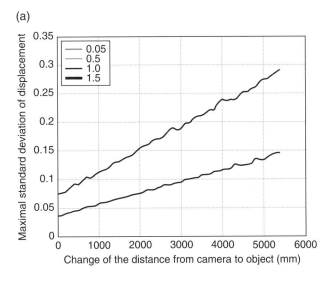

(b)

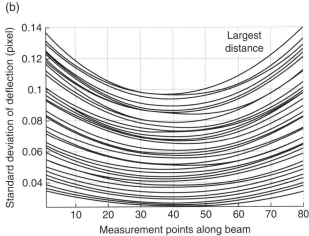

Figure 10.9 The deflection measurement uncertainty in varying distance test: (a) maximal standard deviation of static deflection for reference case; and (b) standard deviation of static deflection measured along beam

standard deviation of 0.5 pixels. The curves with the smallest values of standard deviations are related to the smallest change of camera distance. However, the ratio between these parameters, measured for both the worst and the best case along the beam, decreases with an increase in the distance.

The second analysis deals with the assessment of the uncertainty propagation for different focal lengths being taken from the interval of 12 mm up to 80 mm. The position and orientation of the camera was fixed in the simulation [Figure 10.8(b)]. Again, there are assumed

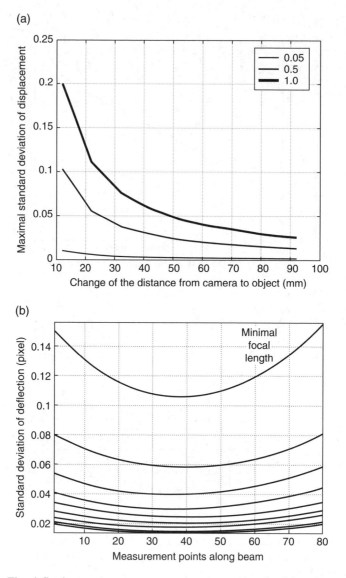

Figure 10.10 The deflection measurement uncertainty in varying focal length test: (a) maximal standard deviation of static deflection for reference case; and (b) standard deviation of static deflection measured along the beam

different standard deviations attached to the localizations of marker corners: from 0.05 up to 1.0 pixel.

Figure 10.10(a) presents the maximal standard deviation of static deflection (found in the endpoints) of the studied beam computed for the examined range of focal lengths. Figure 10.10(b) shows an exemplary group of curves found for all measurement points localized along the beam and defining the standard deviation of the static deformation for an exemplary case with the input standard deviations 0.5 pixels. The curves with the smallest values of standard deviations are related to the greatest focal length. The ratio between these parameters, measured for both the worst and the best case along the beam, decreases as the focal length decreases.

The last numerical test [Figure 10.8(c)] deals with uncertainty propagation in the case with the camera orbiting around the object. The camera was moved on a circular path around the construction and, at the same time, the angle between the optical axis of the camera and direction perpendicular to the construction's plane was varied, so that the optical axis always pointed at the centre of the beam. The internal parameters were fixed during investigation. The assumed standard deviations of the computed localizations of marker corners were taken from a range of 0.05 up to 1.0 pixel.

Figure 10.11 presents the maximal standard deviation of static deflection of the studied beam found for all measurement points for all referential cases.

Figure 10.12 shows groups of curves found for all measurement points localized along the beam and defining the standard deviation of the static deformation for all referential cases with the input standard deviations 0.5 pixels.

Figure 10.13 presents how the output resultant maximal standard deviation of the static deformation evolves while the input standard deviation increases for all referential cases and chosen parallel displacements of the camera. The linear relationships between input and output standard deviations have been found irrespective of the referential case chosen.

The research has indicated that the deflection's standard deviation is smaller than the corner positions' standard deviation for all cases of the input noise level. The largest impact of the input's noise has been observed on the endpoints of the beam. It is reasonable, since these are the smallest displacements of points and their measurement is easily disturbed by noise in the image. The highest value of the deflection's standard deviation has been found for the focal length $f = 12$ mm, irrespective of the chosen point of measurement. However, the effect of the noise imposed on the corresponding points coordinates has been insignificant in all other cases. The standard deviation of deflection in each of the measurement points has been a function of the camera's parallel displacement, for all considered cases of focal lengths. For example, if the camera was moved by a distance of -100 mm with respect to the central position, the standard deviation of points located on the beam's left side was much larger than the right side's standard deviation. However, the maximum variation of the measured deflection, captured on both ends of the beam, was constant, irrespective of the camera's displacement.

10.7.3   Model Based Probability of Damage Detection

In the next test, the model based probability of damage detection was performed. The main goal of the experiment was to prove the ability of the vision measurement method to detect a failure of the construction in the form of a crack, based on the analysis of the deflection curve's change. The investigation involved the numerical model of the construction and camera system described in the previous subsection. The crack was modelled as a local decrease of the

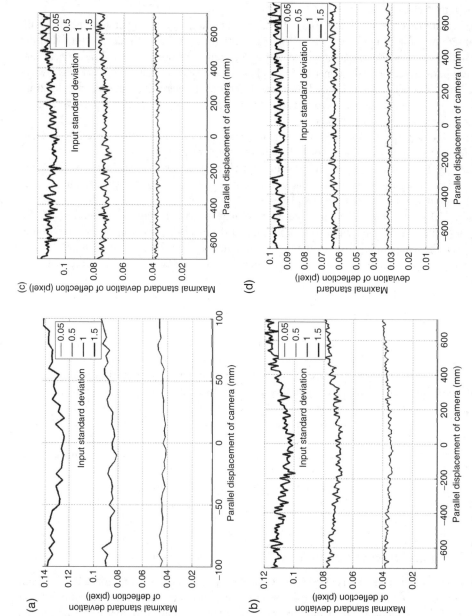

Figure 10.11 The deflection measurement uncertainty in the test with orbiting camera. Maximal standard deviation of static deflection for reference cases: (a) distance 500 mm/focal length 12 mm; (b) 1200 mm/35 mm; (c) 1800 mm/50 mm; and (d) 2500 mm/80 mm

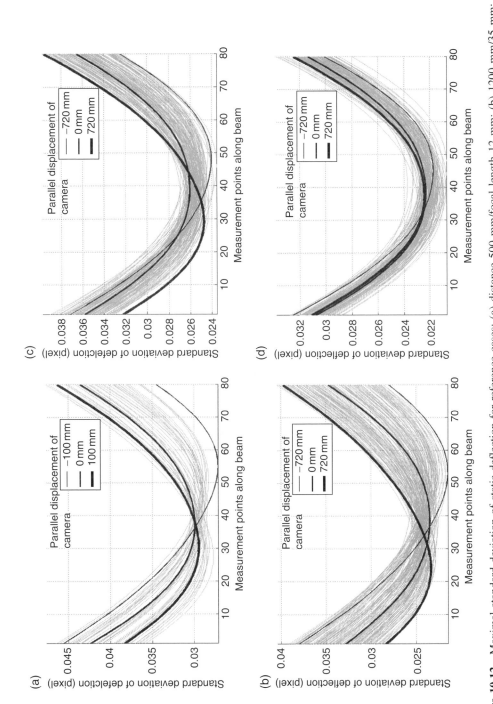

Figure 10.12 Maximal standard deviation of static deflection for reference cases: (a) distance 500 mm/focal length 12 mm; (b) 1200 mm/35 mm; (c) 1800 mm/50 mm; and (d) 2500 mm/80 mm

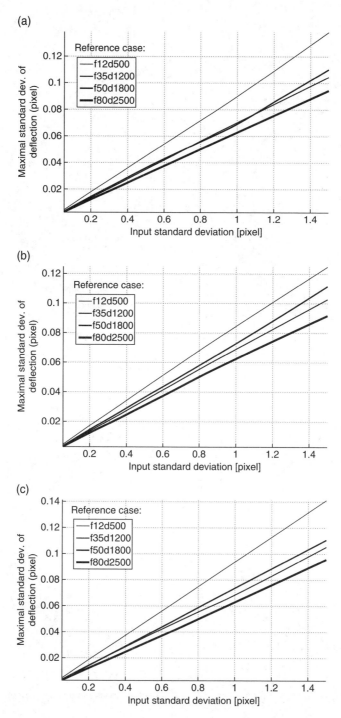

Figure 10.13 Output resultant maximal standard deviation of the static deformation for parallel displacement: (a) 100 mm; (b) 0 mm; and (c) 100 mm

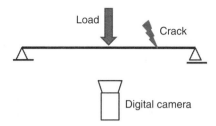

Figure 10.14   The model of the beam used in the numerical investigations

FEs stiffness (Figure 10.14). The beam supported at both ends was investigated. The position of the crack was detected by CWT coefficients' analysis on many levels of decomposition. The numerical investigations involved the noise, the position of the crack and its severity impact on the performance of the algorithm. The POD curves were computed on the basis of the numerical data. A hit and miss method of POD analysis was carried out.

In the preliminary tests, the ability of the system to detect a crack corresponding to different values of stiffness' decrease was examined. The damage detection and localization was performed for the crack at a distance equal to 680 mm from the first beam's support. Three values of crack severity were tested corresponding to 5, 10 and 15% of the stiffness decrease. The point load was acting in the middle of the structure. The standard deviation of the noise level simulating inaccuracy of correlation measurement was equal to 0.05 pixels. In all analysed cases, the damage was correctly detected and localized. The obtained CWT plots are presented in Figure 10.15.

In the next test the influence of the increasing noise was investigated. The test was carried out for one value of decrease in stiffness equal to 15%. The position of the crack as well as the load of the beam were the same as in the previous case. The increase of the standard deviation of the noise from 0.005 to 0.05 pixels did not influence the damage detection algorithm. The values of the noise larger than 0.5 pixels had an adverse effect on the method's performance, not encountered in the measurement with the use of well manufactured optics. The standard deviation of the noise present on the images captured by a Canon EOS 5D Mark II camera with 5616 × 3744 resolution ranged from 0.008 to 0.05 pixels. The results of the CWT transform of the deflection signal for increasing value of the image noise are presented in Figure 10.16.

The POD curves were computed on the basis of the data from numerical simulations. The curves were plotted for two different values of the standard deviation of the image noise equal to 0.005 and 0.05 pixels (Figure 10.17).

10.8   Laboratory Investigation of the Method

The laboratory investigations of the method were carried out to prove the results obtained from the numerical tests using the constructed prototype of the system. The uncertainty analysis of the deflection measurement for images taken from a reference camera as well as the images captured after the application of the rectification was performed. The POD was carried out for a beam specimen with a growing crack modelled as a cut of increasing depth. Additionally,

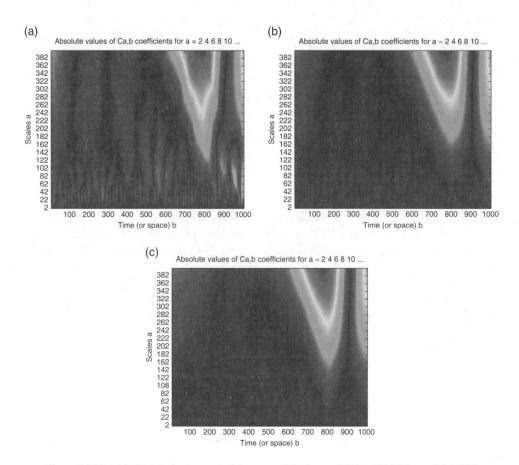

Figure 10.15   CWT plots for increasing damage of the beam: (a) 5%; (b) 10% ; and (c) 15%

the accuracy of the method was calculated by a comparison of the results computed using vision algorithm with the data obtained from a laser tracker device.

The laboratory setup consisted of an aluminium beam of length 1.8 m and cross section 50 × 10 mm [Figure 10.18(c)]. The beam was fixed at both ends and it was loaded with a point weight acting centrally. The random intensity pattern was placed on the measured plane of the construction. The set of rectangular markers necessary for homography matrix computation was placed coplanar with the beam's plane. The photographs of the construction were acquired by a system of two digital Canon 5D Mark II cameras with 21.1 MPix image resolution and lens Canon 24–70 mm f/2.8L with focal length $f = 50$ mm adjusted. The optical axis of the first, reference camera was perpendicular to the plane of the construction. The second camera was translated and rotated with respect to the first one and it was used for calculation of the deflection field from the images after application of rectification [Figure 10.18(a) and (b)].

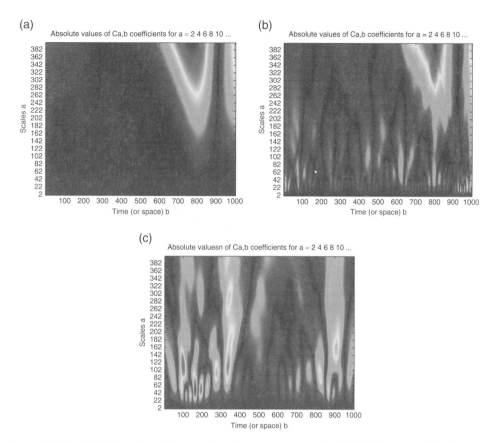

Figure 10.16 CWT plots for different values of measurement noise: (a) 0.005 pixels; (b) 0.05 pixels; and (c) 0.5 pixels

10.8.1 Tests of the Method on the Laboratory Setup

In the first experiment, the camera was moved along an elliptical path around the construction and at the same time, its optical axis was rotated so that it pointed at the centre of the construction. The values of distances between the camera and the object, as well as corresponding angles between the optical axis and direction perpendicular to the plane, are presented in Table 10.2. In the second test, the camera was moved away from the object, but the orientation of its optical axis was not varied (Table 10.2). In both cases, the focal length of the camera was fixed ($f = 50$ mm).

The experiment on the laboratory setup proved the results obtained from the numerical tests. In the case of the camera orbiting around the construction, the measurement of the displacement was characterized by a higher value of standard deviation for the points placed on the left side of the beam. The calculated uncertainty was not only the result of the homography computation from noisy data but came from the phenomena associated with lens construction, such as the depth of the field or the radial and tangential distortions. Concerning the

(a)

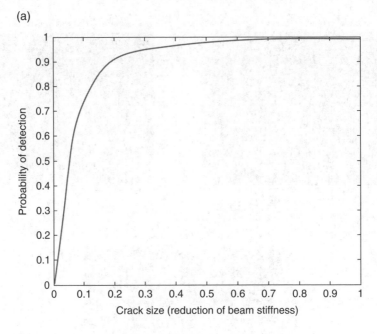

(b)

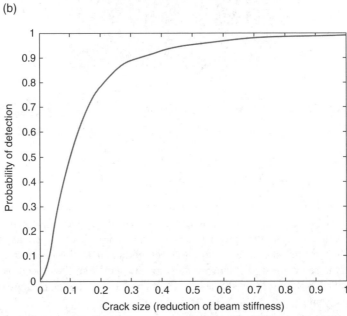

Figure 10.17 Model based POD curves for different values of the standard deviation input image noise: (a) 0.005 pixels; and (b) 0.05 pixels

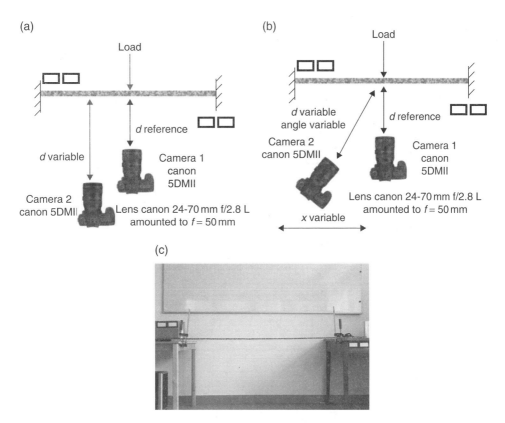

Figure 10.18   The laboratory setup: (a) the first experimental test; (b) the second experimental test; and (c) the investigated beam

Table 10.2   The distances and corresponding angles of the optical axis with respect to the direction perpendicular to the construction's plane for two experimental cases

Laboratory test 1 Distance d and corresponding angle α (mm, deg)	Laboratory test 2 Distance d of the camera (mm)
2800, 0	2800
2750, 5	3000
2750, 10	3400
2700, 15	3700
2600, 20	4000
2500, 30	
2400, 35	
2200, 45	

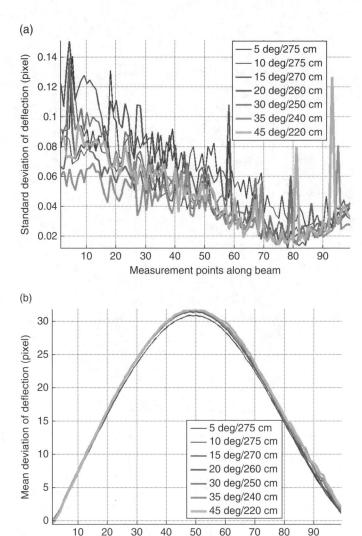

Figure 10.19   Standard deviation (a) and mean values (b) of static deflection measured along the beam for different angles of camera orientation

variable distance between the camera and the construction, the standard deviation increased with the distance, as expected. The highest value of the standard deviation was less than 0.18 pixels. The experiment has confirmed that the mean, maximum and minimum value of the standard deviation, measured along the beam's length, was constant irrespective of the angle between the camera's optical axis and the direction perpendicular to the construction's plane. These parameters have increased with the distance between the camera and the construction (Figure 10.19, Figure 10.20 and Figure 10.21).

(a)

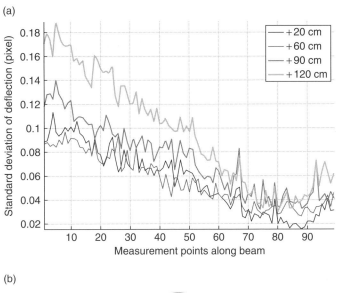

(b)

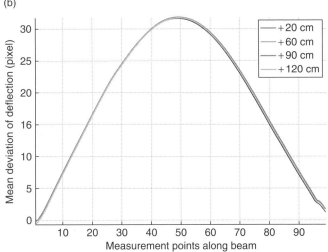

Figure 10.20   Standard deviation (a) and mean values (b) of static deflection measured along the beam for different distances from the camera to the object

10.8.2   The Probability of Detection of the Method in the Laboratory Investigation

The laboratory setup consisted of an aluminium beam of length 800 mm and cross section 40 × 5 mm (Figure 10.22). The beam was supported at both ends and it was loaded with a point weight acting centrally. The damage of the beam was modelled by cutting the beam perpendicular to its largest dimension. The value of the cut's depth increased from 2 to 20 mm. The camera used for image acquisition was the same as in the previous research. The

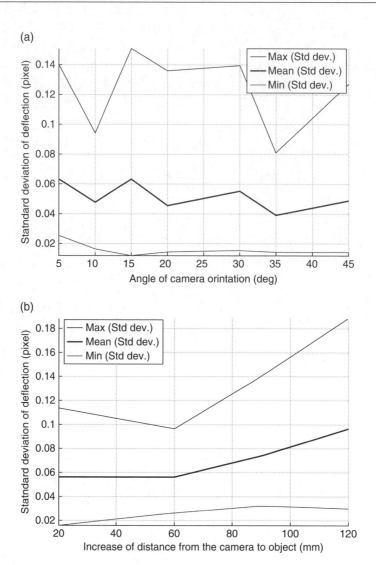

Figure 10.21 The scatter of standard deviation of static deflection: (a) for different camera orientations with respect to the construction; and (b) for different distances from the camera to the object

deflection courses computed by the method were transformed by CWT and used for damage detection and localization (Figure 10.23).

The developed vision system was able to correctly detect and localize the damage with a length that was equal to or greater than 4 mm. The localization of the damage was assumed to be correct if and only if its computed position was different from the real position by less than 5 mm. It was observed that the wavelet coefficient corresponding to the position of the crack was increasing monotonically with the crack's length. Figure 10.24 shows the CWT

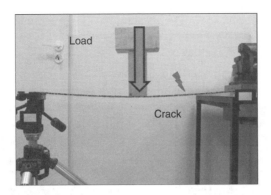

Figure 10.22   The image of the laboratory setup with the position of the load and modelled crack

coefficients' plots for increasing value of the crack's length. It presents the coefficients of CWT for a given scale level. The increase in magnitude of CWT with increasing length of crack can be observed.

The POD curve has been computed on the basis of the data from the laboratory experiment; it is shown in Figure 10.25.

10.8.3   Investigation of the Developed Method's Accuracy

The laboratory setup consisted of an aluminium beam of length 1.8 m and cross section 50×10 mm (Figure 10.26). The beam was fixed at both ends. The random intensity pattern was placed on the measured plane of the beam. The beam was loaded with a point weight acting at the middle of its length. The set of rectangular markers necessary for homography matrix computation was placed coplanar with the beam's plane. Photographs of the construction were acquired by a system of two digital Canon 5D Mark II cameras with 21.1 MPix image resolution and lens Canon 24–70 mm f/2.8 L with focal length $f = 50$ mm adjusted. The positions of the cameras were fixed in space. One of the cameras had its optical axis perpendicular to the plane of the construction. The reference images were acquired by that camera. The second camera was translated and rotated with respect to the first one and it was used for calculation of the deflection field from the images after application of rectification. The distance d_0 between the reference camera and the beam was 2539.307 mm. The distance d_1 between the second camera and the beam was 2527.408 mm and the angle between the optical axes of the first and the second camera was 16.67°. The scene was illuminated by a lighting system with 1 kW power.

In order to prove the measurements performed on the described optical system were correct, they were compared with the results obtained with the use of a Leica LaserTracker. A Leica LTD 860 model was used. This system is used in the Laboratory of Coordinate Metrology at the Kraków University of Technology, which is an accredited calibration laboratory according to ISO 17025 standard. The LaserTracker used is an accredited device and its accuracy is constantly monitored. The maximum permissible error for point measurement was 0.025 mm. The measurements were performed with the use of a self-centring technique. The cone-shaped

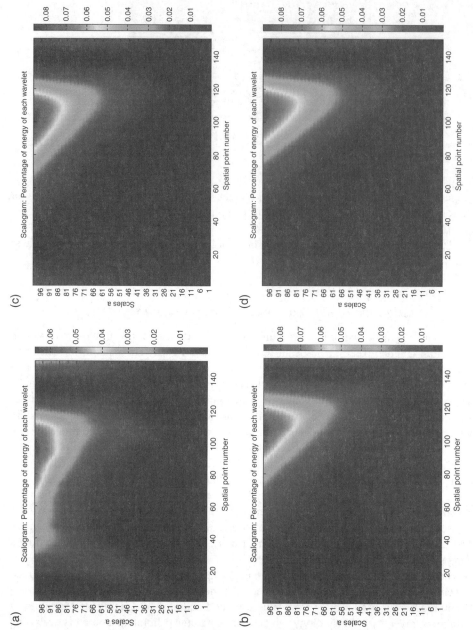

Figure 10.23 The results of the damage detection for different lengths of cracks: (a) 4 mm; (b) 8 mm; (c) 12 mm; and (d) 20 mm

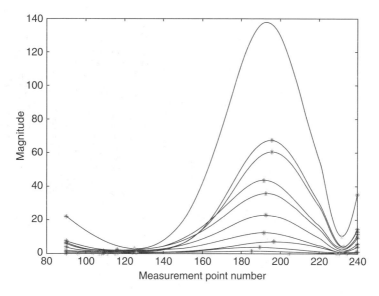

Figure 10.24 Plots of the CWT coefficients for a given decomposition scale. The maxima of the curves correspond to the position of the crack. The crack length was increased from 2 to 20 mm

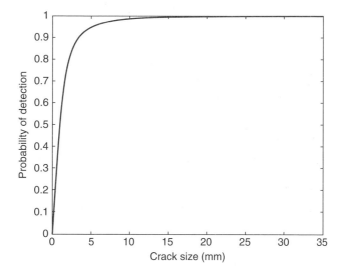

Figure 10.25 POD curve for the developed measurement system

holes of diameter 4 mm were drilled in the front plane of the beam at regular distances. The measurements were first performed on the beam without a load (reference measurements) and then on the loaded one. As a result of measuring a single point, its coordinates x, y, z were obtained. Differences in coordinates of corresponding points before and after deformation gave the value of deflection.

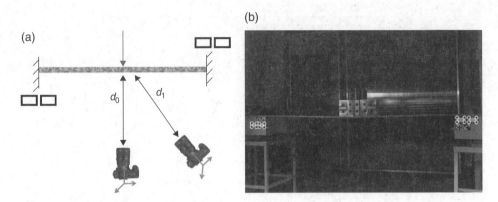

Figure 10.26   The laboratory setup for accuracy analysis: (a) the laboratory setup shown schematically; and (b) the laboratory setup photographed by one of the cameras

Table 10.3   The comparison of results of beam deflection measurements performed by the two systems

Point no.	Optical system		Leica LaserTracker	
	Deflection (mm)	Uncertainty (mm)	Deflection (mm)	Uncertainty (mm)
1	0.027	0.004	0.096	0.015
2	0.682	0.004	0.634	0.021
3	2.162	0.004	2.127	0.007
4	4.259	0.004	4.249	0.007
5	6.622	0.003	6.584	0.008
6	8.965	0.002	9.001	0.014
7	11.047	0.003	11.129	0.012
8	12.672	0.001	12.606	0.027
9	13.505	0.002	13.433	0.009
10	13.494	0.003	13.464	0.026
11	12.615	0.001	12.524	0.018
12	11.006	0.001	11.019	0.008
13	8.802	0.001	8.627	0.019
14	6.213	0.001	6.239	0.004
15	3.762	0.001	3.861	0.022
16	1.730	0.001	1.884	0.009
17	0.424	0.001	0.468	0.021
18	0.000	0.000	0.051	0.009

10.8.3.1   Results

Table 10.3 shows the comparison of results of beam deflection measurements performed by both systems. The results presented for the 'Optical system' are for measurements performed by the camera mounted perpendicular to the front plane of the beam.

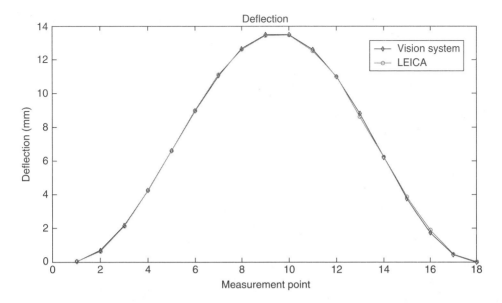

Figure 10.27   Results of deflection measurement (camera perpendicular to the front plane of the beam)

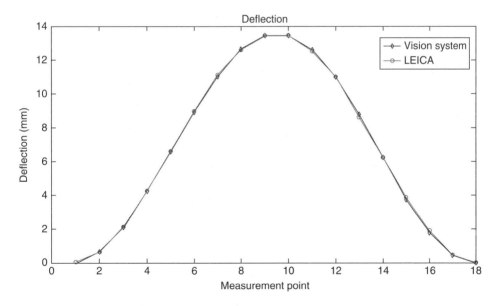

Figure 10.28   Results of deflection measurement (after rectification)

The maximum difference between the optical system and the Leica LaserTracker was 0.174 mm, the minimum 0.010 mm, and the mean was difference 0.063 mm (Figure 10.27).

The results for the second camera (obtained after application of rectification) are similar (Figure 10.28). The maximum difference between the optical system in this configuration and the Leica LaserTracker was 0.149 mm, the minimum 0.002 mm and the mean difference was 0.064 mm.

The results presented in the previous paragraph show that deflection values obtained by the created optical system are comparable with the results from the Leica Laser-Tracker metrological device, whose accuracy is proved and constantly monitored. On that basis, the correctness of the described optical system can be assumed. The results also show that measurements done by a camera with its optical axis not perpendicular to the measured object are almost equal to the case when it is perpendicular. This means that measurements are possible from different positions and angles between the measured object and camera.

10.9   Key Studies and Evaluation of the Method

The results of the developed vision based method for civil engineering constructions' in-plane deflection measurement and state monitoring have been presented in Kohut *et al.* (2012b). To evaluate the mentioned method the deflection of a tram viaduct is investigated and results are compared with the data obtained by interferometric radar.

10.9.1   Tram Viaduct Deflection Monitoring

The measurements involved monitoring a 28-m-long viaduct segment (Figure 10.29). Measurement markers with random shapes as well as three calibration patterns consisting of white circles and black crosses, used for determining the scale coefficient, were attached on the surface of the monitored object. Measurements where performed using two Canon EOS 5D Mark II SLR cameras. The cameras were placed at a distance of 24.6 m from the object. The first camera was equipped with a Canon EF 24–70 mm f/2.8L lens set to work with 70 mm focal length. The field of view for this camera spanned 13.5 m. The second camera was equipped with a telephoto lens (Canon EF 100–400 mm f/4.5–5.6L) set to 400 mm focal length. The camera was used for monitoring a smaller fragment of the construction which was 2.32 m in length. The angle between the monitored object's surface and the optical axis was 7°. In addition, the continuous deflection measurement was performed using an interference radar IBIS-S. The aim of the field test was to register the construction's deflection under the load of passing trams. Figure 10.29 shows the monitored object. Figure 10.30 illustrates the measurement setup diagram with selected system's elements.

Correlation function values were determined with the use of correlation window 120×120 pixels for photographs taken by the 400 mm lens camera and 80×80 pixels for those captured by the 70 mm lens camera.

Displacements were analysed at 7 points on the span, located in the central part of the structure. Points' positions were numbered and estimated relative to the first calibration target (CalT1), taken as a reference for both types of measurements. The measurements were performed by both systems while a tram was passing. Due to the frequency of the radar recording data (100 Hz), information about displacement of the points was averaged for 0.25-s intervals in order to compare results. In the case of the vision based method, the intermediate

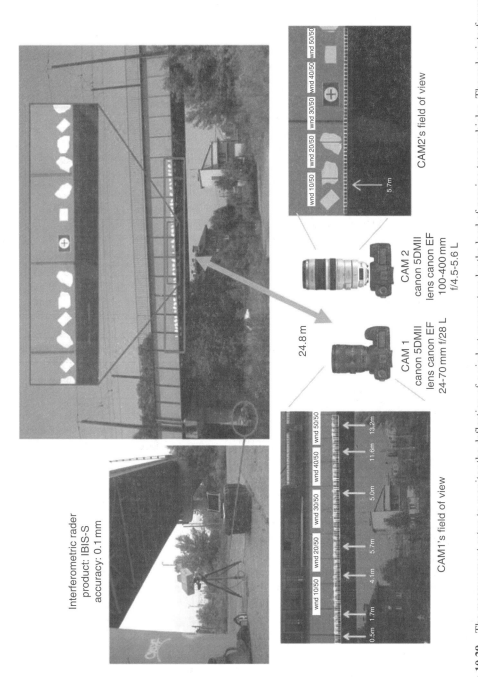

Figure 10.29 The measurement setup to monitor the deflection of a viaduct segment under the load of passing tram vehicles. The radar interferometer IBIS-S and two Canon EOS 5D Mark II digital cameras with different types of lenses and corresponding fields of view are also shown

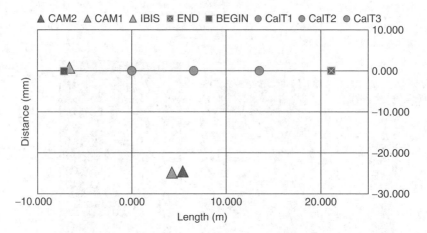

Figure 10.30 The measurement setup diagram with selected system's elements. IBIS, the interferometric radar; CAM1, CAM2, the two cameras of the vision system; CalT1, CalT2, CalT3, calibration markers

Table 10.4 The comparison of results of beam deflection measurements performed by both systems. Deflection difference (mm)

Time (s)			Measurement	points	(m)		
	−0.464	**1.664**	**4.064**	**5.664**	**8.964**	**11.564**	**13.164**
1.750	−0.550	−0.550	−0.665	−0.490	−0.269	−0.122	0.072
2.000	−0.221	−0.041	−0.254	−0.123	−0.151	−0.275	−0.201
2.250	−0.093	0.104	−0.320	−0.057	0.124	−0.168	−0.339
2.500	0.383	0.194	−0.102	0.027	0.175	0.119	−0.261
2.750	0.364	−0.089	0.111	−0.047	0.204	0.141	−0.269
3.000	0.088	−0.127	−0.284	−0.173	0.144	0.026	−0.385

deflection values were interpolated in time. The maximum values of deflection for each point were compared. The observations were collected for 8 tram passages. For each passage the maximum deflection for both methods was calculated, as well as differences between the two methods. Moreover, for each set of displacements the minimum and maximum values were calculated. The minimum difference between the two techniques was 0.08 mm, while the maximum difference reached 0.99 mm. The results for the chosen case of tram passage have been collected in Table 10.4 and Table 10.5. Table 10.4 presents the difference between measurement of displacements for both methods and for all measurement points. In Table 10.5, the mean value and the standard deviation of the difference between displacements obtained by the two systems are shown.

Figure 10.31 shows the shape of the viaduct at the moment of the maximum deflection during a sample tram passing. It presents the results obtained by the two methods, as well as the difference between them. In Figure 10.32(a) and (b) the deflection of the two chosen

Table 10.5 The average and the standard deviation of the differences between measurement by IBIS-S interferometric radar and vision based method. The results were obtained from all analysed tram passages. Deflection difference (mm)

Time (s)		Measurement	points	(m)			
	−0.464	1.664	4.064	5.664	8.964	11.564	13.164
Average (mm)	−0.393	−0.162	−0.082	0.157	0.615	0.99	0.894
Standard deviation (mm)	0.420	0.401	0.279	0.318	0.365	0.455	0.572

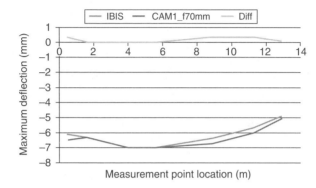

Figure 10.31 The shape of the viaduct at the moment of the maximum deflection during a sample tram passing. The results obtained by the two methods as well as the difference between them are shown

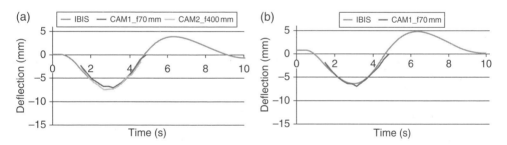

Figure 10.32 The displacement of two chosen points on the span during the same tram passing is presented in the time domain: (a) the deflection of the point at a distance a 5.7 m from the start point on the span measured by three methods; and (b) the deflection of the point at a distance of 9.0 m from the start point of the span

points on the span during the same tram passing is presented in the time domain. These results were also obtained using both techniques.

The performed field tests allowed to determine the applicability of the proposed vision based technique for measuring the deflection of civil engineering constructions. The vision system measurement resulted in data which matched the points' displacement registered by the interferometric radar. The minimum difference between the vision technique and the radar was 0.08 mm, while the maximum difference was 0.99 mm. The method can be applied to structures such as bridges, footbridges, chimneys, viaducts, girders, ceilings, halls, masts, wind turbines, buildings, machines and devices. The main advantages of the proposed methods are the simplicity of measurement, possibility of obtaining dense field of deflection using only one acquisition device, application of a commonly available digital camera, the easy image acquisition process which can be performed from different points in space as well as fast and easy analysis and interpretation of measurement results.

10.10   Conclusions

The vision based measurement system enables structure static states monitoring to be performed by means of analysis of changes in the geometric properties of the structure, such as shape of the deflection curve. The introduction of image registration techniques has improved the flexibility, universality and accuracy of this method. The technique of in-plane deflection measurement presented in this chapter enables images of the construction to be taken from different points of view during examination. The developed techniques of marker detection and matching make it possible to create an application which can be used in automatic vision condition diagnostics systems, in which the construction can be monitored during its everyday use. Furthermore the developed methods can be applied with user-friendly software which can allow the state of the construction to be quickly assessed during inspection. The system uses an easily available camera as the measuring sensor for the measurement of static or quasi-static states of structures. The method can be applied to the structures with a texture in the form of noise or natural texture of the materials as well as in cases when special markers are available. The system provides high measurement density without application of active optical methods.

References

Behrens A, Bommes M, Stehle T, Leonhardt S and Aach T 2011 Real-time image composition of bladder mosaics in fluorescence endoscopy. *Computer Science Research and Development* **26**(102), 51–64.

Benhimane S and Malis E 2007 Homography-based 2D visual tracking and servoing. *International Journal of Robotic Research* **26**(7), 661–676.

Berfield T, Patel J, Shimmin R, Braun P, Lambros J and Sottos N 2007 Micro- and nanoscale deformation measurement of surface and internal planes via digital image correlation. *Experimental Mechanics* **47**, 51–62.

Borkowski S, Riff O and Crowley J 2003 Projecting rectified images in an augmented environment. *IEEE International Workshop on Projector-Camera Systems*. Nice, France, pp. 1–8.

Brown M and Lowe DG 2003 Recognising panoramas. *Proceedings of the 9th IEEE International Conference on Computer Vision*. IEEE Computer Society, vol. 2, pp. 1218–1225.

Budzik B, Grzelka M and Oleksy M 2010 Evaluation of silicone mould accuracy with use of optical measurements. *Measurement Automation & Monitoring* **1**, 18–19.

Cawley P and Adams R 1979 The location of defects in structures from measurements of natural frequencies. *Journal of Strain Analysis* **14**, 49–57.

Chen X, Zhu H and Chen C 2005 Structural damage identification using test static databased on grey system theory. *Journal of Zhejiang University Science* **6**(5), 790–796.

Cheng T, Dai C and Ran R 2007 Viscoelastic properties of human tympanic membrane. *Annals of Biomedical Engineering* **35**, 305–314.

Chu T, Ranson W and Sutton M 1985 Applications of digital-image-correlation techniques to experimental mechanics. *Experimental Mechanics* **25**, 232–244.

Eshel R and Moses Y 2008 Homography based multiple camera detection and tracking of people in a dense crowd *IEEE Conference on Computer Vision and Pattern Recognition 2008*. Anchorage, USA, pp. 1–8.

Fang Y, Dawson D, Dixon W and de Queiroz M 2002 Homography-based visual servoing of wheeled mobile robots. *Proceedings of the 41st IEEE Conference on Decision and Control*. Las Vegas, USA, vol. 3, pp. 2866–2871.

Fujun Y and Xiaoyuan H 2006 Digital speckle projection for vibration measurement by applying digital image correlation method. *Key Engineering Material* **326–328**, 99–102.

Gancarczyk J and Gancarczyk T 2010 Non-invasive museum object inspection by VIS photography, UV fluorescence and IR reflectography. *Measurement, Automation & Monitoring* **3**, 268–271.

Gonzales R and Woods R 2002 *Digital Image Processing*. Prentice Hall.

Guan H and Karrbhari V 2008 Improved damage detection method based on element modal strain damage index using sparse measurement. *Journal of Sound and Vibration* **309**, 455–494.

Guo HJ and Li Z 2011 Structural damage detection based on strain energy and evidence theory. *Journal of Applied Mechanics and Materials* **48–49**, 1122–1125.

Harris C and Stephens M 1988 *Proceedings of the 4th Alvey Vision Conference*. Manchester, UK, pp. 147–151.

Hartley R and Zisserman R 2004 *Multiple View Geometry in Computer Vision*. Cambridge University Press.

Helm J 2008 Digital image correlation for specimens with multiple growing cracks. *Experimental Mechanics* **48**, 753–762.

Infantino I and Chella A 2001 Architectural scenes reconstruction from uncalibrated photos and map based model knowledge. *Proceedings of the 7th Congress of the Italian Association for Artificial Intelligence on Advances in Artificial Intelligence*. Springer pp. 356–361.

Iwase S and Saito H 2002 Tracking soccer players based on homography among multiple views. *Visual Communications and Image Processing* **50–51**, 283–292.

Jaishi B and Rex W 2006 Damage detection by finite element model updating using modal flexibility residual. *Journal of Sound and Vibration* **290**, 369–387.

Jang S, Sim S and Spencer B 2007 Structural damage detection using static strain data. *Proceedings of the World Forum on Smart Materials and Smart Structures Technology*. Chongqing and Nanjing, China, pp. 286–291.

Kohut P, Holak K and Martowicz A 2012a An uncertainty propagation in developed vision based measurement system aided by numerical and experimental tests. *Journal of Theoretical and Applied Mechanics* **50**, 1049–1061.

Kohut P, Holak K and Uhl T 2008 Application of image correlation for SHM of steel structures *Structural Health Monitoring 2008: Proceedings of the Fourth European Workshop*. DEStech Publications, *Inc.*, pp. 1257–1264.

Kohut P, Holak K and Uhl T 2011 Prototype of the vision system for deflection measurements. *Diagnostyka* **4**, 3–11.

Kohut P, Holak K, Uhl T, Krupinski K, Owerko T and Kuras P 2012b Structure's condition monitoring based on optical measurements. *Key Engineering Materials* **518**, 338–349.

Kohut P and Kurowski P 2005 The integration of vision based measurement system and modal analysis for detection and localization of damage. *The International Journal of Ingenium* **4**, 391–398.

Kohut P and Kurowski P 2006 The integration of vision system and modal analysis for SHM applications *Proceeding of IMAC-XXIV*, St Louis, USA, pp. 1–8.

Kosetska Y, Ma Y, Sastry S and Soatto S 2004 *An Invitation to 3D Computer Vision*. Springer.

Li H, Yong H, Ou J and Bao Y 2011 Fractal dimension-based damage detection method for beams with a uniform cross-section. *Computer-Aided Civil and Infrastructure Engineering* **26**(3), 190–206.

Li W, Sutton M, Li X and Schreier H 2008 Full-field thermal deformation measurements in a scanning electron microscope by 2D digital image correlation. *Experimental Mechanics* **48**, 635–646.

Liebowitz D and Zisserman A 1998 Metric rectification for perspective images of planes. *IEEE Conference on Computer Vision and Pattern Recognition*. Santa Barbara, USA, pp. 482–488.

Lieven N and Ewins D 1988 Spatial correlation of mode shapes, the coordinate modal assurance criterion (COMAC). *Proceedings of the Sixth International Modal Analysis Conference*. Kissimmee, USA, pp. 690–695.

Loop C and Zhengyou Z 1999 Computing rectifying homographies for stereo vision. *IEEE Computer Society Conference on Computer Vision and Pattern Recognition*. Fort Collins, USA, vol. 1.

Mei X, Mahesh R and Shaohua K 2005 Video background retrieval using mosaic images. *IEEE International Conference on Acoustics*. Philadelphia, USA, vol. 2, pp. 441–444.

Pandy A, Biswas M and Samman M 1991 Damage detection from changes in curvature mode shapes. *Journal of Sound and Vibration* **145**, 321–332.

Park S and Trivedi M 2007 Homography-based analysis of people and vehicle activities in crowded scenes. *Proceedings of the Eighth IEEE Workshop on Applications of Computer Vision*, IEEE Computer Society, p. 51.

Patsias S and Staszewski WJ 2002 Damage detection using optical measurements and wavelets. *Structural Health Monitoring* **1**(1), 5–22.

Perera R and Ruiz A 2008 A multi-stage FE updating procedure for damage identification in large scale structures based on multi-objective evolutionary optimization. *Mechanical Systems and Signal Processing* **22**, 970–991.

Rahmatalla S and Eun H 2010 A damage detection approach based on the distribution of constraint forces predicted from measured flexural strain. *Smart Materials and Structures* **19**(10), 105–116.

Rethore J, Gravoil A, Morestin F and Combescure A 2005 Estimation of mixed-mode stress intensity factors using digital image correlation and an interaction integral. *International Journal of Fracture* **132**, 65–79.

Robert L, Nazaret F, Cutard T and Orteu J 2007 Use of 3-D digital image correlation to characterize the mechanical behavior of a fiber reinforced refractory castable. *Experimental Mechanics* **47**, 761–773.

Rossi M, Broggiato G and Papalini S 2008 Application of digital image correlation to the study of planar anisotropy of sheet metals at large strains. *Meccanica* **43**, 185–199.

Roux S and Hild F 2006 Stress intensity factor measurements from digital image correlation post-processing and integrated approaches. *International Journal of Fracture* **140**, 141–157.

Rucka M and Wilde K 2006 Crack identification using wavelets on experimental static deflection profiles. *Engineering Structures* **28**, 279–288.

Salehi M, Ziaei-Rad S, Ghayour M and Vaziri-Zanjani M 2010 A structural damage detection technique based on measured frequency response functions. *Contemporary Engineering Sciences* **3**(5–8), 215–226.

Schmidt T, Tyson J and Galanulis K 2003 Full-field dynamic displacement and strain measurement using advanced 3D image correlation photogrammetry. *Experimental Techniques* **27**, 41–44.

Schmidt T, Tyson J, Shahinpoor M and Galanulis K 2002 Biomechanics deformation and strain measurements with 3D image correlation. *Experimental Techniques* **26**, 39–42.

Scoleri T, Chojnacki W and Brooks M 2005 A multiobjective parameter estimator for image mosaicing. *Proceedings of IEEE International Symposium of Signal Processing and its Applications*, pp. 551–554.

Shankar N, Ravi N and Zhong Z 2009 A real-time print-defect detection system for web offset printing. *Measurement* **42**(5), 645–652.

Sierra G, Wattrisse B and Bordreui I C 2008 Structural analysis of steel to aluminium welded overlap joint by digital image correlation. *Experimental Mechanics* **48**, 213–223.

Spagnolo G, Paolett i D, Paoletti A and Ambrosini D 1997 Roughness measurement by electronic speckle correlation and mechanical profilometry. *Measurement* **20**(4), 243–249.

Teo L and Xue M 2007 A novel method to study strain-induced delamination in plastic IC packages using a digital image correlation tool. *9th Electronics Packaging Technology Conference*. Singapore, pp. 576–581.

Tiwari V and Sutton, M. and McNeill S 2007 Assessment of high speed imaging systems for 2D and 3D deformation measurements: Methodology development and validation. *Experimental Mechanics* **47**, 561–579.

Tyson J 2000 Non-contact full-field strain measurement with 3D ESPI. *Sensors* **17**, 62–70.

Uhl T, Kohut P and Holak K 2009 Image correlation and homography mapping in optical deflection measurement. *Structural Health Monitoring 2008: Proceedings of the Fourth European Workshop*. Shaker Publishing pp. 191–200.

Uhl T, Kohut P, Holak K and Krupinski K 2011 Vision based condition assessment of structures. *Journal of Physics Conference Series* **305**, 1–10.

West M 1984 Illustration of the use of modal assurance criterion to detect structural changes in an orbiter test specimen. *Proceedings of the Air Force Conference on Aircraft Structural Integrity*. Palm Springs, USA, pp. 1–6.

Wieczorkowski M, Koteras R and Znaniecki P 2010 Use of optical scanner for car body quality control. *Measurement Automation & Monitoring* **1**, 40–41.

Zhang G, Hu Y and Jian-Min Z 2009 New image analysis-based displacement-measurement system for geotechnical centrifuge modeling tests. *Measurement* **42**(1), 87–96.

Zhang Z and He LW 2007 Whiteboard scanning and image enhancement. *Digital Signal Processing.* **17**(2), 414–432.

Zhu Y, Dai R, Xiao B and Wang C 2008 Perspective rectification of camera-based document images using local linear structure. *Proceedings of the 2008 ACM Symposium on Applied Computing*. New York, USA, pp. 451–452.

Zitova B and Flusser J 2003 Image registration methods: A survey. *Image and Vision Computing* **21**, 977–1000.

Index

Advanced Structural Damage Detection: From Theory to Engineering Applications, First Edition.
Edited by Tadeusz Stepinski, Tadeusz Uhl and Wieslaw Staszewski.
© 2013 John Wiley & Sons, Ltd. Published 2013 by John Wiley & Sons, Ltd.